Interacting Catalytic Feller Diffusions: Finite System Scheme and Renormalisation

Den Naturwissenschaftlichen Fakultäten
der Friedrich-Alexander-Universität Erlangen-Nürnberg
zur
Erlangung des Doktorgrades

vorgelegt von

Christian Penßel
aus Nürnberg

Als Dissertation genehmigt von den Naturwissenschaftlichen Fakultäten der Friedrich-Alexander-Universität Erlangen-Nürnberg

Tag der mündlichen Prüfung: 11. Februar 2004

Vorsitzender der Promotionskommission:	Prof. Dr. L. Dahlenburg
Erstberichterstattung:	Prof. Dr. A. Greven
Zweitberichterstattung:	Prof. Dr. A. Klenke Universität Mainz

Bibliografische Information Der Deutschen Bibliothek

Die Deutsche Bibliothek verzeichnet diese Publikation in der Deutschen Nationalbibliografie; detaillierte bibliografische Daten sind im Internet über http://dnb.ddb.de abrufbar.

ISBN 3-8325-0528-8

Logos Verlag Berlin
Comeniushof, Gubener Str. 47,
10243 Berlin
Tel.: +49 030 42 85 10 90
Fax: +49 030 42 85 10 92
INTERNET: http://www.logos-verlag.de

Danksagung

Ich möchte an dieser Stelle all denen danken, die mich bei dieser Arbeit unterstützt haben und mir den Rücken gestärkt haben.

Als erstes möchte ich hier meine Eltern und meinen Großvater nennen. Wahrscheinlich waren es meine Eltern, die in mir Begeisterung und Freude an der Mathematik geweckt haben. In den letzten Jahren waren sie mir ein wichtiger Rückhalt; es hat mir gut getan, zu wissen, daß sie den Fortschritt der Arbeit mit guten Gedanken begleiten.
Ebenso möchte ich meiner Freundin Renate danken. Von ihr habe ich die Aufmunterung erfahren, die nötig war, auch mathematische Durststrecken durchzustehen.

Weiterhin danke ich dem Betreuer meiner Arbeit, Professor Greven. Er hat es mir ermöglicht, einen Einblick in aktuelle mathematische Forschung zu gewinnen. Die Gespräche mit ihm haben meinen mathematischen Horizont erweitert, so daß es mir eine intellektuelle Bereicherung war, ihn zum Betreuer zu haben.
Die Zusammenarbeit mit den Kollegen der Arbeitsgruppe - in letzter Zeit insbesondere mit Anita, Iljana, Jan, Peter, Alexander und Tobias – empfand ich als außerordentlich angenehm und stimulierend.
Für finanzielle und ideelle Förderung danke ich dem Freistaat Bayern, dessen Stipendium mir das Studiums erleichtert hat, der deutschen Forschungsgemeinschaft, von der das Geld für meine Promotionsstelle stammte, sowie der Studienstiftung des deutschen Volkes, der ich nicht nur einige Euros, sondern insbesondere ein herausragendes allgemeinbildendes Rahmenprogramm für Studium und Promotion verdanke.

Bubenreuth, Oktober 2003 Christian Penßel

Zusammenfassung

In dieser Arbeit untersuchen wir verschiedene Aspekte linear interagierender katalytischer Fellerdiffusionen. Das betrachtete Modell reiht sich ein in die Reihe der Zweitypenmodelle; für eine abelsche Gruppe Λ wird die Evolution des betrachteten Prozesses durch ein $2\,|\Lambda|$-dimensionales System stochastischer Differentialgleichungen beschrieben. Der Prozeß kann in gewisser Weise als Diffusionslimes eines verzweigenden Teilchenprozesses verstanden werden kann; dies ist insofern nützlich, als wir damit zumindest auf der heuristischen Ebene eine anschauliche Charakterisierung der Zeitentwicklung des Prozesses erhalten. Die beiden Teilchentypen werden *Katalysten* und *Reaktanten* genannt. Motiviert ist diese Nomenklatur durch die Zeitentwicklung des Prozesses, die sich folgendermaßen beschreiben läßt:

Sowohl Katalysten als auch Reaktanten seien kritisch verzweigende Irrfahrten in stetiger Zeit auf der zugrundeliegenden Gruppe. Die Verzweigungsrate für die Katalysten werde als zeitlich und räumlich konstant angenommen, während sie im Falle der Reaktanten variiere und lokal durch ein Vielfaches der Anzahl der dort anzutreffenden Katalysten bestimmt sei.

Zunächst diskutieren wir das Langzeitverhalten für $\Lambda = \mathbb{Z}^d$ ($d \in \mathbb{N}$). Es stellt sich heraus, daß abhängig von der Dimension d (genau genommen abhängig davon, ob die Symmetrisierung der zugrundeliegenden Irrfahrt rekurrent oder transient ist) qualitativ unterschiedliche Universalitätsklassen des Langzeitverhaltens beobachtet werden können.

Der zweite Teil der Arbeit ist dem Vergleich endlicher und unendlicher Systeme gewidmet: auf der einen Seite betrachten wir $\Lambda = \mathbb{Z}^d$, auf der anderen Seite wählen wir endliche d-dimensionale Tori als zugrundeliegende Gruppe. Hier wenden wir das sogenannte FINITE SYSTEM SCHEME an. Diese von Cox und Greven [CG90] eingeführte Methode ermöglicht es uns, Zusammenhänge zwischen großen endlichen Systemen einerseits und dem unendlichen Fall andererseits herzustellen. Insbesondere wird deutlich, für welche Zeitskalierungen (von der Größe des Torus abhängig) das unendliche System als Limes großer endlicher Systeme gesehen werden kann.

Schließlich betrachten wir linear interagierende katalytische Fellerdiffusionen für $\Lambda = \Omega_{N,L}$, also auf der hierarchischen Gruppe mit L Hierarchien, die die Eigenschaft besitzt, daß jedes Gruppenelement $N-1$ nächste Nachbarn hat. Gemäß der von Dawson und Greven [DG93a] eingeführten MULTIPLE SPACE TIME SCALE ANALYSIS weisen wir nach, daß für große N eine Mean-Field Approximation sinnvoll ist, also der Einfluß der Nachbarn (und der dort sich entwickelnden Prozesse) eines Elements der hierarchischen Gruppe mehr und mehr deterministisch wird. Dies kann zurückgeführt werden auf das Skalenverhalten von empirischen Mitteln über Kugeln (bezüglich der hierarchischen Metrik). Eine Einbeziehung aller Level der hierarchischen Gruppe in die Betrachtung ermöglicht es, einen hierarchischen Mean-Field Limes zu formulieren. Die Abhängigkeiten der empirischen Mittel verschiedener Hierarchieebenen können durch eine Markovkette, die sogenannte *interaction chain*, wiedergegeben werden. Mit Hilfe dieser Markovkette ist es möglich, das Langzeitverhalten der betrachteten Prozesse zu klassifizieren.

Abstract

We discuss several aspects of linearly interacting catalytic Feller diffusions. This is a two-type model with interacting types living on a countable Abelian group Λ. The evolution of the process can be described in terms of a $2\,|\Lambda|$-dimensional system of stochastic differential equations. Since the interaction between the types is a one-way interaction, the model we discuss can be classified as a stochastic process in a random medium that is itself a stochastic process.

This is the motivation to call particles of the first type *catalysts* and particles of the second type *reactants*. It is possible to characterize the process as the diffusion limit of a particle model: Both, catalyst and reactant particles perform migration and critical branching; the branching rate of the catalyst particles is assumed to be constant in space and time, whereas for a given site the local branching rate of the reactant is given as a multiple of the number of catalyst particles to be found there. Hence, in the diffusion picture the catalyst is given as an autonomously evolving interacting Feller diffusion process, whereas the reactant process is an interacting diffusion process that depends on the catalyst process.

At first, we investigate the long term behavior in the case $\Lambda = \mathbb{Z}^d$. Dependent on dimension (or, more precisely, on the recurrence/transience properties of the symmetization of the underlying random walk), one observes different classes of universality.

The next part is dedicated to the investigation of large finite systems in comparision to infinite systems. In particular, we discuss to what extent it is possible to approximate lattice systems by processes defined on tori. The method we use is the so-called FINITE SYSTEM SCHEME introduced by Cox and Greven in 1990 [CG90]. A main tool of their approach is the discussion of fluctuations of averages after a certain rescaling of time. It turns out that this rescaling depends linearly on the cardinality of Λ.

Besides the investigation of catalytic interacting Feller diffusion on the lattice, we discuss the process on the hierarchical group $\Omega_{N,L}$, (L is the number of hierarchies, $N-1$ the number of next neighbors of each element of $\Omega_{N,L}$). It turns out that a rescaling analysis similar to the Finite System Scheme can be carried out. The method used for processes on the hierarchical group is called MULTIPLE SPACE TIME-SCALE ANALYSIS and was introduced by Dawson and Greven in 1993 ([DG93a], [DG93b], [DG93c]). We find out that for large N a mean-field approximation makes sense: roughly speaking, the influence of neighboring processes can be replaced by their expectation as $N \to \infty$. In a second step, we extend this result considering empirical averages on various levels. We find out that the fluctuations of these averages resemble the original process as $N \to \infty$ with respect to rescaled time. By means of the so-called interaction chain, this gives insight into the interaction of different hierarchies and the long-range dependencies of evolution.

Contents

Chapter 0

Introduction

0.1 The model

In this thesis we investigate various aspects of catalytic interacting Feller diffusions. This process is a two-type interacting Feller diffusion process on a countable Abelian group Λ. One type acts as catalyst and evolves autonomously as interacting Feller diffusion, whereas the second type – the reactant – evolves with respect to the random medium created by the catalyst.

The reactant is an interacting Feller branching diffusion process with a varying branching rate that is determined by the catalyst. In order to clarify what we are talking about, we provide the corresponding evolution equations at this place:

$$\begin{aligned} dX_t(i) &= c_x \sum_{j\in\Lambda} a(i,j)\left(X_t(j) - X_t(i)\right) dt \\ &\quad + \sqrt{2\gamma_x X_t(i)\, Y_t(i)}\, dB_t^x(i) \quad (i \in \Lambda)\,, \qquad (1) \\ dY_t(i) &= c_y \sum_{j\in\Lambda} a(i,j)\left(Y_t(j) - Y_t(i)\right) dt \\ &\quad + \sqrt{2\gamma_y Y_t(i)}\, dB_t^y(i) \quad (i \in \Lambda)\,. \qquad (2) \end{aligned}$$

Remark 1 *We intentionally have not yet introduced the ingredients of this pair of stochastic differential equations; we hope that they are more or less self-explaining. A precise definition of the model can be found in 1.2.1.*

We want to point out two properties of the model:

- *catalytic setting:* One type evolves autonomously, whereas the second type can only be constructed dependent on the first type. Therefore, the second type may be regarded as a stochastic process in a randomly fluctuating medium.
- *branching property*: The process arises as diffusion limit of a catalytic interacting branching particle system. This gives rise to infinite divisibility of the law of any type not influenced by other types. Hence, techniques relying on infinite divisibility can directly be applied when considering the catalyst. It is possible to make this method applicable to the reactant by considering the reactant for fixed catalyst[1].

[1]this is the quenched point of view.

There has already been research on the topic of catalytic branching: The particle model on the lattice and its analogue on $\mathbb{R}^d$ have been considered by Greven, Klenke and Wakolbinger ([GKW99], [GKW02]). Their interest mainly centres on the longtime behavior of the processes. The case of the two-dimensional lattice turned out to be quite delicate. Investigations on it have been carried out by Etheridge and Fleischmann [EF98]. Further references on the topic of longtime behavior are papers by Dawson and Fleischmann ([DF97a], [DF97b]).

0.2 Research on related models

Investigations of catalytic branching are embedded in research on several closely related models. All these models have in common that they are two-type or, more general, multi-type models: each type can be described as measure-valued process on an Abelian group. In any case, there is a competition between migration and branching, or, in the case of diffusions, migration and local fluctuations. The different types only affect the branching rates of one another, not the properties of migration. The branching rate at a given site can possibly be influenced by the presence of other types. Typically, the evolution of related processes (with $n \in \mathbb{N}$ types) $\mathfrak{X} = \left(\mathfrak{X}^{(1)}, ..., \mathfrak{X}^{(n)}\right)$ is of the following form:

$$d\mathfrak{X}_t^{(k)}(i) = \sum_{j \in \Lambda} a\left(i,j\right) \left(\mathfrak{X}_t^{(k)}\left(j\right) - \mathfrak{X}_t^{(k)}\left(i\right)\right) dt \tag{3}$$

$$+ \sqrt{2\mathfrak{X}_t^{(k)} \cdot g^{(k)}\left(\mathfrak{X}_t^{(1)}(i), \ldots, \mathfrak{X}_t^{(n)}(i)\right)}\, dB_t^{(k)}(i) \quad \left(k \in \{1, ..., n\}, i \in \Lambda\right). \tag{4}$$

Migration (heat flow) is described by (3), while (4) describes branching (fluctuations). It is assumed that the random fluctuations (4) are governed by a family of independent Brownian motions, one per site and type. The functions $g^{(k)} : \mathbb{R}_+^n \to \mathbb{R}_+$ describe how the branching of type $k \in \{1, ..., n\}$ at a given site i is affected by the vector describing the population at this site.

In particular, the following models have been objects of research:

$n = 2$	$g^{(1)}\left(x_1, x_2\right) = x_2,$ $g^{(2)}\left(x_1, x_2\right) = const.$	catalytic branching
$n = 2$	$g^{(1)}\left(x_1, x_2\right) = x_2,$ $g^{(2)}\left(x_1, x_2\right) = x_1.$	mutually catalytic branching
n arbitrary	$g^{(k)}\left(x_1, ..., x_n\right) = x_{(k+1) \bmod n}$ $(k \in \{1, ..., n\})$	cyclically catalytic branching
n arbitrary	$g^{(k)}\left(x_1, ..., x_n\right) = h(\sum_{i=1}^n x_i)$ $(k \in \{1, ..., n\})$	state dependent branching

with $h \in \mathcal{C}\left(\mathbb{R}_+, \mathbb{R}_+\right)$ locally Lipschitz, $h(x) = o\left(1 + x\right)$.

Mutually catalytic branching: This model was introduced by Carl Mueller. It has been studied by Dawson and Perkins [DP98]. The peculiarity of this two-type model is that the types act both as catalyst and as reactant. As the interaction of both types

is mutual, methods relying on infinite divisibility cannot be applied. An important feature of this model is that there is a self-duality relation. This was discovered by Mytnik [Myt98], who applies this fact to prove uniqueness.
The investigations of Dawson and Perkins center on the longtime behavior of the process. They have obtained that there is a dichotomy between coexistence of the two types going along with the existence of a stationary measure on the one hand, and separation of the two types on the other hand. This dichotomy is related to recurrence/transience properties of the underlying random walk: if dimension is small, separation of types is observed, whereas there is stationarity and coexistence of types in the high-dimensional case. The process has been investigated from the rescaling point of view by Cox, Dawson and Greven [CDG03], who have applied the Finite System Scheme to the process. Furthermore, an investigation of the hierarchical mean-field limit is carried out.
There is a continuous space analogue of mutually catalytic branching. This process was introduced in the one-dimensional case by Dawson and Perkins [DP98] in terms of stochastic partial differential equations. If dimension exceeds one, the process has to be discussed as solution of a martingale problem. It turned out to be non-trivial to prove that the process exists: this was only known for $d = 1$ at first; recent research has revealed a partial result for the two-dimensional case: Provided that the collision rate between types is sufficiently small, existence can be proven for $d = 2$, too. This result can be found in [DEF+02a].
A survey on the current state of knowledge on this topic is provided by a trilogy of papers by Dawson, Etheridge, Fleischmann, Mytnik, Perkins and Xiong: [DEF+02a], [DEF+02b], [DFM+02].

Cyclically catalytic branching: It is possible to generate a cycle from many "one way" interactions of catalyst-reactant type as described in 0.1. In this model, each type acts both as catalyst and as reactant. It has been considered by Fleischmann and Xiong [FX01]. Since so far no correspondence of Mytnik's argument has been found, uniqueness remains an open problem.

State dependent multitype branching: It is characteristic for *state dependent multitype branching* models that the branching rate of an individuum depends on the total population at its location. The important point is that this dependency is the same for all types. Therefore, multi-type state-dependent branching models can be regarded as branching models with the additional feature that particles are colored such that the pedigrees of single particles are made visible. After all, this is the main objective of the discussion of these models: the aim is to obtain insight into the family structure of interacting diffusions and their historical processes. Often the type space is chosen to be continuous in generalization of the setting of the table provided in 0.2. State dependent multitype branching has been investigated by Dawson and Greven [DG03] and Pfaffelhuber [Pf03]. The main purpose of the paper of Dawson and Greven is to construct the processes as measure-valued diffusions, to characterize them by means of martingale problems and to give results concerning the longtime behavior. In particular, they discuss the question whether the processes locally tend to be mono- or multitype in the long run considering relative weights of types as a stochastic process in the medium given by the total mass process. This approach was also applied by Pfaffelhuber, who focusses on the rescaling analysis of this class of processes carrying out the Finite System Scheme.

0.3 Focus of interest

In this thesis we investigate different aspects of catalytic interacting branching diffusions. In particular, we focus on three topics, namely

- the longtime behavior,
- finite and infinite systems and rescaling: FINITE SYSTEM SCHEME in the lattice case, MULTIPLE SPACE TIME SCALE ANALYSIS when considering the process on the hierarchical group,
- and – this is more or less a by-product – a discussion of duality as a method of proving convergence of processes with the branching property.

0.3.1 Longtime behavior

We consider the limit distributions of the considered processes as time tends to infinity.

It is a common feature to many stochastic processes that there occurs a dichotomy between persistence and local extinction. Which of these regimes typically occurs depends on whether the underlying random walk[2] is transient or recurrent. In fact, this dichotomy occurs in all cases.

Things get more complicated if we discuss processes consisting of two (or even more) interacting types. What makes things interesting is the fact that the reactant needs the catalyst to be able to create branching events; roughly speaking, only if the catalyst dies out quickly we can be sure that the reactant persists.

We will provide a theorem which

- gives a sufficient condition for non-extinction of the reactant in the recurrent cases (dimension $d = 1, 2$)
- guarantees that the process has an equilibrium it converges to in the transient cases (dimension $d \geq 3$)

0.3.2 Rescaling analysis in space and time

Lattice case

As we have just mentioned, there are different regimes of longtime behavior. These regimes correspond to properties of the underlying random walks: in general, the long-term asymptotics of the Green function of the symmetrized random walk give insight which universality class of longtime behavior the considered process belongs to. Since these asymptotics are closely related to the dimension of space, there are different regimes of longtime behavior for different dimensions.

[2]to be more precise: its symmetrization

Processes on the lattice can be approximated by processes on tori. Though this approximation certainly works for fixed time, it cannot be expected that the longtime limits of a sequence of finite systems converge to those of the approximated infinite system. A familiar example of this non-exchangeability of limits – namely the limit of letting time to infinity and the limit of letting the size of the system to infinity – is the interacting Feller diffusion process with $d \geq 3$: This process dies out on finite systems almost surely, but it persists when considered on the lattice. As the longtime limits of finite and infinite model differ qualitatively, it is a challenge to pose the question to what extent finite systems could still be regarded as approximations of infinite systems.

To let system size approach infinity and to let time approach infinity – these limits cannot be exchanged. Hence, it is quite natural to combine them, letting them approach infinity jointly. This procedure, moreover, can be motivated by the fact that empirical averages over growing boxes fluctuate at time-scales which depend on the size of the box. This fluctuation of averages gives rise to a rescaling analysis: The approach of Finite System Scheme (compare [CG90]) allows to discuss both fluctuation of averages and local similarity between finite and infinite models. A key tool of this approach is the introduction of two different timescales for fluctuations on a macroscopic scale and fluctuations on a microscopic scale.

Hierarchical group case

Processes on the hierarchical group can be discussed using the Multiple Space Time Scale Analysis (compare [DG93a]). This method allows to discuss properties of large systems (more precisely: a hierarchical mean-field limit is carried out) such as longtime behavior and cluster growth. The analysis relies on a consideration of space-time renormalized systems, combining slow and fast timescales in order to investigate averages over different levels. The qualitative behavior of a process can be characterized by the so-called interaction chain: this Markov chain describes how low-level fluctuations are influenced by high-level averages. The entrance laws of the interaction chain play a key role: according to their limit behavior, the longtime behavior can be classified making it possible to collect different processes with similar properties in classes of universality.

0.3.3 Duality and convergence

It is due to the catalyst-reactant setting of the processes discussed here that methods based on infinite divisibility can be used: The catalyst is infinitely divisible by definition; as to the reactant, infinite divisibility can be applied, too: the key idea is to use the quenched point of view, that means considering the reactant for a given (typical) realization of the catalyst process. One of the advantages of infinite divisibility is that one obtains multiplicativity of Laplace-transforms, hence additivity of log-Laplace transforms. We will be able to describe the considered Laplace transforms in terms of certain dual processes. In particular, we will lay out how the concept of dual processes can be used to describe finite dimensional distributions. This leads to a new strategy of proving convergence of finite dimensional distributions: finite dimensional distributions can be regarded as some kind of random occupation times. These random occupation times can again be described by dual processes. It turns out that convergence of these dual processes is sufficient for convergence (with respect to finite dimensional distributions) of the processes originally considered.

Chapter 1

Catalytic interacting Feller diffusions: Definition of the model

In this chapter we introduce and define the object of interest of this thesis, namely *Catalytic Interacting Feller Diffusions.* At first, we provide a short discussion of the particle analogue, then we make clear what exactly we will talk about. The description of the process consists of the following parts

- the evolution equations of the process;
- assumptions on the initial conditions of the process necessary to make the process well-defined on a state space, which is preserved.

1.1 Particle picture

A detailed discussion of the ergodic properties of the particle analogue of the process we discuss here has been carried out by Greven, Klenke and Wakolbinger [GKW99]. We describe the particle model discussed by them. Afterwards, we make clear how this particle model is related to the diffusion model we discuss here. We give a description of the mechanisms of migration and branching of both, particles of catalyst and particles of reactant type.

1.1.1 Particles of catalyst type

The catalyst process evolves in the following way:

- *Migration:* Assume that to each particle there corresponds an exponentially distributed waiting time with constant expectation. After this waiting time, the particle jumps somewhere else. This migration mechanism is described by a stochastic $\Lambda \times \Lambda$-matrix $\bar{a}$ (Λ denotes an Abelian group equipped with a translation invariant metric d). It is assumed that $\bar{a}$ has the following properties:

  - The transition rates are homogenous in space:

$$\bar{a}(i,j) = \bar{a}(0, j-i) \quad (i, j \in \Lambda). \tag{1.1}$$

– The second moment of the jump distance is finite

$$\sum_{i\in\Lambda} (d(0,i))^2 \bar{a}(0,i) < \infty \quad . \tag{1.2}$$

– The random walk kernel is supposed to be irreducible:

$$\forall t > 0, i \in \Lambda : \bar{a}_t(0,i) > 0 \quad , \tag{1.3}$$

where the transition rates after time t are given by

$$\bar{a}_t(\cdot,\cdot) = \sum_{k=0}^{\infty} \exp(-t) \frac{t^k}{k!} \bar{a}^{(k)}(\cdot,\cdot) \quad . \tag{1.4}$$

Let a be defined by

$$a(i,j) := \bar{a}(j,i) \quad (i,j \in \Lambda) \quad . \tag{1.5}$$

With the homogeneity of space provided by (1.1), a is a stochastic matrix, too.

- *Critical binary branching:* After another exponential distributed waiting time with constant expectation, a particle is affected by a branching event: with probability $\frac{1}{2}$, it dies; with probability $\frac{1}{2}$, it is replaced by two particles. These new-born particles migrate and branch just like their parent particle, but independently from each other and independently from all other particles.

  This mechanism yields, that the total number of particles is a martingale.

- *Independence*: All particles are assumed to evolve independently. In particular, initial condition, migration and branching are assumed to be independent.

1.1.2 Particles of reactant type

The reactant process evolves in the following way:

- *Migration:* Each particle is assigned to an exponentially distributed waiting time with constant expectation. After this waiting time, the particle moves to another site. The jump probabilities are given by a translation invariant random walk kernel. The transition probabilities are given by the transition matrix $\bar{a}$.

- *Critical binary Branching with fluctuating rate:* Again, we have a binary branching mechanism, but the local rate of branching (=inverse expected waiting time) is chosen proportional to the number of catalysts at the corresponding site.

- *All particles evolve independently:* In particular, migration and branching are assumed to be independent. In addition to that, we assume that the evolution of the process and its initial condition are independent.

Remark 2

1. *Since the waiting times are exponentially distributed, the process has the Markov property.*

2. *In most cases, one is interested in models started with an infinite number of particles. This forces us to impose certain requirements on the initial law. We want to avoid "explosions" caused by particles coming in from infinity. This problem has been discussed by Liggett and Spitzer in 1981 [LS81]. They give conditions that the initial law of interacting processes should satisfy to avoid such explosions. More details will be given in the context of the diffusion limit.* *(see page 11)*

3. *Throughout this thesis, we will assume that the law of initial conditions is ergodic and translation invariant. We assume that the intensity is finite.*

1.2 Diffusion process

1.2.1 Diffusion limit and definition of the diffusion model

The interacting diffusion process considered in this thesis can be obtained by taking the diffusion limit of the particle process: the point is that instead of particles one observes a continuously fluctuating fluid. In the following, we describe how the diffusion process can be obtained from the particle process. In general, one has to carry out two rescaling procedures simultanously to obtain the diffusion limit:

- We convert the particles into a fluid by cutting them into pieces. Introducing the notion of mass, this is a procedure which preserves total mass. This way, we increase successively the number of particles.
- The branching rate per particle has to be increased proportional to the number of particles. Without changing the branching rate per particle, fluctuations caused by branching would be smudged away as a consequence of a law of large numbers. Hence, the branching rate per particle has to be increased.

The construction of the diffusion limit of the particle process consists of two steps. First, we carry out the diffusion limit of the catalyst. The diffusion limit we obtain is used as medium of the reactant processes. In particular, note that the catalyst diffusion process serves as catalytic medium for the reactant particle processes.

The second step is then the diffusion limit of the reactant processes for given diffusive catalyst.

1. *Diffusion limit of the catalyst:*
   An interacting diffusion process describing the diffusion limit of the catalyst particle process can be obtained by
   - increasing the number of particles: Every particle (with mass 1) is replaced by N particles of mass $\frac{1}{N}$.
   - increasing the branching rate: the branching rate per catalyst particle is substituted by its N-th multiple.

2. *Diffusion limit of the reactant for given catalyst:*
   Given the catalyst diffusion process, a reactant particle process can be defined with the catalyst diffusion process as local branching rate. In the manner we have just described, it is possible to construct a diffusion limit of the reactant particle model.

The evolution of the limit process can be described by a system of stochastic differential equations:

$$\begin{aligned} dX_t(i) &= c_x \sum_{j\in\Lambda} a(i,j)\left(X_t(j) - X_t(i)\right) dt \\ &\quad + \sqrt{2\gamma_x X_t(i)\, Y_t(i)}\, dB_t^x(i) \quad (i \in \Lambda), \qquad (1.6) \\ dY_t(i) &= c_y \sum_{j\in\Lambda} a(i,j)\left(Y_t(j) - Y_t(i)\right) dt \\ &\quad + \sqrt{2\gamma_y Y_t(i)}\, dB_t^y(i) \quad (i \in \Lambda). \qquad (1.7) \end{aligned}$$

1.2.2 Ingredients

The evolution equations contain the following ingredients:

rate of catalyst migration:	$c_y > 0$	
rate of reactant migration:	$c_x > 0$	
rate of catalyst branching:	$\gamma_y > 0$	
rate of reactant branching:	$\gamma_x > 0$	
driving Brownian motions:	$\left\{(B_t^y(i))_{t\geq 0}\right\}_{i\in\Lambda}$, $\left\{(B_t^x(i))_{t\geq 0}\right\}_{i\in\Lambda}$	independent families of independent BM
stochastic migration matrix:	$(a(i,j))_{(i,j)\in\Lambda\times\Lambda}$ with properties (1.1), (1.2), (1.3), (1.4)	

1.2.3 Remarks on the choice of the state space – Well-definedness of the process

Before discussing the evolution mechanism provided by (1.6) and (1.7) it should be made clear what initial conditions we shall allow. In other words, we have to specify on what state space S it is possible to define processes solving (1.6) and (1.7). This provides the widest possible frame of initial conditions. Afterwards we will line out what initial conditions we are going to focus at.

The state space S has to satisfy the following:

1. $S \subset \mathcal{M}(\Lambda) \times \mathcal{M}(\Lambda)$.[1]

2. S is preserved under the evolution given by (1.6) and (1.7). (The law of the process at any time t should be re-usable as initial law.)

3. The process characterized by the evolution equations exists and it is unique. (In particular, no "explosions" occur in finite time)

[1] We use the measure-valued notation for $(\mathbb{R}_+ \times \mathbb{R}_+)^\Lambda$.

Definition 1 *The state space $S \subset \mathcal{M}(\Lambda) \times \mathcal{M}(\Lambda)$ fulfills a* Liggett-Spitzer condition *if the following holds:*
There exists $\alpha : \Lambda \to \mathbb{R}_+ \backslash \{0\}$ with

$$\sum_{j \in \Lambda} a(\cdot, j)\, \alpha(j) \leq M \alpha(\cdot) \tag{1.8}$$

for some $M > 0$ s. t. $S \subseteq S_\alpha$, which is defined by

$$S_\alpha := \{(\mathfrak{R}, \mathfrak{T}) \in \mathcal{M}(\Lambda) \times \mathcal{M}(\Lambda) : \langle \mathfrak{R}, \alpha \rangle + \langle \mathfrak{T}, \alpha \rangle < \infty\} \quad . \tag{1.9}$$

Remark 3 *An $\alpha \in (\mathbb{R}_+ \backslash \{0\})^\Lambda$ for which (1.8) holds can be constructed as follows:*
Choose $M > 1$ and let

$$\alpha(i) = \sum_{n=0}^{\infty} M^{-n} \sum_{j \in \Lambda} a^{(n)}(i, j)\, \beta(j) \tag{1.10}$$

with $\beta \in (\mathbb{R}_+ \backslash \{0\})^\Lambda$ and $\sum_{x \in \Lambda} \beta(x) < \infty$.

1.2.4 Assumptions on the initial law

Now and henceforth we assume that the following holds for $\mathcal{L}\{X_0, Y_0\}$:

Independence We assume that the initial configuration (X_0, Y_0) and the family of independent Brownian motions $\left\{(B_t^x(i), B_t^y(i))_{t \geq 0}\right\}_{i \in \Lambda}$ driving the evolution of the process are independent.

Translation invariance We assume that for arbitrary $y \in \Lambda$ we have

$$\mathcal{L}\left\{(X_0(i), Y_0(i))_{i \in \Lambda}\right\} = \mathcal{L}\left\{(X_0(i+y), Y_0(i+y))_{i \in \Lambda}\right\} \quad . \tag{1.11}$$

Finite intensity We assume that first moments are finite in the beginning:

$$\theta_x := \mathbf{E}[X_0(i)] < \infty \quad , \tag{1.12}$$
$$\theta_y := \mathbf{E}[Y_0(i)] < \infty \quad (i \in \Lambda). \tag{1.13}$$

Remark 4 *The evolution equations (1.6) and (1.7) yield that these finite intensities are preserved quantities of evolution:*

$$\forall t \geq 0 :$$

$$\theta_x := \mathbf{E}[X_t(i)] < \infty \quad , \tag{1.14}$$
$$\theta_y := \mathbf{E}[Y_t(i)] < \infty \quad (i \in \Lambda). \tag{1.15}$$

Moments We assume that there exists $p > 2$ with

$$\mathbf{E}\left[(X_0(0))^p + (Y_0(0))^p\right] < \infty \quad . \tag{1.16}$$

Remark 5 *In the appendix, we provide*

- *a result about uniform boundedness of moments for processes living on tori (Lemma 14, p. 247).*
- *a result about boundedness of moments for processes living on the high-dimensional ($d \geq 3$) lattice (Lemma 16, p. 257).*

X_0 and Y_0 not correlated We assume:

$$\forall i \in \Lambda : \mathbf{E}\left[X_0(i) \,|\sigma(Y_0)\right] = \mathbf{E}\left[X_0(i)\right] \tag{1.17}$$

Remark 6 *An immediate consequence of (1.17) and the evolution equations (1.6), (1.7) is*

$$\forall t \geq 0, (i,j) \in \Lambda^2 : \mathbf{E}\left[X_t(i)\, Y_t(j)\right] = \theta_x \theta_y \quad . \tag{1.18}$$

Non-negative autocorrelation We assume that for arbitrary $(i,j) \in \Lambda^2$ we have

$$\mathbf{E}\left[X_0(i)\, X_0(j)\right] \geq (\theta_x)^2 \tag{1.19}$$

and

$$\mathbf{E}\left[Y_0(i)\, Y_0(j)\right] \geq (\theta_y)^2 \quad . \tag{1.20}$$

Lemma 1 *With non-negative autocorrelation in the beginning as given by (1.19) and (1.20), we also have non-negative autocorrelation forever:*

$$\forall t \geq 0 :$$
$$\mathbf{E}\left[X_t(i)\, X_t(j)\right] \geq (\theta_x)^2 \quad , \tag{1.21}$$
$$\mathbf{E}\left[Y_t(i)\, Y_t(j)\right] \geq (\theta_y)^2 \quad . \tag{1.22}$$

Proof. Since the driving Brownian motions are independent at different sites, we can conclude from (1.6) and (1.7) that $\mathbf{E}\left[X_t(i)\, X_t(j)\right]$ and $\mathbf{E}\left[Y_t(i)\, Y_t(j)\right]$ can be described by the following differential equations:

$$\begin{aligned} &\frac{\partial}{\partial t}\mathbf{E}\left[X_t(i)\, X_t(j)\right] \\ &= c_x \sum_{k \in \Lambda} a(j,k)\, \mathbf{E}\left[X_t(k)\, X_t(i)\right] + a(i,k)\, \mathbf{E}\left[X_t(k)\, X_t(j)\right] \\ &\quad - 2c_x \mathbf{E}\left[X_t(i)\, X_t(j)\right] \quad . \end{aligned} \tag{1.23}$$

With the abbreviation

$$E_t^{i,j} := \mathbf{E}\left[X_t(i)\, X_t(j)\right] - (\theta_x)^2 \quad (i, j \in \Lambda) \tag{1.24}$$

we can formulate the claim (1.21) as follows:

$$\forall t \geq 0, \forall \varepsilon > 0 : E_t^{i,j} \geq -\varepsilon \quad . \tag{1.25}$$

Equation (1.23) becomes

$$\frac{\partial}{\partial t} E_t^{i,j} = c_x \left(\sum_{k \in \Lambda} \left(a(j,k)\, E_t^{i,k} + a(i,k)\, E_t^{j,k} \right) - 2E_t^{i,j} \right) \quad . \tag{1.26}$$

Define $\tau \geq 0$ by

$$\tau = \inf\left\{ t : \exists\, (i_0, j_0) \in \Lambda^2 \text{ with } E_t^{i_0,j_0} = -\varepsilon \right\} \quad . \tag{1.27}$$

Note that we have $\tau \geq \frac{1}{2c_x}$, as the following consideration shows:

Assume that there exist $\tau' > 0, i_0', j_0' \in \Lambda^2$ with $\tau' = \frac{1}{2c_x} - \delta$, $\delta > 0$ and $E_{\tau'}^{i_0',j_0'} = -\varepsilon$ Furthermore, let

$$\sigma := 0 \vee \sup\left\{ t \in \left[0, \frac{1}{2c_x} - \delta\right] : E_t^{i_0',j_0'} > 0 \right\} \quad . \tag{1.28}$$

With continuity of $t \longmapsto E_t^{i_0',j_0'}$, the set $(0, a)$ is not empty. By definition of τ' and with (1.28) we have

$$\forall t \in \left[\sigma, \tau'\right] : c_x \left(\sum_{k \in \Lambda} \left(a(j_0', k)\, E_t^{i_0',k} + a(i_0, k)\, E_t^{j_0',k} \right) - 2E_t^{i_0',j_0'} \right) \geq -2c_x\varepsilon \quad . \tag{1.29}$$

This yields

$$E_\tau^{i_0',j_0'} \geq (\tau - \sigma) \cdot (-2c_x\varepsilon) \geq \tau \cdot (-2c_x\varepsilon) = -\varepsilon + 2c_x\varepsilon\delta > -\varepsilon \quad . \tag{1.30}$$

This is a contradiction to $E_\tau^{i_0',j_0'} = -\varepsilon$; we can conclude that $\tau \geq \frac{1}{2c_x}$ is true.
Since – applying continuity of solutions –

$$-\varepsilon = \min_{t \in [0,\tau]} \left\{ E_t^{i_0,j_0} \right\} \tag{1.31}$$

we obtain:

$$\frac{\partial}{\partial t} E_t^{i_0,j_0} \leq 0 \quad \text{for } t = \tau. \tag{1.32}$$

Hence, by (1.26) we get

$$\sum_{k \in \Lambda} \left(a(j_0, k)\, E_\tau^{i_0,k} + a(i_0, k)\, E_\tau^{j_0,k} \right) - 2E_\tau^{i_0,j_0} \leq 0 \quad . \tag{1.33}$$

By the definition of τ, this yields

$$\sum_{k \in \Lambda} \left(a(j_0, k)\, E_\tau^{i_0,k} + a(i_0, k)\, E_\tau^{j_0,k} \right) \leq 2\varepsilon \quad . \tag{1.34}$$

This is only possible if for all k with $a\left(j_0', k\right) > 0$ we have

$$E_\tau^{i_0,k} = -\varepsilon \quad . \tag{1.35}$$

Since a is irreducible, we get

$$\forall k \in \Lambda : E_\tau^{i_0,k} = -\varepsilon \quad . \tag{1.36}$$

Together with the assumption of homogeneity (1.11), this yields

$$\forall \left(k, \ell\right) \in \Lambda^2 : E_\tau^{k,\ell} = -\varepsilon \quad .$$

Taking (1.26) into account, we get

$$E_t^{k,\ell} = -\varepsilon \quad \forall t \geq \tau,$$

and with the definition of τ

$$E_t^{k,\ell} \geq -\varepsilon \quad \forall t \geq 0. \tag{1.37}$$

Letting $\varepsilon \to 0$, we finally obtain our claim (1.19).

The second claim (1.20) follows from analogous calculations. ■

Spatial ergodic properties We assume that the following spatial ergodic properties hold:

$$\lim_{t\to\infty} \mathbf{E}\left[\left(\sum_{i\in\Lambda} a_t\left(0,i\right)\left(X_0\left(i\right), Y_0\left(i\right)\right) - \left(\theta_x, \theta_y\right)\right)^2\right] = 0 \quad . \tag{1.38}$$

$$\lim_{t\to\infty} \mathbf{E}\left[\left(\sum_{i\in\Lambda} \bar{a}_t\left(0,i\right)\left(X_0\left(i\right), Y_0\left(i\right)\right) - \left(\theta_x, \theta_y\right)\right)^2\right] = 0 \quad . \tag{1.39}$$

Definition 2

1. *Allowed initial laws:*
   Let $\mathcal{S} \subset \mathcal{M}^1(\mathcal{M}(\Lambda) \times \mathcal{M}(\Lambda))$ *be the set of all those probability distributions on* $\mathcal{M}(\Lambda) \times \mathcal{M}(\Lambda)$ *which are laws of random variables* $(X_0(i), Y_0(i))_{i \in \Lambda}$ *for which the assumptions mentioned above hold; namely, these assumptions are:*

   - $(X_0(i), Y_0(i))_{i \in \Lambda}$ *almost surely fulfills a Liggett-Spitzer condition*
   - *Independence of initial law and driving Brownian motions*
   - *Translation invariance*
   - *Finite intensity*
   - *Finiteness of moments up to a* $p > 2$.
   - X_0 *and* Y_0 *not correlated*
   - *Non-negative autocorrelation*
   - *Spatial ergodic properties*

2. *Initial laws on the lattice with given intensity:*
   For $(\theta_x, \theta_y) \in \mathbb{R}_+ \times \mathbb{R}_+$ *let* $S^{(\theta_x, \theta_y)} \subset \mathcal{M}^1(\mathcal{M}(\mathbb{Z}^d) \times \mathcal{M}(\mathbb{Z}^d))$ *be the set of those probability laws* $P_0 \in \mathcal{M}^1(\mathcal{M}(\mathbb{Z}^d) \times \mathcal{M}(\mathbb{Z}^d))$ *for which we have:*

   - $P_0 \in \mathcal{S}$.
   - $\int (X, Y)\, P_0(d(X, Y)) = (\theta_x, \theta_y)$.

Remark 7

1. *A closer inspection of the equations governing evolution yields that the intensity* (θ_x, θ_y) *is a preserved quantity of the evolution of our process.*

2. *When discussing the Finite System Scheme, we will compare families of processes* $(X^N(i), Y^N(i))$ *on tori of increasing size. In this context, we will assume that (1.38) holds uniformly in N (this assumption can be found in 3.1.3)).*

Conclusion 1 *The process under consideration is well-defined (i. e. existence and uniqueness) for the following reasons:*

- *One problem could be an "explosion" arising from mass cascading in from infinity. This is avoided, since almost surely a Liggett-Spitzer condition is fulfilled. (compare "allowed initial laws" (Definition 2, p. 15 and the definition of the Liggett Spitzer condition (Definition 1.8, p. 11))*

- *Uniqueness could be in danger if one of the coordinates of the process approaches zero. Taking only the catalyst into account the well-known results of Le Gall [LG83] can be applied. In our case in is important to know that the catalyst evolves autonomously. Therefore, the reactant is an interacting diffusion in a random medium, namely the catalyst. For this class of processes, the problem of proving existence and uniqueness has been solved by Greven, Klenke and Wakolbinger ([GKW02], Theorem 1). They show that a unique strong solution always exists.*

Chapter 2

The models - elementary properties

2.1 Finite models

With the expression "finite models" we describe processes satisfying (1.6) and (1.7) with finite Λ. The total mass processes of the catalyst,

$$Y_t(\Lambda) = \sum_{i\in\Lambda} Y_t(i) \tag{2.1}$$

with initial condition $Y_0(\Lambda) = \sum_{i\in\Lambda} Y_0(i)$ can be described as strong solution of the stochastic differential equation

$$dY_t(\Lambda) = \sum_{i\in\Lambda} \sqrt{2\gamma_y Y_t(i)}dB_t^y(i) \quad . \tag{2.2}$$

It is a rather well-known fact[1] that the following process $\tilde{Y}$ is equivalent in law: Let $\left(\tilde{Y}_t\right)_{t\geq 0}$ be defined as strong solution of

$$\begin{aligned} \tilde{Y}_0 \; &: = Y_0(\Lambda) \quad , &(2.3)\\ d\tilde{Y}_t &= \sqrt{2\gamma_y \tilde{Y}_t dB_t^y} \quad . &(2.4) \end{aligned}$$

Hence the total mass process of the catalyst can be regarded as Feller's branching diffusion. As a consequence of the martingale convergence theorem, this process converges almost surely, since it is a positive martingale with finite expectation. The limit point is zero – only at zero the diffusion term vanishes.
The total mass process of the reactant, $X_t(\Lambda)$, solves the stochastic differential equation

$$X_t(\Lambda) = \sum_{i\in\Lambda} \sqrt{2\gamma_y X_t(i)\, Y_t(i)}dB_t^x(i) \quad . \tag{2.5}$$

This process converges almost surely: it can fluctuate at most as long as the total mass process of the catalyst does not vanish. Since the catalyst dies out almost surely, the total

[1] Note that the generators of both processes coincide. Furthermore, the branching property of the approximating particle systems is reflected by the mentioned equivalence.

mass process stops fluctuating at almost surely finite time. For $\Lambda = \{0\}$, the reactant can be represented as a Feller diffusion with randomly changed timescale. It may happen that the reactant dies out before the catalyst vanishes. Therefore, the reactant dies out with non-vanishing probability.
We take a closer look at three examples of finite models:

1. the *zero-dimensional model* where Λ consists of just one element

2. the *"no geometry"-model (Mean-Field-Model)*, where

   - Λ consists of a finite number N of elements:

     $$\Lambda = \{0, ..., N-1\}$$

   - $a(\cdot,\cdot)$ only generates the discrete topology (hence there is not really a underlying geometrical structure):

     $$a(i,j) = \frac{1}{N}$$

     for all $i, j \in \{0, ..., N-1\}$.

   By construction, the processes $(X_t(i), Y_t(i))$ $(i \in \{0, ..., N-1\})$ are exchangeable in distribution.

3. the *torus-model:*

- Here, Λ is the d-dimensional torus $\{-N, ..., N-1\}^d$.

2.1.1 Zero-dimensional model - archetype of random survival

We consider the zero-dimensional model ($\Lambda = \{0\}$). We have to consider a $\mathbb{R}_+ \times \mathbb{R}_+$-valued diffusion $(X_t, Y_t)_{t\geq 0}$ given by

$$dX_t = \sqrt{2\gamma_x X_t Y_t}\, dB_t^x \quad , \tag{2.6}$$

$$dY_t = \sqrt{2\gamma_y Y_t}\, dB_t^y \quad . \tag{2.7}$$

The catalyst Y converges almost surely due to the martingale convergence theorem; the limit point is zero. The reactant fluctuates at most as long as the catalyst survives, hence, in the limit, the reactant may have died out or not. To visualize how this typically can look like, we provide simulations[2] of the corresponding particle model, drawing the path of the process. In one sample the reactant is extinct, in the other it survives.

[2]The source code of the PERL programs used to generate the following pictures can be found in Appendix D.

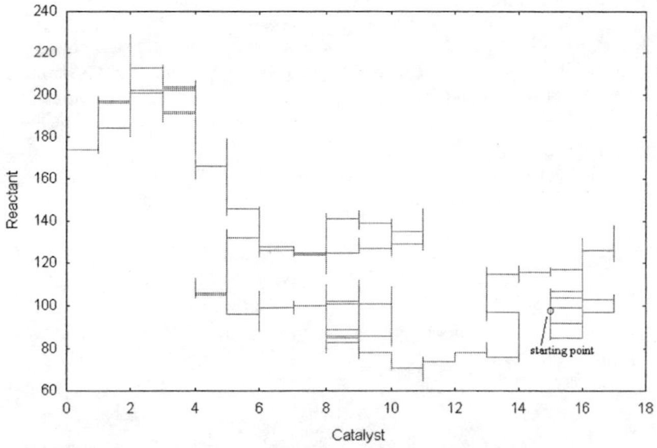

The reactant survives

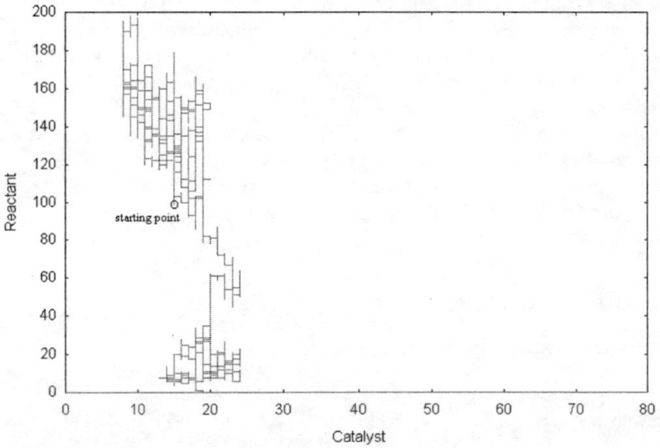

The reactant dies out

It is possible to calculate explicitely the probability that the reactant dies out. Namely, we can prove

$$\lim_{t\to\infty} \mathbf{P}\left[X_t = 0\right] = \frac{1}{\sqrt{\frac{4\gamma_y}{\gamma_x}\frac{X_0}{Y_0^2} + 1}} \quad . \tag{2.8}$$

We do not give a proof at this point, since we have not yet discussed the methods used in the proof. More on this topic can be found in Appendix D, where (2.8) is proven. Appendix D also contains the programming code used for simulation.

2.1.2 "No geometry" - model

Normally, when considering interacting particle systems, one assumes that these particles are located in some colonies which are related one to another by a certain geometrical structure. This geometrical structure is reflected by transition probabilities between the colonies. A common simplification is to neglect these geometrical structures: then the probability kernel a has the simple form

$$a(i,j) = \frac{1}{N}, \quad (i,j \in \{0,...,N-1\}) \quad . \tag{2.9}$$

This way, all transition probabilities have become equal. The focus of interest lies on the question whether the influence the other $N-1$ colonies have on one specific colony could be approximated using some kind of law of large number (a so-called mean-field limit) for large N. The "no geometry" model is used to approximate the local behavior of processes on the hierarchical group.

2.1.3 Torus model

We will soon see that the case $\Lambda = \{-N,...,N-1\}^d =: \mathbb{T}_N^d$ – we equip this set with torus topology – can be recognized as the intermediate step between finite and infinite systems.

On the one hand, one certainly observes the trapping properties that are characteristic to finite systems: the catalyst process dies out almost surely.

On the other hand, we can approximate infinite lattices by letting N to infinity. We observe persistence of the process on the infinite lattice in the case $d \geq 3$.

It is a challenging project to try to cross the borderline between the two regimes of long term behavior, namely the finite case, for which local extinction of the catalyst occurs, and the infinite case $\Lambda = \mathbb{Z}^d$, for which persistence is observed.

This investigation is made by following the agenda of the so-called Finite System Scheme. The crucial point is that both limits, the limit of time ($t \to \infty$) and system size ($N \to \infty$) are carried out simultaneously.

2.2 Infinite models

2.2.1 Processes on the infinite lattice

We consider catalytic interacting Feller diffusions on the d-dimensional lattice $\mathbb{Z}^d$. We have introduced this process in 1.2.1.

Longterm behavior of the catalyst

The catalyst's behavior is well understood: Corresponding to the recurrence / transience properties of symmetric random walk with finite variance transition kernel, there is a dichotomy between local extinction ($d = 1,2$) and persistence ($d \geq 3$):

- *local extinction of the catalyst*

$$\forall \varepsilon > 0, A \subset \mathbb{Z}^d \text{ bounded}: \lim_{t\to\infty} \mathbf{P}[Y_t(A) > \varepsilon] = 0 \quad . \tag{2.10}$$

  for $d = 1,2$.

- *survival of the catalyst*

$$\exists \varepsilon > 0, A \subset \mathbb{Z}^d \text{ bounded} : \limsup_{t \to \infty} \mathbf{P}\left[Y_t(A) > \varepsilon\right] > 0 \quad . \tag{2.11}$$

  for $d = 3$.
  (The catalyst does not only survive, but converges in distribution to a non-vanishing stationary distribution.)

The first of these two regimes, the regime of local extinction can again be split up. This is due to the different types of clustering one can find for $d = 1$ resp. $d = 2$.

Remarks on clustering

We have mentioned that for $d = 1$ and $d = 2$ we have local extinction of the catalyst. Since we have persistence of intensity, this local extinction goes along with the phenomenon that mass concentrates in clusters, which grow in size and height and become more and more distinct in space. We describe two possibilities to gain control over more details over the growth of these clusters, namely the consideration of

- random occupation times
- laws of averages of growing boxes

Random occupation times One of the questions arising when discussing clustering is whether sites are typically visited by clusters infinitely often or only a finite number of times.

For this purpose, one considers the average occupation of a finite set $A \subset \mathbb{Z}^d$ until time t, namely

$$O_t(A) := \frac{1}{t} \int_0^t Y_s(A)\, ds \quad . \tag{2.12}$$

Under the condition that there exists a probability law $\Pi \in \mathcal{M}^1(\mathbb{R}_+)$ with

$$\mathcal{L}\{O_t(A)\} \overset{t \to \infty}{\Longrightarrow} \Pi \quad .$$

we can distinguish the following two cases:

1. *Non-recurrence of clusters:*
$$\Pi = \delta_{\{0\}} \quad . \tag{2.13}$$
2. *Recurrence of clusters:*
$$\Pi \neq \delta_{\{0\}} \quad . \tag{2.14}$$

Believing that Interacting Feller Diffusions behave similar to Branching Brownian Motion[3], non-recurrence of clusters takes place for $d = 1$, while recurrence of clusters can be observed for $d = 2$. For Branching Brownian motion, this result can be found in [CGr85].

[3] A process and its diffusion limit have obviously similar ergodic properties.

Averages over growing boxes Another possibility to discuss the growth of clusters is to discuss averages over growing boxes. To be more precise, it is the longtime behavior of

$$H_\alpha(t) := \frac{1}{\#\left(B_{t^{\frac{\alpha}{2}}}\right)} Y_s\left(B_{t^{\frac{\alpha}{2}}}\right) \tag{2.15}$$

one is interested in for $\alpha \in \mathbb{R}_+$, where $B_{t^{\frac{\alpha}{2}}}$ is given by

$$B_{t^{\frac{\alpha}{2}}} := \left\{i \in \mathbb{Z}^d : |i| < t^{\frac{\alpha}{2}}\right\} \quad . \tag{2.16}$$

The idea behind is the following. If α is too big the averaging effect wins: The number of clusters contained in $B_{t^{\frac{\alpha}{2}}}$ increases in time, and as limit of $H_\alpha(t)$ we observe the initial intensity for $t \to \infty$. If, on the other hand, α is chosen too small, we observe local extinction and $H_\alpha(t)$ tends to zero almost surely. It turns out that there are two types of clustering, depending whether there is only one α for which a non-trivial limit can be expected or whether a non-trivial behavior is to be found if α is contained in some non-empty interval. In the latter case the limits can be obtained by means of a diffusion process (therefore one speaks of *diffusive clustering*).

Effects on the reactant

For $d = 1$ and $d \geq 3$ it is not hard to guess what long time behavior has to be expected: As soon as at some site the catalyst dies out, the evolution of the reactant becomes deterministic mass flow.

In the case $d \geq 3$, the catalyst persists, converging in law to a stationary process. Qualitatively the catalyst influences the reactant like a multiple of the constant process.

We summarize what we have obtained using the heuristic arguments we have just described:

high-dimensional case For $d \geq 3$ the catalyst is persistent. Qualitatively, this has the same effect on the reactant as a constant catalyst with some effective intensity: The whole catalyst-reactant system converges to a stationary process characterized by the intensities given by the initial condition.

low-dimensional case For $d = 1$ the catalyst dies out very quickly. Hence, the averaging effect wins globally against any clustering coming from branching - there is no branching in quickly growing areas. Hence, the limit is simply a multiple of counting measure. The initial intensity is preserved.

critical case: It is not obvious what happens for $d = 2$. This problem has been solved for the particle case [GKW99] as well as for the continuous space case of catalytic interacting super Brownian motion (CSBM). ([DF97b], Proposition 13). The key to the long-time behavior of the continuous space case lies in its self-similarity: It turns out that the law of CSBM can be obtained from itself by

- speeding up time $t \mapsto Kt$
- rescaling space $(x, y) \mapsto \left(\sqrt{K}x, \sqrt{K}y\right)$
- adjusting mass $\cdot \mapsto \frac{\cdot}{K}$.

This leads to the conclusion that in the critical case the limit law of the reactant is a non-trivial mixture of counting measure of different intensities. In some sense, this is similar to the zero-dimensional model introduced above.

2.2.2 Processes on the hierarchical group

We choose the hierarchical group as space of sites. It can be visualized as the set of the leaves of a tree carrying words of an "alphabet" $\{0, ..., N-1\}$ as labels.

Trees are characterized by their height $L \in \mathbb{N} \cup \{\infty\}$ and the branching parameter N. We distinguish

- trees of finite height $(L < \infty)$:

$$\Omega_{N,L} := \left\{(\omega_0, ..., \omega_{L-1}) \in \{0, ..., N-1\}^L\right\} \tag{2.17}$$

- trees of infinite height $(L = \infty)$:

$$\Omega_{N,\infty} := \left\{(\omega_0, \omega_1, ...) \in \{0, ..., N-1\}^{\mathbb{N}} \,|\exists k : \omega_j = 0 \; \forall j > k\right\} \tag{2.18}$$

In both cases, $L < \infty$ and $L = \infty$, the group contains a countable number of elements. The hierarchical group can be made a metric space by introducing the distance function

$$d: \begin{array}{ccc} \Omega_{N,L} \times \Omega_{N,L} & \longrightarrow & \mathbb{N}_0 \\ (\omega_0, \omega_1, ...), (\tilde{\omega}_0, \tilde{\omega}_1, ...) & \longmapsto & \min\{k|\omega_j = \tilde{\omega}_j \,\forall j \geq k\} \end{array} \tag{2.19}$$

There is a somehow geographical interpretation of the hierarchical group as a population model: Consider a population which is subdivided into colonies, which themselves contain a certain number of clans, which are made up from families...

A member of the population can then be characterized by its family, which clan the family belongs to and so on. The hierarchical distance of two members of the population then indicates the level of subdivision needed to distinguish them. Another interpretation could of course be a genealogical one . This is not the picture we shall have in mind, as we want to avoid to confuse the genealogical interpretation of trees with the genealogy of branching particles.

It is natural to define the interaction kernel on the hierarchical group in such a way that the probability of a jump from site i to site j depends as well on their genealogical distance as on N. It turns out that the choice

$$a^N(i,j) = \frac{c_{d(i,j)}}{N^{d(i,j)-1}} \frac{1}{N^{d(i,j)}} \tag{2.20}$$

(where $(c_k)_{k\in\mathbb{N}_0} \in \mathbb{R}_+^{\mathbb{N}_0}$) leads to interesting rescaling results if we let $N \to \infty$. This motivates us to consider the process given by the following finite system of stochastic differential equations:

$$dX_t^N(i) = \sum_{k=1}^{L} \frac{c_{k-1}}{N^{k-1}} \left(\frac{1}{N^k} \sum_{j \in \Omega_{N,L}: d(i,j) \leq k} X_t^N(j) - X_t^N(i) \right) dt$$
$$+ \sqrt{2\gamma_x X_t^N(i) Y_t^N(i)}\, dB_t^x(i) \quad ; \tag{2.21}$$
$$dY_t^N(i) = \sum_{k=1}^{L} \frac{c_{k-1}}{N^{k-1}} \left(\frac{1}{N^k} \sum_{j \in \Omega_{N,L}: d(i,j) \leq k} Y_t^N(j) - Y_t^N(i) \right) dt$$
$$+ \sqrt{2\gamma_y Y_t^N(i)}\, dB_i^y(i), \quad (i \in \Omega_{N,L}) \quad . \tag{2.22}$$

In order to be able to discuss what happens in the limit $L \to \infty$, we have to assume

$$\sum_{k=0}^{\infty} \frac{c_k}{N^k} < \infty \tag{2.23}$$

if only N is large enough. The most common approach to systems of this kind is the so-called Multiple Space-Time Scale Analysis. It was carried out for the catalyst system in [DG96]. One of the key questions of this approach is: How do rescaled averages behave in their timescale in the limit $N \to \infty$? Is there a non-trivial entrance law of the so-called interaction chain (the Markov chain describing what we can know about averages on low levels if we know an average on a higher level) and what can be concluded about the dichotomy stability versus clustering?

Chapter 3

Discussion and Results: Lattice models

3.0 Longtime behavior (Theorem 0)

3.0.0 Preliminary remarks – Regimes of longtime behavior

This section is intended to give an overview of the longtime behavior of the lattice models. We investigate the weak limit of $\mathcal{L}\{(X_t, Y_t)\}$ as $t \to \infty$.

Corresponding to the different regimes of longterm behavior of the catalyst, we get different results for the cases $d = 1, 2$ (recurrence of the underlying random walk) and $d \geq 3$ (transience of the underlying random walk) respectively. We prove that in the low-dimensional case, the reactant is not extinct if a certain condition on the transition kernels is fulfilled; it converges weakly to a (possibly degenerated) mixture of counting measure.

As the comparison of infinite high-dimensional[1] lattice models with their torus approximations is one of the main themes of this thesis, it is important to know about the long term behavior on the infinite lattice for $d \geq 3$. We prove that for given initial intensities of reactant and catalyst (we assume that a. s. the initial state belongs to a set of "allowed" initial conditions) there exists a uniquely defined stationary measure the marginal laws of the process converge to. In particular, there is preservation of both, catalyst intensity and reactant intensity in the limit $t \to \infty$.

Remark 8

We will use the following notations[2]:

- $P^Y_{reac}, \mathbf{E}^Y_{reac}, var^Y_{reac}$ *refer to probability measure, expectation resp. variance of the reactant for given catalyst.*

- P_{cat} *denotes the law of the catalyst*

We consider $\mathcal{M}\left(\mathbb{Z}^d\right) \times \mathcal{M}\left(\mathbb{Z}^d\right)$-valued processes $(X_t, Y_t)_{t \geq 0}$, the evolution of which given by (1.6) and (1.7). The initial condition satisfies the regularity criteria formulated in 1.2.4.

[1] $d \geq 3$

[2] More details about the catalytic setting can be found in Chapter 5 ("Methods and Tools").

Theorem 0

1. $d \in \{1,2\}$

   We already know that the catalyst dies out in this case. We will show the following:

   (a) *The reactant is* locally flat *in the limit* $t \to \infty$:

   $$\forall \alpha, \beta \in \mathbb{Z}^d \forall \varepsilon > 0 : \lim_{t \to \infty} P\left[|X_t(\alpha) - X_t(\beta)| \geq \varepsilon\right] = 0 \quad . \tag{3.1}$$

   (b) *Assume that there exists* $K < \infty$ *with*

   $$\forall t \geq 0 : K a_{c_y t}(0, \cdot) \geq a_{c_x t}(0, \cdot) \quad . \tag{3.2}$$

   Then the reactant is not extinct for almost all realizations of the catalyst process: For any $\varrho \in (0, \theta_x)$

   $$\limsup_{t \to \infty} P^Y_{reac}\left[X_t(0) > \varrho\right] > 0 \quad P_{cat}\text{-a.s.}. \tag{3.3}$$

2. $d \geq 3$

   Suppose

   $$\mathcal{L}\{(X_0, Y_0)\} \in S^{(\theta_x, \theta_y)} \quad . \tag{3.4}$$

   Then the laws $\mathcal{L}\left\{(X_{t+s}, Y_{t+s})_{s \geq 0}\right\}$ $(t \geq 0)$ *converge weakly to the law* $\nu_{(\theta_x, \theta_y)}$ *of a stationary process with full intensity* (θ_x, θ_y)

   $$\mathcal{L}\left\{(X_{t+s}, Y_{t+s})_{s \geq 0}\right\} \overset{t \to \infty}{\Longrightarrow} \nu_{(\theta_x, \theta_y)} \quad . \tag{3.5}$$

3.1 Finite System Scheme (Theorem 1)

We want to achieve a better understanding of the dichotomy between the two regimes we observe when discussing lattice systems. These two regimes are specified in terms of the long-time behavior of the catalyst: There is either persistence or local extinction of the catalyst.

We aim to focus on the borderline between the two regimes; we want to cross this borderline on the way from finite torus systems – for these systems there is local extinction of the catalyst – to the infinite lattice case, for which we have persistence for $d \geq 3$. Approximating the lattice by a sequence of tori of increasing size has the advantage that as well tori as lattices can be made a group in a canonical way. This facilitates comparison.

Notation: Whenever thinking of $\{-N, ..., N-1\}^d$ equipped with the torus topology, we denote this set by $\mathbb{T}^d_N$.

Having replaced $\mathbb{Z}^d$ by a torus, we need to adjust the interaction kernel $a(\cdot, \cdot)$ to this new situation. Let

$$a^N(i,j) := \sum_{k \in (2\mathbb{Z})^d} a(i, j + kN); \qquad \left(i, j \in \mathbb{T}^d_N\right). \tag{3.6}$$

In the long run, the catalytic interacting Feller diffusion process $\left(X^N, Y^N\right)$ on the torus $\mathbb{T}^d_N$ will certainly exhibit its finite character:

- the catalyst will disappear at an a.s. finite random time.
- the reactant will converge to a possibly random multiple of counting measure.

In contrast to the intensity θ of infinite systems, the corresponding microscopic variable of finite systems – the empirical averages – fluctuate. We use the following notation for the processes of empirical averages:

$$\bar{X}_t^N := \frac{1}{|\mathbb{T}_N^d|} X_t^N\left(\mathbb{T}_N^d\right), \quad \bar{Y}_t^N := \frac{1}{|\mathbb{T}_N^d|} Y_t^N\left(\mathbb{T}_N^d\right) \tag{3.7}$$

It will turn out that the speed of these fluctuations depends on the system size. This gives rise to a rescaling approach.

3.1.1 Rescaling of catalyst averages

The process of empirical averages, $\left(\bar{Y}_t^N\right)_{t\geq 0}$ can be described itself as a Feller diffusion: There is a Brownian motion B such that

$$\bar{Y}_t^N = \bar{Y}_0^N + \int_0^t \frac{1}{\sqrt{|\mathbb{T}_N^d|}} \sqrt{2\gamma_y \bar{Y}_s^N}\, dB_s \tag{3.8}$$

holds. In the spirit of rescaling, we can write

$$\bar{Y}_{|\mathbb{T}_N^d|t}^N = \bar{Y}_0^N + \int_0^t \sqrt{2\gamma_y \bar{Y}_{|\mathbb{T}_N^d|s}^N}\, dB_s \quad . \tag{3.9}$$

As $N \to \infty$, $\bar{Y}_0^N$ a. s. converges to θ_y. Therefore, the rescaled averages of the catalyst converge for larger and larger systems to Feller's branching diffusion with original branching rate γ_y and point of initialization θ_y.

3.1.2 Idea of the Finite System Scheme

The idea of *Finite System Scheme* was introduced by Cox and Greven in 1990 [CG90]. It has been applied to a broad spectrum of interacting diffusion processes (applications to multi-type branching models: e. g. [CDG03], [Pf03]). Especially, it has been carried out for interacting Feller diffusion by Cox, Greven and Shiga in 1995 [CGS95].

Hence, all work has been done for the catalyst process. It is our aim to show that the Finite System Scheme can also be applied to the multi-type system we are discussing here. We give a short survey of the machinery of Finite System Scheme.

Macroscopic variables - rescaling of time

As already mentioned above, we have in mind a correspondence between the intensity (θ_x, θ_y) of the infinite system and the empirical intensities (averages) $\left(\bar{X}^N, \bar{Y}^N\right)$ of finite systems. We understand these variables as macroscopic variables qualitatively describing the state of the system. Choosing empirical averages as macroscopic variables, we find out that after rescaling a weak convergence property holds. The rescaling consists in replacing t by $\left|\mathbb{T}_N^d\right| t$. Note that the system size is given by

$$\left|\mathbb{T}_N^d\right| = (2N)^d \quad . \tag{3.10}$$

The laws of the rescaled processes of empirical averages,

$$\left\{\left(\bar{X}^N_{|\mathbb{T}_N^d|t}, \bar{Y}^N_{|\mathbb{T}_N^d|t}\right)_{t\geq 0}\right\}_{N\in\mathbb{N}} \tag{3.11}$$

converge weakly to the law of the zero-dimensional process. We have introduced this limit process in 2.1.1. This convergence property is the content of the first part of the theorem formulated below (in 3.1.3).

In the second part of this theorem we consider the microscopic point of view.

Microscopic variables

In the case of infinite, high-dimensional lattices we observe that the local behavior of the equilibrium is given by the intensity of the whole system. In the finite case, there is no equilibrium. Nevertheless, there is no reason not to expect that the local behavior of large but finite systems differs much from that of infinite systems as long as we consider times that are small in comparison with the size of the system.

3.1.3 Formulation of the theorem

Ingredients:

- Law $\Psi \in \mathcal{M}^1\left(\mathcal{C}\left(\mathbb{R}_+, \mathbb{R}_+ \times \mathbb{R}_+\right)\right)$ of the limit of rescaled empirical averages:

  $$\Psi := \mathcal{L}\left\{\left(\hat{X}_t, \hat{Y}_t\right)_{t\geq 0}\right\} \tag{3.12}$$

  where $\left(\hat{X}_t, \hat{Y}_t\right)_{t\geq 0}$ is given as strong solution of the system of stochastic differential equations

  $$\hat{X}_0 = \theta_x \quad , \tag{3.13}$$
  $$\hat{Y}_0 = \theta_y \quad , \tag{3.14}$$
  $$d\hat{X}_t = \sqrt{2\gamma_x \hat{X}_t \hat{Y}_t} dB_t^x \quad , \tag{3.15}$$
  $$d\hat{Y}_t = \sqrt{2\gamma_y \hat{Y}_t} dB_t^y \quad . \tag{3.16}$$

  with marginals $\Psi_t \in \mathcal{M}^1\left(\mathbb{R}_+ \times \mathbb{R}_+\right)$ $(t \geq 0)$. $(B^x, B^y$ are independent Wiener processes)

- Law of the stationary process associated with the equilibrium of the d-dimensional infinite system with intensity (θ_x, θ_y) – the existence of this law is guaranteed by Theorem 0.

  $$\nu^d_{(\theta_x,\theta_y)} \in \mathcal{M}^1\left(\mathcal{C}\left(\mathbb{R}_+, \mathcal{M}\left(\mathbb{Z}^d\right) \times \mathcal{M}\left(\mathbb{Z}^d\right)\right)\right) \tag{3.17}$$

- processes to be compared:

  We compare catalytic interacting Feller diffusion on the lattice $\mathbb{Z}^d$ with its approximations on the torus.

  initial conditions (lattice): Let (X_0, Y_0) be a random variable with values in $\mathcal{M}(\mathbb{Z}^d) \times \mathcal{M}(\mathbb{Z}^d)$, satisfying the conditions formulated in 1.2.4.

  initial conditions (torus): Let (X_0^N, Y_0^N) $(N \in \mathbb{N})$ be $\mathcal{M}(\mathbb{T}_N^d) \times \mathcal{M}(\mathbb{T}_N^d)$-valued random variables on the probability space on which (X_0, Y_0) is defined already.

  We assume:

  - *convergence of intensities:* For

  $$\left(\theta_x^N, \theta_y^N\right) := \frac{1}{(2N)^d} \sum_{i \in \mathbb{T}_N^d} \left(X_0^N(i), Y_0^N(i)\right) \tag{3.18}$$

    we demand

  $$\lim_{N \to \infty} \left(\theta_x^N, \theta_y^N\right) = (\theta_x, \theta_y) \tag{3.19}$$

    to hold almost surely.

  - *higher than second moments uniformly bounded:*[3]

  $$\exists p > 2 : \sup_{N \in \mathbb{N}} \left(\mathbf{E}\left[\left(X_0^N(0)\right)^p\right] + \mathbf{E}\left[\left(Y_0^N(0)\right)^p\right]\right) < \infty \quad . \tag{3.20}$$

  - *uniform convergence:*

  $$\lim_{t \to \infty} \mathbf{E}\left[\sup_{N \in \mathbb{N}} \left| \sum_{i \in \mathbb{T}_N^d} a_t^N(0,i) \left(X_0^N(i), Y_0^N(i)\right) - \left(\theta_x^N, \theta_y^N\right) \right| \right] = 0 \quad . \tag{3.21}$$

Remark 9 *The following counterexample is intended to make clear intuitively what we want to avoid by (3.21).*

$$X_0^N(i_1, \dots, i_d) = \prod_{n=1}^{d} \cos\left(\frac{i_n}{N}\pi\right) + 1 \quad . \tag{3.22}$$

shall not be allowed.

evolution equations: We once again recall the evolution equations (1.6) and (1.7):

$$\begin{aligned} dX_t(i) &= c_x \sum_{j \in \mathbb{Z}^d} a(i,j)\left(X_t(j) - X_t(i)\right) dt \\ &\quad + \sqrt{2\gamma_x X_t(i) Y_t(i)}\, dB_t^x(i) \quad , \end{aligned} \tag{3.23}$$

$$\begin{aligned} dY_t(i) &= c_y \sum_{j \in \mathbb{Z}^d} a(i,j)\left(Y_t(j) - Y_t(i)\right) dt \\ &\quad + \sqrt{2\gamma_y Y_t(i)}\, dB_t^y(i) \quad (i \in \mathbb{Z}^d) \quad . \end{aligned} \tag{3.24}$$

[3]Note that by assumption we have translation invariance. Therefore, the moment property has to be fulfilled for any site, not only at site "0".

and

$$\begin{aligned} dX_t^N(i) &= c_x \sum_{j \in \mathbb{T}_N^d} a^N(i,j)\left(X_t^N(j) - X_t^N(i)\right) dt \\ &\quad + \sqrt{2\gamma_x X_t^N(i)\, Y_t^N(i)}\, dB_t^x(i) \quad , \end{aligned} \tag{3.25}$$

$$\begin{aligned} dY_t^N(i) &= c_y \sum_{j \in \mathbb{T}_N^d} a^N(i,j)\left(Y_t^N(j) - Y_t^N(i)\right) dt \\ &\quad + \sqrt{2\gamma_y Y_t^N(i)}\, dB_t^y(i) \quad (i \in \mathbb{T}_N^d) \quad . \end{aligned} \tag{3.26}$$

Theorem 1 *For the processes just defined the following holds:*

1. *(Fluctuation of averages)*
   The limit behavior of rescaled averages is given by

$$\mathcal{L}\left\{\left(\bar{X}^N_{|\mathbb{T}_N^d|t}, \bar{Y}^N_{|\mathbb{T}_N^d|t}\right)_{t \geq 0}\right\} \overset{N \to \infty}{\Longrightarrow} \Psi \quad . \tag{3.27}$$

2. *(Local behavior)*
   For $s \geq 0$ we have

$$\mathcal{L}\left\{\left(X^N_{|\mathbb{T}_N^d|s+t}, Y^N_{|\mathbb{T}_N^d|s+t}\right)_{t \geq 0}\right\} \overset{N \to \infty}{\Longrightarrow} \int \nu^d_{(\tilde{\theta}_x, \tilde{\theta}_y)} \quad \Psi_s\left(d\left(\tilde{\theta}_x, \tilde{\theta}_y\right)\right) \quad . \tag{3.28}$$

Chapter 4

Discussion and Results: Processes on the hierarchical group

In this chapter, we will give results about catalytic interacting Feller diffusions on the hierarchical group. There are several reasons why it makes sense to choose the hierarchical group as site space:

- Discussing models motivated by biology, there is the following application of the hierarchical group:

  Members of a population can be addressed by the family, clan, colony etc. they belong to. Each colony being made up by a certain number of clans and each clan consisting of families with a certain number of members, it seems reasonable to use the hierarchical group as population model.

- Furthermore, there is a certain correspondance between interacting (finite variance) branching processes on the infinite hierarchical group and interacting branching processes on the two-dimensional lattice.

  This result is due to Dawson, Gorostiza and Wakolbinger [DGW01]. They describe the asymptotics of occupation time fluctuations of certain branching systems (in fact, they even discuss multi-level branching) in terms of the Green potential operator and the Green's function considering the limit $t \to \infty$.

  This way, they find a correspondance between

  - (possibly multi-level) branching systems on $\mathbb{R}^d$, where the underlying motion is an α-stable symmetric Levy-process,
  - (possibly multi-level) branching systems on the infinite hierarchical group; as underlying motion they choose the so-called c-hierarchical random walk.

  They show that the asymptotics of the Green's function depends on a parameter γ, which is given by

$$\gamma = \frac{d}{\alpha} - 1 \tag{4.1}$$

  for the case $\mathbb{R}^d$ and by

$$\gamma = \frac{\log c}{\log \frac{N}{c}} \tag{4.2}$$

for the hierarchical group case.

If Brownian motion on $\mathbb{R}^2$ is considered, one has $\gamma = 0$ (compare (4.1)). This can be approximated in the hierarchical group case by letting $N \to \infty$ in (4.2).

As already mentioned, the finite-level hierarchical group is defined by

$$\Omega_{N,L} := \left\{ (\omega_0, ..., \omega_{L-1}) \in \{0, ..., N-1\}^L \right\} \quad , L \in \mathbb{N}. \tag{4.3}$$

We want to discuss how large systems behave, where "large" is meant in the sense that we let N increase to infinity. It will turn out that the way the process fluctuates locally is governed by the average of the process taken over the directly neighboring sites. This average itself fluctuates, but on a different (size-dependent) time-scale.

In summary, one can say that low-level averages fluctuate quickly, governed successively by averages on higher levels, which fluctuate more slowly. It turns out that fluctuations of averages have to be rescaled (w.r.t. time) by factors which depend on the level of the concerned average in order to obtain non-trivial limit processes letting $N \to \infty$.

4.1 Mean Field Limit (Theorem 2)

We start with a calculation of the typical behavior on a given branch of the tree for large N. Neglecting any interaction with higher levels we can restrict the discussion to the case $L = 1$. Since the multi-level case is discussed in Theorem 3, Theorem 2 is a special case of Theorem 3. Nevertheless, the case $L = 1$ is discussed seperately in order to clarify the methods used to prove Theorem 3.

4.1.1 Notation

The site space we consider is simply the set $\{0, ..., N-1\}$ with discrete topology. We have introduced the interacting diffusion process considered on this space earlier, calling it the "no geometry"-model. We get the following system of stochastic differential equations, choosing $\left(X_t^N, Y_t^N\right)_{t \geq 0}$ as notation for the process on $\{0, ..., N-1\}$. Let

$$Z^N := \left(X^N, Y^N\right) \quad . \tag{4.4}$$

We consider the following system of stochastic differential equations:

$$\begin{aligned} dX_t^N(i) &= c_x \left(\frac{1}{N} \sum_{j=0}^{N-1} X_t^N(j) - X_t^N(i) \right) dt \\ &\quad + \sqrt{2\gamma_x X_t^N(i)\, Y_t^N(i)}\, dB_t^x(i), \end{aligned} \tag{4.5}$$

$$\begin{aligned} dY_t^N(i) &= c_y \left(\frac{1}{N} \sum_{j=0}^{N-1} Y_t^N(j) - Y_t^N(i) \right) dt \\ &\quad + \sqrt{2\gamma_y Y_t^N(i)}\, dB_t^y(i) \quad . \end{aligned} \tag{4.6}$$

4.1.2 Initial conditions

As we let N approach infinity, our state space

$$\mathcal{M}(\{0,...,N-1\}) \times \mathcal{M}(\{0,...,N-1\})$$

blows up, too. It be regarded as embedded canonically in $\mathcal{M}(\mathbb{N}_0) \times \mathcal{M}(\mathbb{N}_0)$. The latter can therefore be chosen as common state space.

We assume that the initial law of the process is a product measure: Namely, let $(Z_0(i))_{i\in\mathbb{N}_0} = (X_0(i), Y_0(i))_{i\in\mathbb{N}_0}$ be a family of iid $(\mathbb{R}_+ \times \mathbb{R}_+)$-valued random variables. In addition to that, we assume that $X_0(i)$ and $Y_0(i)$ are not correlated $(i \in \mathbb{N}_0)$.

We assume that the initial conditions are integrable

$$\mathbf{E}[X_0(i)] =: \theta_x < \infty \quad , \tag{4.7}$$
$$\mathbf{E}[Y_0(i)] =: \theta_y < \infty \quad . \tag{4.8}$$

Let

$$\theta := (\theta_x, \theta_y) \quad . \tag{4.9}$$

In addition to that, we assume that there exists $p > 2$ with

$$\mathbf{E}[(X_0(i))^p + (Y_0(i))^p] < \infty \quad . \tag{4.10}$$

Let

$$\mathcal{L}\left\{\left(Z_0^N(i)\right)_{i\in\{0,...,N-1\}}\right\} := \mathcal{L}\left\{(Z_0(i))_{i\in\{0,...,N-1\}}\right\} \quad . \tag{4.11}$$

4.1.3 Limit law for $N \to \infty$

We want to prove that the laws of the processes Z^N $(N \in \mathbb{N})$ converge weakly as N increases to infinity. It is rather obvious that the empirical averages converge:

$$\mathcal{L}\left\{\frac{1}{N}\sum_{i=0}^{N-1} X_t^N(i)\right\} \overset{N\to\infty}{\Longrightarrow} \delta_{\{\theta_x\}} \quad , \tag{4.12}$$
$$\mathcal{L}\left\{\frac{1}{N}\sum_{i=0}^{N-1} Y_t^N(i)\right\} \overset{N\to\infty}{\Longrightarrow} \delta_{\{\theta_y\}} \quad . \tag{4.13}$$

Therefore, we have a canonical candidate for the weak limit of the processes Z^N $(N \in \mathbb{N})$: Consider the $\mathcal{M}(\mathbb{N}_0) \times \mathcal{M}(\mathbb{N}_0)$-valued process $(Z_t^\infty)_{t\geq 0} = (X_t^\infty, Y_t^\infty)_{t\geq 0}$ defined by

$$X_0^\infty(i) = \theta_x, \tag{4.14}$$
$$Y_0^\infty(i) = \theta_y, \tag{4.15}$$
$$dX_t^\infty(i) = c_x(\theta_x - X_t^\infty(i))\,dt + \sqrt{2\gamma_x X_t^\infty(i)\,Y_t^\infty(i)}\,dB_t^x(i), \tag{4.16}$$
$$dY_t^\infty(i) = c_y(\theta_y - Y_t^\infty(i))\,dt + \sqrt{2\gamma_y Y_t^\infty(i)}\,dB_t^y(i) \quad , i \in \mathbb{N}_0 \tag{4.17}$$

where $\left\{(B_t^x(i), B_t^y(i))_{t\geq 0}\right\}_{i\in\mathbb{N}_0}$ is a family of independent planar Brownian motions.

The processes $\left\{(Z_t^\infty(i))_{t\geq 0}\right\}_{i\in\mathbb{N}_0}$ are independent: There is no interaction between processes living on different sites, and the initial measure is assumed to be a product measure.

In fact, $\mathcal{L}\left\{(Z_t^\infty)_{t\geq 0}\right\}$ turns out to be the weak limit of the sequence $\left(\mathcal{L}\left\{(Z_t^N)_{t\geq 0}\right\}\right)_{N\in\mathbb{N}}$. This limiting property is called McKean-Vlasov limit.

Theorem 2 *With the assumptions, definitions and notations given above, the following holds:*

$$\mathcal{L}\{Z^N\} \quad \overset{N\to\infty}{\Longrightarrow} \quad \mathcal{L}\{Z^\infty\} \quad . \tag{4.18}$$

The following is needed to formulate Theorem 3.

Notation and definitions: For $c_x = c_y = c$ and $\theta = (\theta_x, \theta_y)$ we define

$$\Psi_c^\theta := \mathcal{L}\left\{(Z_t^\infty(0))_{t\geq 0}\right\} \quad . \tag{4.19}$$

In Appendix E we prove the following: For any $c > 0, \gamma_x, \gamma_y, \theta_x, \theta_y \geq 0$ there is an unique stationary process $\hat{\Psi}_c^\theta$ with

$$\mathcal{L}\left\{(Z_{s+t}^\infty(0))_{t\geq 0}\right\} \quad \overset{s\to\infty}{\Longrightarrow} \quad \hat{\Psi}_c^\theta \quad . \tag{4.20}$$

Let Γ_c^θ denote the marginal law of $\hat{\Psi}_c^\theta$. In particular

$$\mathcal{L}\{Z_t^\infty(0)\} \quad \overset{t\to\infty}{\Longrightarrow} \quad \Gamma_c^\theta \tag{4.21}$$

holds.

Later on, we will need notations corresponding to (4.19), (4.20) and (4.21) only taking the catalyst into regard:

$$\Psi\!\!\!\!Y_c^\theta := \mathcal{L}\left\{(Y_t^\infty(0))_{t\geq 0}\right\} \quad , \tag{4.22}$$

$$\mathcal{L}\left\{(Y_{s+t}^\infty(0))_{t\geq 0}\right\} \quad \overset{s\to\infty}{\Longrightarrow} \quad \hat{\Psi\!\!\!\!Y}_c^\theta \quad , \tag{4.23}$$

$$\mathcal{L}\{Y_t^\infty(0)\} \quad \overset{t\to\infty}{\Longrightarrow} \quad \Gamma\!\!\!\!Y_c^\theta \quad . \tag{4.24}$$

4.2 Behavior on various levels – Multiple Space-Time scale analysis (Theorem 3)

In this section we want to discuss the behavior of our system with the hierarchical group $\Omega_{N,L}$ as site space. We have defined this process in 2.2.2 (p. 23) – please note that we assume that reactant and catalyst coincide with respect to deterministic heat flow since the sequence (c_k) is used for both processes. This is not a necessary assumption for the theorems we are going to formulate, as a closer inspection of the proof yields, but it simplifies notation.

We investigate two limits: we can either let N to infinity or discuss what happens if the number of levels L increases to infinity. By Theorem 2, a mean field limit holds for $L = 1$. Taking various levels into account, we have to use different timescales corresponding to the empirical intensities of different levels. What we obtain is similar to what we have seen when considering the lattice model:

The theorem we are going to prove informs about the following:

- Theorem 3 provides a limit statement about the convergence of the processes of empirical averages $\left(\bar{X}^{N,k}, \bar{Y}^{N,k}\right)$ speeded up by the factor N^k.
- Theorem 3 gives insight into the local behavior: locally, the process can be described by a mean field limit as discussed above (Theorem 2). The limit law of the intensity of a given level is governed by the empirical intensities of the next higher level. This leads to a Markov chain that describes the interdependencies of various levels.

An important part of the work has beed done already: In [DG96], Dawson and Greven investigate the catalyst process by carrying out the program of Multiple Space Time Scale Analysis.

4.2.1 Ingredients

We need the following ingredients to formulate the rescaling results we observe when considering the process on the hierarchical group. In the following, let $k, j, L \in \mathbb{N}$ with $k \le j < L$.

- *empirical averages:* For $i \in \Omega_{N,L}$, let

$$\bar{X}_t^{N,k}(i) := \frac{1}{N^k} \sum_{j:d(j,i)\le k} X_t^N(j) \tag{4.25}$$

$$\bar{Y}_t^{N,k}(i) := \frac{1}{N^k} \sum_{j:d(j,i)\le k} Y_t^N(j) \tag{4.26}$$

- *Law of catalytic one-level process Ψ_c^θ, law of stationary catalytic one-level process $\hat{\Psi}_c^\theta$, equilibrium measure Γ_c^θ, as introduced in 4.1.3* – the existence of these ingredients is verified in Appendix E.
- *interaction chain :* For $k \le j+1$ define the following measure $\mu_\theta^{j,k} \in \mathcal{M}^1(\mathbb{R}_+ \times \mathbb{R}_+)$:

$$\mu_\theta^{j,k}(\cdot) = \int_{\mathbb{R}_+\times\mathbb{R}_+} \cdots \int_{\mathbb{R}_+\times\mathbb{R}_+} \Gamma_{c_k}^{\theta_{j-k}}(\cdot) \cdots \Gamma_{c_{j-1}}^{\theta_1}(d\theta_2)\, \Gamma_{c_j}^{\theta}(d\theta_1) \quad (k \le j+1) \tag{4.27}$$

$$\mu_\theta^{j,j+1}(\cdot) = \delta_{\{\theta\}} \quad . \tag{4.28}$$

In particular

$$\mu_{\theta}^{j,j}(\cdot) = \Gamma_{c_j}^{\theta}(\cdot) \quad .$$

Notation 1 *We also need the interaction chain of the catalyst. Let*

$$\mu_{\theta_y}^{j,k}(\cdot) = \int_{\mathbb{R}_+} \cdots \int_{\mathbb{R}_+} \mathbb{F}_{c_k}^{\theta_{j-k}}(\cdot) \cdots \mathbb{F}_{c_{j-1}}^{\theta_1}(d\theta_2)\, \mathbb{F}_{c_j}^{\theta_y}(d\theta_1) \quad . \tag{4.29}$$

4.2.2 Formulation of the theorem

Theorem 3 *We consider the processes* $\left(X_t^N, Y_t^N\right)_{t\geq 0}$ *started in a product measure – catalyst and reactant not correlated – with first moments given by* $\theta_0 = (\theta_0^x, \theta_0^y)$. *In addition to that, we assume that at time zero there exists* $p > 2$ *s. t. p-th moments exist.*

Let $i \in \Omega_{N,L}$. *We claim:*

1.
$$\mathcal{L}\left\{\left(\bar{X}_{N^k t}^{N,k}(i), \bar{Y}_{N^k t}^{N,k}(i)\right)_{t\geq 0}\right\} \overset{N\to\infty}{\Longrightarrow} \Psi_{c_k}^{\theta_0} \quad . \tag{4.30}$$

2. *Let* $k \leq j$. *For* $s : \mathbb{N} \to \mathbb{R}_+$ *with* $\lim_{N\to\infty} s(N) = \infty$, $\lim_{N\to\infty} \frac{s(N)}{N} = 0$ *we have*

$$\mathcal{L}\left\{\left(\bar{X}_{N^j s(N)+N^k t}^{N,k}(i), \bar{Y}_{N^j s(N)+N^k t}^{N,k}(i)\right)_{t\geq 0}\right\} \overset{N\to\infty}{\Longrightarrow} \int \hat{\Psi}_{c_k}^{\theta^*} \;\mu_{\theta_0}^{j,k+1}(d\theta^*) \quad . \tag{4.31}$$

Chapter 5

Methods and Tools

The intention of this chapter is to collect some of the tools frequently used in the proofs. We want to provide theoretical background and examples of application. Especially, we focus on the following topics:

Topological issues

We discuss the topological concepts needed to give a notion of convergence of the considered measure valued processes. In particular, the following topics are discussed:

- The space of càdlàg paths as a metric space
  We intoduce and discuss the Skorohod metric, which enables us to formulate convergence of càdlàg paths.
- The space of probability measures as a metric space
  We give a definition of the Prohorov metric. In general, this metric describes weak convergence of probability measures on metric spaces. In our case, the metric space is the space of càdlàg paths occuring as realizations of the stochastic processes we discuss. The Prohorov metric gives us a quantitative notion of convergence of laws of stochastic processes.
- Skorohod representation
  We give a formulation of the Skorohod representation of a sequence of weakly converging laws of paths.

Catalytic setting

The processes we consider can be thought of as realizations of a two-step experiment: The reactant can be constructed for given catalyst, which serves as random medium. With the quenched approach we will use a concept from the theory of stochastic processes in random media allowing us to simplify the analysis by using the fact that the interaction between the two types of particles is one-way.

Duality and random occupation times

In many cases, our proofs are based on a dual approach. For this reason, we give a description of the concept of duality. We mention some examples of application. Applying the concept of duality to the processes discussed here we explain how a generalization of random occupation times of the catalyst respectively the reactant for given catalyst can be characterized by means of duality. What makes random occupation times important for us is that we can use them to describe the laws of finite dimensional distributions of the processes.

Considering the dual picture, a certain type of integro-differential equation occurs. We discuss the properties of their solutions.

Clustering and Kallenberg criterion

In this part, we give some theoretical background about the formation of clusters for infinitely divisible processes, such as branching processes. The Kallenberg criterion allows us to decide whether there is persistence or local extinction.

5.1 Topological issues

Naturally, a discussion of convergence properties of stochastic processes - in fact, we do nothing else - is a topological problem. Therefore we give a short survey on the topic of convergence on the space of multi-dimensional càdlàg processes. We discuss processes with state space embedded in $\mathbb{R}_+^\Lambda$ $[= \mathcal{M}(\Lambda)]$, where Λ is assumed to be countable.

We introduce two notions of distance:

- Distance of paths:
  The *Skorohod* metric d_{Sk} measures the distance of càdlàg paths.

- Distance of probability measures on the space of paths:
  The *Prohorov* metric ϱ describes weak convergence of processes.

5.1.1 Distance of paths

Though we will only consider processes with continuous paths, we provide the theory for a wider class of processes, namely processes with càdlàg paths. The motivation to use this general setting is that we do not know a priori that the limit points of continuous processes are continuous – in fact they are in the cases we consider here.

First, we discuss $\mathbb{R}_+$-valued càdlàg processes with state space $\mathcal{D}(\mathbb{R}_+, \mathbb{R}_+) \subsetneq \mathbb{R}_+^{\mathbb{R}_+}$.

The space $\mathcal{D}(\mathbb{R}_+, \mathbb{R}_+)$ becomes a metric space if we equip it with the so-called Skorohod metric d_{Sk}. Before we provide the definition of the Skorohod metric, we want to mention two properties of it in order to motivate the way it is defined.

1. Any sequence of continuous paths which is converging with respect to the topology introduced by compact convergence[1] shall be convergent with repect to d_{Sk}.

2. Consider the following sequence $\left\{(f_n(t))_{t\geq 0}\right\}_{n\in\mathbb{N}}$ of pure jump paths

$$f_n(t) = \sum_{k=1}^{\infty} \mathbb{1}_{[t_{n,k},\infty)}(t) \quad . \tag{5.2}$$

For all $n \in \mathbb{N}$, f_n is increasing with sequence of discontinuities $(t_{n,k})_{k\in\mathbb{N}}$. The sequence (f_n) converges (with respect to the topology induced by d_{Sk}) if there exists a sequence $(t_{\infty,k})_{k\in\mathbb{N}}$ with

$$\forall k \in \mathbb{N} : \lim_{n\to\infty} t_{n,k} = t_{\infty,k} \quad . \tag{5.3}$$

We hope that the latter helps to understand why in the definition of Skorohod metric there has to occur the notion of time-change.

---

[1] A sequence $\left((f_t^n)_{t\in\mathbb{R}_+}\right)_{n\in\mathbb{N}} \in \left(\mathbb{R}_+^{\mathbb{R}_+}\right)^{\mathbb{N}}$ is said to converge compactly with limit $\left((f_t^\infty)_{t\in\mathbb{R}_+}\right)_{n\in\mathbb{N}}$ if for each compact set $K \subset \mathbb{R}_+$ the following is true:

$$\lim_{n\to\infty} \sup_{t\in K} |f_t^n - f_t^\infty| = 0 \quad . \tag{5.1}$$

Definition of the Skorohod metric d_{Sk} on $\mathcal{D}(\mathbb{R}_+, \mathbb{R}_+)$

We need the following ingredients:

- Metric on $\mathbb{R}_+$:
  Let d be a bounded metric on $\mathbb{R}_+$ with $d \leq 1$.
- Notion of time change:
  Let Λ' be the collection of strictly increasing functions mapping $\mathbb{R}_+$ onto itself. Each such $\ell \in \Lambda'$ defines a measure $\tilde{\lambda} \in \mathcal{M}(\mathbb{R}_+)$ by

  $$\tilde{\lambda}_\ell([0,a]) = \ell(a)$$

  Let Λ consist of those $\ell \in \Lambda'$ for which the measures $\tilde{\lambda}_\ell$ and λ (Lebesgue-measure) are equivalent with bounded density functions, namely

  $$\ell \in \Lambda \quad \Longleftrightarrow \quad \gamma(\ell) := \sup_{t \geq 0} \left(\frac{d\lambda}{d\tilde{\lambda}_\ell}(t) \vee \frac{d\tilde{\lambda}_\ell}{d\lambda}(t) \right) < \infty \quad . \tag{5.4}$$

Definition 3 *For two paths $X, \tilde{X} \in \mathcal{D}(\mathbb{R}_+, \mathbb{R}_+)$ the Skorohod metric d_{Sk} is defined by*

$$d_{Sk}\left(X, \tilde{X}\right) := \inf_{\ell \in \Lambda} \left[\gamma(\ell) \vee \int_0^\infty e^{-u} d^*\left(X, \tilde{X}, \ell, u\right) du \right], \tag{5.5}$$

where

$$d^*\left(X, \tilde{X}, \ell, u\right) := \sup_{t \geq 0} d\left(X_{t \wedge u}, \tilde{X}_{\ell(t) \wedge u}\right) \quad . \tag{5.6}$$

This way the Skorohod distance quantifies the effort necessary to transform X into $\tilde{X}$ by deformation of X and change of time.

Remark 10 *Note that the definition of Skorohod metric as a metric on $\mathcal{D}(\mathbb{R}_+, \mathbb{R}_+)$ is equivalent to the definition given in [EK86] (Section III.5). We have modified the notation given there in order to enlighten the concept of time change.*

Remark 11 *By [EK86], Theorem 5.5.6, the pair $(\mathcal{D}(\mathbb{R}_+, \mathbb{R}_+), d_{Sk})$ is a complete seperable metric space, hence a Polish space.*

Definition of the Skorohod metric on $(\mathcal{D}(\mathbb{R}_+, \mathbb{R}_+))^\Lambda$ (Λ countable)

The definition of the Skorohod metric can be canonically extended to the countable dimensional case where the set of paths is $(\mathcal{D}(\mathbb{R}_+, \mathbb{R}_+))^\Lambda$.

Definition 4 *Let $\left((X_t(i))_{t \geq 0}\right)_{i \in \Lambda}, \left(\left(\tilde{X}_t(i)\right)_{t \geq 0}\right)_{i \in \Lambda} \in (\mathcal{D}(\mathbb{R}_+, \mathbb{R}_+))^\Lambda$. Let $\beth : \Lambda \to \mathbb{N}$ denote an injective[2] mapping from Λ to $\mathbb{N}$. The* Skorohod distance *between the two paths is*

[2] if Λ is an infinite set, it is natural to choose a bijective mapping

then defined by

$$d_{Sk}^{\beth}\left(\left((X_t(i))_{t\geq 0}\right)_{i\in\Lambda},\left(\left(\tilde{X}_t(i)\right)_{t\geq 0}\right)_{i\in\Lambda}\right)$$
$$:=\sum_{i\in\Lambda}\frac{1}{2^{\beth(i)}}d_{Sk}\left((X_t(i))_{t\geq 0},\left(\tilde{X}_t(i)\right)_{t\geq 0}\right) \quad . \tag{5.7}$$

Remark 12

1. *Note that $d_{Sk}^{\beth}$ are equivalent for all injective $\beth\in\mathbb{N}^{\Lambda}$. We hence drop the superscript $\beth$.*

2. *Again it turns out that $\left((\mathcal{D}(\mathbb{R}_+,\mathbb{R}_+))^{\Lambda}, d_{Sk}\right)$ is a Polish space.*

5.1.2 Distance of probability measures

Throughout this thesis, we consider probability measures on Polish spaces. It is a quite well-known fact (compare [EK86], Theorem 3.1.7) that $\mathcal{M}^1(S)$ can be made a Polish space $(\mathcal{M}^1(S), \varrho)$ if (S,d) is Polish. The corresponding metric ϱ is called *Prohorov metric.* It is defined as follows: (taken from [EK86], III.1)

Definition 5 *(Prohorov metric)*
Let (S,d) be a metric space and $P,Q\in\mathcal{M}^1(S)$. The Prohorov metric ϱ is defined by

$$\varrho(P,Q)=\inf\{\varepsilon>0: P(F)\leq Q(F^{\varepsilon})+\varepsilon \quad \forall F\in\mathfrak{C}\} \quad , \tag{5.8}$$

where

- *$\mathfrak{C}$ denotes the set of all closed subsets of S.*
- *F^{ε} is defined by*
$$F^{\varepsilon}:=\{x\in S: d(x,F)<\varepsilon\} \quad . \tag{5.9}$$

By [EK86], Corollary 3.1.6., ϱ induces the weak topology on $\mathcal{M}^1(S)$.

5.1.3 Skorohod representation

We will make frequent use of the following theorem (taken from [EK86], Theorem 3.1.8.)

Theorem *Let (S,d) be separable. Suppose $P_n, n=1,2,\dots$, and P in $\mathcal{M}^1(S)$ satisfy*

$$\lim_{n\to\infty}\varrho(P_n,P)=0 \tag{5.10}$$

[this is equivalent to $P_n \overset{n\to\infty}{\Longrightarrow} P$]. *Then there exists a probability space $(\Omega,\mathcal{F},\nu)$ on which are defined S-valued random variables $X_n, n=1,2,\dots$, and X with distributions $P_n, n=1,2,\dots$, and P, respectively, such that*

$$\lim_{n\to\infty}X_n=X \quad a.s. \tag{5.11}$$

5.2 Catalytic setting

The process we consider is catalytic: it can be decomposed into two parts; between these parts there is a one-way interaction: one of the parts evolves autonomously, the other part is influenced by the first part. This fact is reflected by the evolution equations (1.6) and (1.7):

- Catalyst (random medium)

  The catalyst $(Y_t)_{t\geq 0}$ is given by its initial law $\mathcal{L}\{Y_0\}$ and the system of stochastic differential equations

$$dY_t(i) = c_y \sum_{j\in\Lambda} a(i,j)\left[Y_t(j) - Y_t(i)\right] dt + \sqrt{2\gamma_y Y_t(i)}\, dB_t^y(i) \quad (i \in \Lambda). \tag{5.12}$$

- Reactant

The reactant's evolution is given by

$$\begin{aligned} dX_t(i) = {} & c_x \sum_{j\in\Lambda} a(i,j)\left[X_t(j) - X_t(i)\right] dt \\ & + \sqrt{2\gamma_x X_t(i)\, Y_t(i)}\, dB_t^x(i) \quad (i \in \Lambda). \end{aligned} \tag{5.13}$$

Obviously, the process $(X_t, Y_t)_{t\geq 0}$ can be sampled by means of a two-step experiment. Therefore we have the following inclusion of the sigmaalgebras generated by the processes

$$\sigma(Y) \subset \sigma((X,Y)) \quad . \tag{5.14}$$

This fact is reflected by the possibility to decompose the probability measure P (the probability space the process lives on is denoted by $(\Omega, \mathcal{A}, P)$) according to the catalytic setting.

The catalyst evolves autonomously. Its law, namely

$$P_{cat} := \mathcal{L}\left\{(Y_t)_{t\geq 0}\right\} \tag{5.15}$$

is the starting point of our construction. By (5.13) and (5.14), the reactant can be constructed conditioned on the catalyst. We choose

$$P_{reac}^Y := \mathcal{L}\left\{(X_t, Y_t)_{t\geq 0} \middle| (Y_t)_{t\geq 0}\right\} \tag{5.16}$$

as notation of the law of the process that can be constructed for given catalyst Y by as strong solution of the stochastic differential equations (1.6) resp. (1.7). Since the mapping $Y \mapsto P_{reac}^Y$ is measurable[3], it is possible to decompose P as follows:

$$P[\cdot] = \int P_{reac}^Y[\cdot]\, P_{cat}(dY) \quad . \tag{5.17}$$

In many cases, we do not consider only one catalyst-reactant process, but several of them defined on a common probability space. Then, (5.15) becomes

$$P_{cat} := \mathcal{L}\left\{\left((Y_t^n)_{t\geq 0}\right)_{n\in\mathbb{N}}\right\} \quad . \tag{5.18}$$

[3]Measurability can be proven similar to Lemma 6, p. 109.

Similarly we have to replace (5.16) by

$$P^Y_{reac} := \mathcal{L}\left\{\left((X^n_t, Y^n_t)_{t\geq 0}\right)_{n\in\mathbb{N}} \middle| \left((Y^n_t)_{t\geq 0}\right)_{n\in\mathbb{N}}\right\} \quad . \tag{5.19}$$

Again, we have

$$P[\cdot] = \int P^Y_{reac}[\cdot]\, P_{cat}(dY) \quad , \tag{5.20}$$

where Y abbreviates the vector of processes $\left(Y^1, Y^2, \ldots\right)$.

Typical line of argumentation

We often have to verify that a sequence of catalyst/reactant processes converges weakly. Typically, this looks as follows

$$\mathcal{L}\left\{(X^n_t, Y^n_t)_{t\geq 0}\right\} \overset{n\to\infty}{\Longrightarrow} \mathcal{L}\left\{(X^\infty_t, Y^\infty_t)_{t\geq 0}\right\} \quad . \tag{5.21}$$

In all cases that we consider we already know that the catalyst converges weakly

$$\mathcal{L}\left\{(Y^n_t)_{t\geq 0}\right\} \overset{n\to\infty}{\Longrightarrow} \mathcal{L}\left\{(Y^\infty_t)_{t\geq 0}\right\} \quad . \tag{5.22}$$

The proof of a statement like (5.21) contains two parts:

1. By Skorohod's representation (compare page 41) we may assume[4]

$$\lim_{n\to\infty} d_{Sk}\left((Y^n_t)_{t\geq 0}, (Y^\infty_t)_{t\geq 0}\right) = 0 \quad Y\text{-a.s.} \tag{5.23}$$

   For this purpose, the probability space we work with has to be modified. At this point, it is important that the catalyst-reactant structure is not destroyed. Therefore it remains possible to apply (5.17).

2. We have $\mathcal{L}\left\{(X^n_t, Y^n_t)_{t\geq 0}\right\} \in \mathcal{M}^1\left(\mathcal{C}\left(\mathbb{R}_+, \mathcal{M}(\Lambda) \times \mathcal{M}(\Lambda)\right)\right)$. This space is a Polish space equipped with the Prohorov metric ϱ (defined on page 41). The issue of metrizability has been discussed when considering the topological background. With the metric ϱ, we can reformulate the claim (5.21) as follows:

$$\lim_{n\to\infty} \varrho\left(\mathcal{L}\left\{(X^n_t, Y^n_t)_{t\geq 0}\right\}, \mathcal{L}\left\{(X^\infty_t, Y^\infty_t)_{t\geq 0}\right\}\right) = 0 \quad . \tag{5.24}$$

   Using the decomposition (5.17), we can equivalently write:

$$\lim_{n\to\infty} \varrho\left(\int \mathcal{L}^{reac}_Y\left\{(X^n_t, Y^n_t)_{t\geq 0}\right\} P_{cat}(dY) \quad , \right.$$
$$\left. \int \mathcal{L}^{reac}_Y\left\{(X^\infty_t, Y^\infty_t)_{t\geq 0}\right\} P_{cat}(dY)\right) = 0 \quad . \tag{5.25}$$

   This can be bounded above, as the following lemma shows:

[4] Recall that Y abbreviates the vector of catalyst processes $(Y^1, Y^2, \ldots)$.

Lemma 2 *In the considered situation, the following holds:*
If

$$\lim_{n\to\infty}\int \varrho\left(\mathcal{L}_Y^{reac}\left\{(X_t^n,Y_t^n)_{t\geq 0}\right\}\ ,\ \mathcal{L}_Y^{reac}\left\{(X_t^\infty,Y_t^\infty)_{t\geq 0}\right\}\right) P_{cat}(dY)=0 \quad . \tag{5.26}$$

then the convergence (5.25) is true as well, namely

$$\lim_{n\to\infty}\varrho\left(\int \mathcal{L}_Y^{reac}\left\{(X_t^n,Y_t^n)_{t\geq 0}\right\} P_{cat}(dY)\quad ,\right.$$
$$\left.\int \mathcal{L}_Y^{reac}\left\{(X_t^\infty,Y_t^\infty)_{t\geq 0}\right\} P_{cat}(dY)\right)=0 \quad . \tag{5.27}$$

Proof. First of all, recall the definition of the Prohorov metric (see Definition 5, p. 41). Let S be a Polish space and $P,Q\in\mathcal{M}^1(S)$ .Then the distance of P and Q with respect to the Prohorov metric is given by

$$\varrho(P,Q)=\inf\{\varepsilon>0: P(F)\leq Q(F^\varepsilon)+\varepsilon \quad \forall F\in\mathfrak{C}\} \tag{5.28}$$

where

- $\mathfrak{C}$ denotes the set of compact sets in S
- F^ε is defined as the open ε-environment of F :

$$F^\varepsilon := \{x\in S: d(F,x)<\varepsilon\} \quad . \tag{5.29}$$

In our case, we have $S=\mathcal{M}^1(\mathcal{C}(\mathbb{R}_+,\mathcal{M}(\Lambda)\times\mathcal{M}(\Lambda)))$, which is a Polish space, as we have explained when discussing the topogical framework.
For $F\in\mathfrak{C}$, let

$$\varepsilon_n^{F,Y} := \inf\{\varepsilon>0:$$
$$\left[\mathcal{L}_Y^{reac}\left\{(X_t^n,Y_t^n)_{t\geq 0}\right\}\right](F)\leq\left[\mathcal{L}_Y^{reac}\left\{(X_t^\infty,Y_t^\infty)_{t\geq 0}\right\}\right](F^\varepsilon)+\varepsilon\right\} \quad . \tag{5.30}$$

We know from (5.26) that $\sup_{F\in\mathfrak{C}}\varepsilon_n^{F,Y}$ converges to zero P_{cat}-stochastically. In particular, there exists an increasing sequence $(n_m)_{m\in\mathbb{N}}$ with $\lim_{m\to\infty} n_m=\infty$ such that

$$\forall m\in\mathbb{N},\forall n\geq n_m: P_{cat}\left[\varepsilon_n^{F,Y}>\frac{1}{m}\right]<\frac{1}{m} \quad . \tag{5.31}$$

To show (5.27) we prove the following:

$$\forall n\geq n_m\forall F\in\mathfrak{C}:$$

$$\int\left[\mathcal{L}_Y^{reac}\left\{(X_t^n,Y_t^n)_{t\geq 0}\right\}\right](F)\ P_{cat}(dY)$$
$$\leq\int\left[\mathcal{L}_Y^{reac}\left\{(X_t^\infty,Y_t^\infty)_{t\geq 0}\right\}\right]\left(F^{\frac{2}{m}}\right) P_{cat}(dY)+\frac{2}{m} \quad . \tag{5.32}$$

This is verified as follows: With (5.30) we have

$$
\begin{aligned}
&\int \left[\mathcal{L}_Y^{reac}\left\{(X_t^n, Y_t^n)_{t\geq 0}\right\}\right](F)\ P_{cat}(dY) \\
&\leq \int \left(\left[\mathcal{L}_Y^{reac}\left\{(X_t^\infty, Y_t^\infty)_{t\geq 0}\right\}\right]\left(F^{\varepsilon_n^{F,Y}}\right) + \varepsilon_n^{F,Y}\right) P_{cat}(dY) \quad .
\end{aligned}
$$

Splitting up the integral considering the twoevents

$$
\left[\varepsilon_n^{F,Y} > \frac{1}{m}\right], \ \left[\varepsilon_n^{F,Y} \leq \frac{1}{m}\right]
$$

seperately, we get

$$
\begin{aligned}
&\leq \int \left(\left[\mathcal{L}_Y^{reac}\left\{(X_t^\infty, Y_t^\infty)_{t\geq 0}\right\}\right]\left(F^{\frac{1}{m}}\right) + \frac{1}{m}\right) P_{cat}(dY) \\
&\quad + P_{cat}\left[\varepsilon_n^{F,Y} > \frac{1}{m}\right] \\
&\leq \int \left[\mathcal{L}_Y^{reac}\left\{(X_t^\infty, Y_t^\infty)_{t\geq 0}\right\}\right]\left(F^{\frac{1}{m}}\right) P_{cat}(dY) + \frac{1}{m} + \frac{1}{m} \\
&\leq \int \left[\mathcal{L}_Y^{reac}\left\{(X_t^\infty, Y_t^\infty)_{t\geq 0}\right\}\right]\left(F^{\frac{2}{m}}\right) P_{cat}(dY) + \frac{2}{m} \quad . \qquad (5.33)
\end{aligned}
$$

This yields (still assuming $n \geq n_m$):

$$
\begin{aligned}
\varrho\Bigg(&\int \mathcal{L}_Y^{reac}\left\{(X_t^n, Y_t^n)_{t\geq 0}\right\} P_{cat}(dY) \quad , \\
&\int \mathcal{L}_Y^{reac}\left\{(X_t^\infty, Y_t^\infty)_{t\geq 0}\right\} P_{cat}(dY)\Bigg) \leq \frac{2}{m} \quad , \qquad (5.34)
\end{aligned}
$$

completing the proof of the lemma. ■

This lemma allows us to prove convergence of the considered catalyst-reactant system in two steps:

1. convergence of the catalyst
2. convergence of the catalyst/reactant processes for given catalyst.

5.3 Duality and random occupation times

In our proofs, we frequently apply a dual point of view to characterize the processes we consider. This approach allows us to describe – and to compare – finite dimensional distributions.

At first, we mention some well-known examples of duality before formulating what we need for our investigations.

5.3.1 Some words about duality

First of all, let us describe heuristically what we mean by duality:[5]

The crucial point about duality is that we change our point of view. We have in mind the following picture: The considered stochastic process – say $(\Xi_t)_{t\geq 0}$ – evolves in time. The state Ξ_t of the process at time t can be observed using a measuring device with output $\Phi\left(\langle f, \Xi_t\rangle\right)$, with $f \in \mathcal{F}$, where $\mathcal{F}$ is some non-empty set of functions on the state space of the process and Φ is a function on $\mathbb{R}$. The output of the measuring device is hence a function Φ of a linear functional $\langle f, \cdot\rangle$ of the state of the system at time t. In a typical situation of application, $\mathcal{F}$ is chosen such that the distribution of Ξ_t can be characterized uniquely by the family $\{\mathbf{E}\left[\Phi\left(\langle f, \Xi_t\rangle\right)\right]\}_{f\in\mathcal{F}}$.

In the dual picture, one changes the measuring device instead of letting the process evolve with time: the focus lies on finding a process $(u_t)_{t\geq 0}$ with

$$\mathbf{E}\left[\Phi\left(\langle u_t, \Xi_0\rangle\right)\right] = \mathbf{E}\left[\Phi\left(\langle f, \Xi_t\rangle\right)\right] \quad . \tag{5.35}$$

Remark 13 *More details about the dual point of view can be found in [E00], Theorem 1.23., p. 16.*

5.3.2 Examples

- Dawson-Watanabe process

  From interacting Feller diffusions one gets a non-trivial limit by a time-space-mass rescaling procedure. The process arising as limit has values in $\mathcal{M}\left(\mathbb{R}^d\right)$. It is called Dawson-Watanabe process - in the following denoted by Ξ. It can be characterized by its log-Laplace equation

  $$\mathbf{E}\left[\exp\left(-\langle f, \Xi_t\rangle\right)\right] = \mathbf{E}\left[\exp\left(-u_t, \Xi_0\right)\right] \tag{5.36}$$

  with

  $$u_0 = f \in \mathcal{C}_b\left(\mathbb{R}^d, \mathbb{R}_+\right) \tag{5.37}$$

  $$\frac{\partial}{\partial t} u_t = \frac{1}{2}\Delta u_t - \gamma u_t^2 \quad . \tag{5.38}$$

[5] We only want to make clear what we mean when speaking of the dual point of view. A precise definition of duality is not needed for the proofs, therefore we feel free to omit some details.

- Duality of Voter Model and Coalescent

  By considering the graphical representation of the voter model, it leaps to the eye that there is duality between the voter model and a coalescing process (roughly speaking, the branches of the ancestral tree coalesce when followed back in time). Details on this topic can be found in [Lig85], Section III.4.

- Self-duality of Mutually Catalytic Branching diffusion

  A process which is closely related to the interacting catalytic branching diffusions discussed in this thesis is the mutually catalytic branching diffusion (MCBD). This process is a two-type process as well, but with symmetric interaction between the types. Its state-space $\mathfrak{X}$ is a subspace of $(\mathbb{R}_+ \times \mathbb{R}_+)^{\mathbb{Z}^d}$ satisfying a sub-exponential growth condition. Further details can be found in [K00] or [DP98], for example. The evolution of the MCBD

$$\left\{(\mathfrak{Z}_t(i))_{i\in\mathbb{Z}^d}\right\}_{t\geq 0} = \left\{(V_t(i), W_t(i))_{i\in\mathbb{Z}^d}\right\}_{t\geq 0}$$

  is given by the following pair of stochastic differential equations:

$$dV_t(i) = \sum_{j\in\mathbb{Z}^d} a(i,j)\left[V_t(j) - V_t(i)\right] dt + \sqrt{2\gamma V_t(i) W_t(i)} dB_t^V(i) \quad , \tag{5.39}$$

$$dW_t(i) = \sum_{j\in\mathbb{Z}^d} a(i,j)\left[W_t(j) - W_t(i)\right] dt + \sqrt{2\gamma V_t(i) W_t(i)} dB_t^W(i) \quad , \tag{5.40}$$

  where $\left(B^V(i), B^W(i)\right)_{i\in\mathbb{Z}^d}$ is assumed to be an independent family of planar Brownian motions. As usual, $(a(i,j))_{i,j\in\mathbb{Z}^d}$ denotes a non-degenerate, symmetric, translation invariant stochastic matrix.

  Let

$$\left\{(\mathfrak{H}_t(i))_{i\in\mathbb{Z}^d}\right\}_{t\geq 0}$$

  be another version of MCBD defined on the same probability space.

  Let f be defined by

$$f: \begin{array}{ccc} (\mathbb{R}_+^d)^2 \times (\mathbb{R}_+^d)^2 & \to & \mathbb{C} \\ ((x_1,x_2),(y_1,y_2)) & \mapsto & \exp\left(-\langle x_1+x_2, y_1+y_2\rangle + i\langle x_1-x_2, y_1-y_2\rangle\right) \end{array} . \tag{5.41}$$

  Then we have the following duality equation

$$\mathbf{E}\left[f(\mathfrak{H}_0, \mathfrak{Z}_t)\right] = \mathbf{E}\left[f(\mathfrak{H}_t, \mathfrak{Z}_0)\right] \quad . \tag{5.42}$$

  This duality goes back to Mytnik [Myt96].

5.3.3 Proposition 1: Dual representation of generalized random occupation times of catalytic interacting Feller diffusion

Next, we line out how the concept of duality can be applied to the processes considered here. We are interested to describe the finite dimensional distributions of the reactant process for given reactant. By the Cramér-Wald device, the considered finite-dimensional distributions can be uniquely characterized by distributions of random sums of the following shape

$$\sum_{k=1}^{n} \langle \psi_n, X_{t_n} \rangle \quad , \tag{5.43}$$

where $(\psi_n)_{n\in\mathbb{N}} \in (\mathbb{R}_+)^{\mathcal{M}(\Lambda)}$ is a sequence of test functions and $t_1, ..., t_n$ a sequence of times.[6]

In generalization of this framework, we will provide a proposition on the topic of characterizing generalized random occupation times. This characterization makes extensive use of the dual picture.

It should be mentioned that the considered problem turns out to be closely related to the investigation of random occupation times carried out by Iscoe [Isc86].

Ingredients

First, we collect the ingredients we need to formulate the Proposition.

Let $\varphi : \begin{array}{ccc} \mathbb{R}_+ \times \mathbb{Z}^d & \longrightarrow & \mathbb{R}_+ \\ (t,i) & \longmapsto & \varphi_t(i) \end{array}$ have the following properties:

- φ is bounded
- for all $i \in \mathbb{Z}^d$, let $t \longmapsto \varphi_t(i)$ be Borel-measurable.
- integrability: for all $t > 0$, assume

$$\int_0^t \sum_{i\in\mathbb{Z}^d} \varphi_s(i)\, ds < \infty \quad . \tag{5.44}$$

Let $n \in \mathbb{N}$.
Consider $(t_m)_{m=1,...,n} \in \mathbb{R}_+^{\mathbb{N}}$ strictly increasing with $t_1 > 0$.
Let $t_0 := 0$.
For $m \in \{0, ..., n\}$, let $\psi_m : \begin{array}{ccc} \mathbb{R}_+ & \longrightarrow & \mathbb{R}_+ \\ i & \longmapsto & \psi_m(i) \end{array}$ be bounded.
Given φ and (ψ_m), define

$$S_t(i) := \int_0^t \varphi_s(i)\, ds + \sum_{m=0}^{n} \mathbb{1}_{\{t\geq t_m\}}(t)\, \psi_m(i) \quad \left(i \in \mathbb{Z}^d\right). \tag{5.45}$$

Let $\mathfrak{S}$ denote the set of all $(S_t)_{t\geq 0}$ definable by (5.45) for $\varphi, (\psi_m)$ satisfying the mentioned boundedness conditions.

[6] Notation is made precise below.

Now we are ready to formulate Proposition 1.

Let $S \in \mathfrak{S}$.
Let (X, Y) be interacting catalytic Feller diffusion on $\mathbb{Z}^d$ (compare (1.6),(1.7)).
Let $T > 0$.
Let $(u_t^{rc})_{t\in[0,T]} \in \left(\mathbb{R}_+^{\mathbb{Z}^d}\right)^{[0,T]}$ be defined by

$$u_0^{rc} \equiv 0 \quad . \tag{5.46}$$

$$\begin{aligned} du_t^{rc}(i) = {} & c_x \sum_{j\in\mathbb{Z}^d} a(j,i)\left[u_t^{rc}(j) - u_t^{rc}(i)\right] dt \\ & - \gamma_x Y_{T-t}(i)\left(u_t^{rc}(i)\right)^2 \\ & + dS_t(i) \quad , \left(i \in \mathbb{Z}^d\right). \end{aligned} \tag{5.47}$$

By construction, $(u_t^{rc})_{t\in[0,T]}$ is càdlàg. Let $(u_t)_{t\in[0,T]}$ be the uniquely defined process being "càglàd"[a] and coinciding with $(u_t^{rc})_{t\in[0,T]}$ at its points of continuity.

Proposition 1 *We claim that for almost all Y the following holds:*

$$\mathbf{E}_{reac}^{Y}\left[\exp\left(-\sum_{i\in\mathbb{Z}^d}\int_0^T X_t(i)\, dS_t(i)\right)\right] = \mathbf{E}_{reac}^{Y}\left[\exp\left(-\langle u_T, X_0\rangle\right)\right] \quad . \tag{5.48}$$

[a]limit exists from the right, continous from the left

Proof. We introduce an additional state "Σ". Let $(X_t(\Sigma))_{t\in[0,T]}$ be defined by

$$X_t(\Sigma) := \sum_{i\in\mathbb{Z}^d}\int_0^T X_t(i)\, dS_t(i) \quad . \tag{5.49}$$

Our strategy is to prove a duality equation (namely (5.53)) implying (5.48).
Consider $(\check{u}_t(i))_{t\in[0,T]}$ $\left(i \in \mathbb{Z}^d \cup \{\Sigma\}\right)$ defined by

- initial condition:
$$\check{u}_0(i) = \begin{cases} 0 & \text{for} \quad i \in \mathbb{Z}^d \\ 1 & \text{for} \quad i = \Sigma \end{cases} \tag{5.50}$$
- evolution equation:
  - for $i \in \mathbb{Z}^d$:
$$\begin{aligned} d\check{u}_t(i) = {} & c_x \sum_{j\in\mathbb{Z}^d} a(j,i)\left[\check{u}_t(j) - \check{u}_t(i)\right] dt \\ & - \gamma_x Y_{T-t}(i)\left(\check{u}_t(i)\right)^2 dt \\ & + dS_{T-t}(i) \quad . \end{aligned} \tag{5.51}$$

– for $i = \Sigma$:

$$\forall t \in [0,T] : \breve{u}_t(\Sigma) = 1 \quad . \tag{5.52}$$

By construction, $(\breve{u}_t)_{t\geq 0}$ is càdlàg. To this càdlàg process there corresponds a "càglàd" process $(\tilde{u}_t)_{t\geq 0}$ only differing from $\breve{u}$ at the finite number of times at which $(\breve{u}_t)_{t\geq 0}$ is not continuous.

Notation 2 *We define*

$$\langle a,b\rangle_\Sigma := \sum_{i\in\mathbb{Z}^d\cup\{\Sigma\}} a(i)\cdot b(i)$$

We claim that the following holds for $t\in[0,T]$

$$\mathbf{E}^Y_{reac}\left[\exp\left(-\langle\tilde{u}_t, X_{T-t}\rangle_\Sigma\right)\right] = const. \tag{5.53}$$

In particular, this yields

$$\begin{aligned}\mathbf{E}^Y_{reac}\left[\exp\left(-\langle u_T, X_0\rangle_\Sigma\right)\right] &= \mathbf{E}^Y_{reac}\left[\exp\left(-\langle\tilde{u}_T, X_0\rangle_\Sigma\right)\right]\\ &= \mathbf{E}^Y_{reac}\left[\exp\left(-\langle\tilde{u}_0, X_T\rangle_\Sigma\right)\right]\\ &= \mathbf{E}^Y_{reac}\left[\exp\left(-\sum_{i\in\mathbb{Z}^d}\int_0^T X_s(i)dS_s(i)\right)\right]\end{aligned} \tag{5.54}$$

Hence (5.53) implies (5.48).

We prove (5.53) by differentiating: (since we have assumed (5.44), we can interchange summation and differentiation)

$$\begin{aligned}d\exp\left(-\langle\tilde{u}_{T-t},X_t\rangle_\Sigma\right) &= -\exp\left(-\langle\tilde{u}_{T-t},X_t\rangle_\Sigma\right)\cdot\left(\langle d\tilde{u}_{T-t},X_t\rangle_\Sigma + \langle\tilde{u}_{T-t},dX_t\rangle_\Sigma\right)\\ &\quad+\frac{1}{2}\exp\left(-\langle\tilde{u}_{T-t},X_t\rangle_\Sigma\right)\cdot\left(d\langle\tilde{u}_{T-t},X_t\rangle_\Sigma\right)^2\end{aligned}$$

$$\begin{aligned}&=\exp\left(-\langle\tilde{u}_{T-t},X_t\rangle_\Sigma\right)\\ &\quad\cdot\Bigg[-(-1)\Bigg(\sum_{i\in\mathbb{Z}^d}c_x\sum_{j\in\mathbb{Z}^d}a(j,i)\left[\tilde{u}_{T-t}(j)-\tilde{u}_{T-t}(i)\right]dt\\ &\qquad\qquad-\gamma_x Y_t(i)\left(\tilde{u}_{T-t}(i)\right)^2dt + dS_t(i)\Bigg)X_t(i)\Bigg]\\ &\quad+\exp\left(-\langle\tilde{u}_{T-t},X_t\rangle_\Sigma\right)\\ &\quad\cdot(-1)\cdot\Bigg[\sum_{i\in\mathbb{Z}^d}\tilde{u}_{T-t}(i)\Bigg(c_x\sum_{j\in\mathbb{Z}^d}a(i,j)\left[X_t(j)-X_t(i)\right]dt+...dB_t(i)\Bigg)\\ &\qquad+\underbrace{\tilde{u}_{T-t}(\Sigma)}_{=1}\sum_{i\in\mathbb{Z}^d}X_t(i)\,dS_t(i)\Bigg]\\ &\quad+\frac{1}{2}\exp\left(-\langle\tilde{u}_{T-t},X_t\rangle_\Sigma\right)\Bigg[\sum_{i\in\mathbb{Z}^d}\left(\tilde{u}_{T-t}(i)\right)^2\cdot 2\gamma_x X_t(i)\,Y_t(i)\,dt\Bigg]\quad .\end{aligned} \tag{5.55}$$

Comparison of terms yields that only the martingale term remains:

$$d\exp\left(-\langle \tilde{u}_{T-t}, X_t\rangle_\Sigma\right) = \exp\left(-\langle \tilde{u}_{T-t}, X_t\rangle_\Sigma\right)\cdot(-1)\cdot\left[\sum_{i\in\mathbb{Z}^d}\tilde{u}_{T-t}(i)\cdot(...dB_t(i))\right] \quad , \tag{5.56}$$

hence we get

$$\mathbf{E}^Y_{reac}\left[\exp\left(-\langle \tilde{u}_{T-t}, X_t\rangle_\Sigma\right)\right] = const. \tag{5.57}$$

This completes the proof of Proposition 1. ■

5.3.4 Properties of the cumulant equation

We want to discuss the properties of the countable system of differential equations coming into play as cumulant equation. We need the following notation:

Notation 3 *Let $S^1, S^2 \in \mathfrak{S}$. We write*

$$S^1 \trianglerighteq S^2 \tag{5.58}$$

if for all s_1, s_2 with $0 \le s_1 < s_2 < \infty$ and all $i \in \mathbb{Z}^d$ we have

$$\int_{s_1}^{s_2} dS^1(i) \ge \int_{s_1}^{s_2} dS^2(i) \quad . \tag{5.59}$$

(Lebesgue-Stiltjes integration)

Lemma 3 *For $S \in \mathfrak{S}$, let $\left(v_t^S\right)_{t\ge 0}$ be defined by*

$$\begin{aligned} v_0^S &= \psi_0 \\ dv_t^S(i) &= \sum_{j\in\mathbb{Z}^d} a(j,i)\left[v_t^S(j) - v_t^S(i)\right]dt \\ &\quad - \left(v_t^S(i)\right)^2 dt + dS_t(i) \qquad \left(i\in\mathbb{Z}^d\right). \end{aligned} \tag{5.60}$$

We claim that the following holds:

0. *For $S \in \mathfrak{S}$, there exists a unique solution of (5.60). The solution is contained in $\left(\mathbb{R}_+^{\mathbb{Z}^d}\right)^{\mathbb{R}_+}$.*

1. *Monotonicity: Let $s \ge 0$, $S^1, S^2 \in \mathfrak{S}$, $S^1 \trianglerighteq S^2$. Then we have*

$$v_s^{S^1} \ge v_s^{S^2} \quad . \tag{5.61}$$

2. *Subadditivity: Let $s \ge 0$, $S^1, S^2, S \in \mathfrak{S}$ with $S = S^1 + S^2$. Then we have*

$$v_s^S \le v_s^{S^1} + v_s^{S^2} \quad . \tag{5.62}$$

3. *Finite input vanishes: Let $d \leq 2$.*
   If

$$\sum_{i \in \mathbb{Z}^d} \int_0^\infty dS(i) < \infty \tag{5.63}$$

then we have

$$\lim_{t \to \infty} \langle \mathfrak{n}, v_t^S \rangle = 0 \quad . \tag{5.64}$$

Proof.

0. General properties:

Since the evolution equation (5.60) is homogenous in time, we can get rid of the jumps by restarting $\left(v_t^S\right)_{t \geq 0}$ at the jump times. Therefore, we can assume that there are no jumps (i.e. $n = 0$). We can write down (5.60) as differential equation, namely

$$\begin{aligned} \frac{\partial}{\partial t} v_t^{\psi_0, \varphi}(i) = & \sum_{j \in \mathbb{Z}^d} a(j,i) \left[v_t^{\psi_0,\varphi}(j) - v_t^{\psi_0,\varphi}(i) \right] \\ & - \left(v_t^{\psi_0,\varphi}(i) \right)^2 \\ & + \varphi_t(i) \qquad \left(i \in \mathbb{Z}^d \right). \end{aligned} \tag{5.65}$$

(a) Existence and uniqueness:
We prove existence and uniqueness up to time $T \in (0, \infty)$. Let

$$\alpha_T := T \sup_{t \in [0,T]} \sup_{i \in \mathbb{Z}^d} \varphi_t(i) + \sup_{i \in \mathbb{Z}^d} \psi_0(i) \tag{5.66}$$

By assumption (5.44), α_T is finite. Consider the equation

$$\mathring{v}_0^{\psi_0,\varphi} = \psi_0 \tag{5.67}$$

$$\begin{aligned} \frac{\partial}{\partial t} \mathring{v}_t^{\psi_0,\varphi}(i) = & \sum_{j \in \mathbb{Z}^d} a(j,i) \left(\mathring{v}_t^{\psi_0,\varphi}(j) \wedge \alpha_T \right) - \left(\mathring{v}_t^{\psi_0,\varphi}(i) \wedge \alpha_T \right) \\ & - \left(\left(\mathring{v}_t^{\psi_0,\varphi}(i) \right)^2 \wedge (\alpha_T)^2 \right) + \varphi_t(i) \qquad \left(i \in \mathbb{Z}^d \right). \end{aligned} \tag{5.68}$$

This equation is a differential equation with respect to $\mathbb{R}^{\mathbb{Z}^d}$. Equipped with the supremum norm, this space is a Banach space.

The quantity α_T is finite by assumption. Hence, the right hand side is almost surely bounded and satisfies a Lipschitz condition. Namely, comparing the r.h.s of (5.68) for $x, y \in \mathbb{R}^{\mathbb{Z}^d}$ with $\|x - y\| =: \delta$ we get for arbitrary $i \in \mathbb{Z}^d$

$$\begin{aligned} &\sum_{j\in\mathbb{Z}^d} a(j,i)\,(x(j) \wedge \alpha_T) - (x(i) \wedge \alpha_T) \\ &- \left[\sum_{j\in\mathbb{Z}^d} a(j,i)\,(y(j) \wedge \alpha_T) - (y(i) \wedge \alpha_T)\right] \\ &- \left((x(i))^2 \wedge (\alpha_T)^2\right) - \left((y(i))^2 \wedge (\alpha_T)^2\right) && (5.69) \\ &\leq \delta + 2\alpha_T\delta = (1 + 2\alpha_T)\cdot\delta \quad . && (5.70) \end{aligned}$$

This is sufficient for existence and uniqueness, as the well-known result of Picard and Lindelöf can easily be generalized to arbitrary Banach spaces.
We can conclude that for $t \in [0, T]$, $\mathring{v}_t^{\psi_0,\varphi}$ is well-defined.
Furthermore, we can conclude that solutions $\mathring{v}_t^{\psi_0,\varphi}$ of (5.68) and $v_t^{\psi_0,\varphi}$ of (5.60) coincide and are non-negative for $t \in [0, T_\infty \wedge T_0]$ where T_0, T_∞ are given by

$$T_\infty := \sup\left\{\vartheta \in \mathbb{R}_+ : \mathring{v}_\vartheta^{\psi_0,\varphi}(i) \leq \alpha_T \,\forall i \in \mathbb{Z}^d\right\}, \tag{5.71}$$

$$T_0 := \sup\left\{\vartheta \in \mathbb{R}_+ : 0 \leq \mathring{v}_\vartheta^{\psi_0,\varphi}(i) \;\forall i \in \mathbb{Z}^d\right\}, \tag{5.72}$$

From the evolution equation (5.60), we can conclude[7]

$$\limsup_{h\to 0+} \frac{\sup_{i\in\mathbb{Z}^d} \mathring{v}_{t+h}^{\psi_0,\varphi}(i) - \sup_{i\in\mathbb{Z}^d} \mathring{v}_t^{\psi_0,\varphi}(i)}{h} \leq \sup_{s\in[t,t+h]} \sup_{i\in\mathbb{Z}^d} \varphi_s(i) \quad . \tag{5.73}$$

Hence we get

$$\begin{aligned} \sup_{i\in\mathbb{Z}^d} \mathring{v}_t^{\psi_0,\varphi}(i) &\leq \sup_{i\in\mathbb{Z}^d} \psi_0(i) + \int_0^T \sup_{i\in\mathbb{Z}^d} \varphi_t(i)\,dt \\ &\leq \alpha_T && (5.74) \end{aligned}$$

and $T_\infty \geq T$.

(b) Positivity:
Suppose $T_0 < T$. There is continuity; hence there exists $i_0 \in \mathbb{Z}^d$ with $v_{T_0}^{\psi_0,\varphi}(i_0) = 0$ and

$$\frac{\partial}{\partial t} v_t^{\psi_0,\varphi}(i_0)\Big|_{t=T_0-} \leq 0 \quad . \tag{5.75}$$

This yields (with equation (5.65))

$$0 \geq \sum_{j\in\mathbb{Z}^d} a(j,i_0)\, v_{T_0}^{\psi_0,\varphi}(j) + \varphi_{T_0}(i_0) \quad . \tag{5.76}$$

Since a is irreducible by assumption, this is only possible if

$$\forall j \in \mathbb{Z}^d : v_{T_0}^{\psi_0,\varphi}(j) = 0 \quad . \tag{5.77}$$

[7]Note that differentiability of $\mathring{v}_t^{\psi_0,\varphi}(i)$ with respect to t cannot be taken for given.

Hence negative values can only occur after passing through the zero state. On the other hand, if the system is in the zero state and there is no immigration, it will remain in the zero state forever. By uniqueness of solutions, a non-positive state is not attained.

1. Monotonicity

We consider v^{S^1}, v^{S^2} defined by

$$\begin{aligned} dv_t^{S^1}(i) = &\sum_{j\in\mathbb{Z}^d} a(j,i)\left[v_t^{S^1}(j) - v_t^{S^1}(i)\right] dt \\ &-\left(v_t^{S^1}(i)\right)^2 dt \\ &+ dS_t^1 \quad \left(i\in\mathbb{Z}^d\right), \end{aligned} \tag{5.78}$$

$$\begin{aligned} dv_t^{S^2}(i) = &\sum_{j\in\mathbb{Z}^d} a(j,i)\left[v_t^{S^2}(j) - v_t^{S^2}(i)\right] dt \\ &-\left(v_t^{S^2}(i)\right)^2 dt \\ &+ dS_t^2 \quad \left(i\in\mathbb{Z}^d\right), \end{aligned} \tag{5.79}$$

with initial conditions

$$v_t^{S^1} = \psi_0^1, \quad v_t^{S^2} = \psi_0^2 \quad . \tag{5.80}$$

By the considerations carried out in 0., v^{S^1} and v^{S^2} are well-defined and non-negative. We assume

$$S^1 \trianglerighteq S^2 \quad . \tag{5.81}$$

It is our aim to prove that

$$D := v^{S^1} - v^{S^2} \tag{5.82}$$

is non-negative.

We get from (5.78) and (5.79):

$$\begin{aligned} dD_t(i) \quad = \quad &\sum_{j\in\mathbb{Z}^d} a(j,i)\left[D_t(j) - D_t(i)\right] dt \\ &-\left(\left(v_t^{S^1}(i)\right)^2 - \left(v_t^{S^2}(i)\right)^2\right) dt \\ &+dS^1 - dS^2 \quad . \end{aligned} \tag{5.83}$$

Hence, for given v^{S^1} and v^{S^2}, the difference D solves the following equation.

$$\begin{aligned} dD_t(i) \quad = \quad &\sum_{j\in\mathbb{Z}^d} a(j,i)\left[D_t(j) - D_t(i)\right] dt \\ &-\left(v_t^{S^1}(i) + v_t^{S^1}(i)\right) D_t(i)\, dt \\ &+dS^1(i) - dS^2(i) \quad . \end{aligned} \tag{5.84}$$

Fix $\tilde{T} > 0$. We want to prove

$$D_{\tilde{T}}(i) \geq 0 \quad \forall i \in \mathbb{Z}^d \quad . \tag{5.85}$$

For this purpose, consider $\tilde{D} \in \left(\mathbb{R}^{\mathbb{Z}^d}\right)^{[0,\tilde{T}]}$ defined by

$$\tilde{D}_t(i) := \sum_{j \in \mathbb{Z}^d} a_{\tilde{T}-t}(i,j)\, D_t(j) \quad . \tag{5.86}$$

Note that the r.h.s. of (5.86) converges absolutely by (5.44). Therefore, we can interchange differentiation and summation as follows:

$$d\tilde{D}_t(i) = \sum_{j \in \mathbb{Z}^d} \left(da_{\tilde{T}-t}(i,j)\right) D_t(j) + \sum_{j \in \mathbb{Z}^d} a_{\tilde{T}-t}(i,j)\left(dD_t(j)\right) \quad . \tag{5.87}$$

We get:

$$\begin{aligned} d\tilde{D}_t(i) = & \sum_{j \in \mathbb{Z}^d} \left((-1) \cdot \sum_{k \in \mathbb{Z}^d} a(j,k) \left[a_{\tilde{T}-t}(i,k) - a_{\tilde{T}-t}(i,j)\right]\right) D_t(j)\, dt \\ & + \sum_{j \in \mathbb{Z}^d} a_{\tilde{T}-t}(i,j) \left(\sum_{j \in \mathbb{Z}^d} a(k,j)\left[D_t(k) - D_t(j)\right]\right) dt \\ & + \boxed{\cdots} \quad . \end{aligned} \tag{5.88}$$

where $\boxed{\cdots}$ abbreviates the remaining terms (compare (5.84) and (5.90)). We get:

$$\begin{aligned} d\tilde{D}_t(i) = & (-1) \cdot \sum_{j \in \mathbb{Z}^d} \sum_{k \in \mathbb{Z}^d} a(j,k)\, a_{\tilde{T}-t}(i,k)\, D_t(j)\, dt \\ & - (-1) \cdot \sum_{j \in \mathbb{Z}^d} a_{\tilde{T}-t}(i,j)\, D_t(j)\, dt \\ & + \sum_{j \in \mathbb{Z}^d} a_{\tilde{T}-t}(i,j) \sum_{k \in \mathbb{Z}^d} a(k,j)\, D_t(k)\, dt \\ & - \sum_{j \in \mathbb{Z}^d} a_{\tilde{T}-t}(i,j)\, D_t(j)\, dt \\ & + \boxed{\cdots} \\ = & \boxed{\cdots} \quad . \end{aligned} \tag{5.89}$$

The remaining terms $\boxed{\cdots}$ are given by

$$\boxed{\cdots} = \sum_{j \in \mathbb{Z}^d} a_{\tilde{T}-t}(i,j) \left(\left[-\left(v_t^1(j) + v_t^2(j)\right) D_t(j)\right] dt + dS^1(j) - dS^2(j)\right) \tag{5.90}$$

By (5.44), there exists $K_{\tilde{T}}$ with

$$\forall t \in \left[0, \tilde{T}\right], j \in \mathbb{Z}^d : v_t^1(j) + v_t^2(j) \leq K_{\tilde{T}} \quad . \tag{5.91}$$

Hence we have

$$\begin{aligned}
d\tilde{D}_t(i) &\geq \sum_{j\in\mathbb{Z}^d} a_{\tilde{T}-t}(i,j)\left(\left[-\left(v_t^1(j)+v_t^2(j)\right)D_t(j)\right]dt + dS^1(j) - dS^2(j)\right)\\
&\geq \sum_{j\in\mathbb{Z}^d} a_{\tilde{T}-t}(i,j)\left[-K_{\tilde{T}}D_t(j)\right]dt\\
&\geq -K_{\tilde{T}}\tilde{D}_t(i)\,dt \quad .
\end{aligned} \tag{5.92}$$

In particular, we have made use of $S^1 \trianglerighteq S^2$. We can conclude:

$$\tilde{D}_{\tilde{T}}(i) \geq \tilde{D}_0(i)\exp\left(-K_{\tilde{T}}\cdot\tilde{T}\right) > 0 \quad . \tag{5.93}$$

This completes the proof.

2. Subadditivity
Define

$$\begin{aligned}
\hat{S}_t(i) := &\int_0^t dS_s^1(i) + \int_0^t dS_s^2(i)\\
&+ 2\int_0^t v_s^{S^1}(i)\,v_s^{S^2}(i)\,ds \quad \left(i\in\mathbb{Z}^d, t\geq 0\right).
\end{aligned} \tag{5.94}$$

Evidently, we have $\hat{S}\in\mathfrak{S}$. Therefore $v^{\hat{S}}$ is well-defined.
We claim that the following equation holds:

$$v^{S^1} + v^{S^2} = v^{\hat{S}} \tag{5.95}$$

The r. h. s. of (5.95) exceeds $v^{S^1+S^2}$ by monotonicity. Therefore, it is sufficient for the claim (5.62) to prove equation (5.95). This follows from

$$\begin{aligned}
&d\left(v_t^{S^1}(i) + v_t^{S^2}(i)\right)\\
&= \sum_{j\in\mathbb{Z}^d} a(j,i)\left[v_t^{S^1}(j) + v_t^{S^2}(j) - \left(v_t^{S^1}(i) + v_t^{S^2}(i)\right)\right]dt \qquad (5.96)\\
&- \left(\left(v^{S^1}(i)\right)^2 + \left(v^{S^2}(i)\right)^2\right)dt\\
&+ dS_t^1(i) + dS_t^2(i)\\
&= \sum_{j\in\mathbb{Z}^d} a(j,i)\left[v_t^{S^1}(j) + v_t^{S^2}(j) - \left(v_t^{S^1}(i) + v_t^{S^2}(i)\right)\right]dt\\
&- \left(v^{S^1}(i) + v^{S^2}(i)\right)^2 dt\\
&+ dS_t^1(i) + dS_t^2(i) + 2v^{S^1}(i)\,v^{S^2}(i)\,dt \quad . \qquad (5.97)
\end{aligned}$$

This yields (5.95).

3. Finite input vanishes:
   The proof contains two steps:

   I. If $S_t^I = \psi_0$ for all $t \geq 0$ (only immigration at the beginning) with

$$\sum_{i\in\mathbb{Z}^d} \psi_0(i) < \infty \tag{5.98}$$

   we have

$$\lim_{t\to\infty} \left\langle \mathfrak{n}, v_t^{S^I} \right\rangle = 0 \quad . \tag{5.99}$$

   II. In the second step of the proof, we verify (5.64) for $S^{II} \in \mathfrak{S}$ with

$$\int_0^\infty \sum_{i\in\mathbb{Z}^d} dS_t^{II}(i) < \infty \quad . \tag{5.100}$$

Proof of I.

Let $\varepsilon > 0$. We can conclude from (5.98) that there exists $R_\varepsilon < \infty$ s. t.

$$\sum_{\substack{i\in\mathbb{Z}^d\\ |i|\geq R_\varepsilon}} \psi_0(i) \leq \varepsilon \quad . \tag{5.101}$$

Let $\psi_0^{int}, \psi_0^{ext}$ be defined by

$$\psi_0^{int}(i) = \psi_0(i) \cdot \mathbb{I}_{\{|i|<R_\varepsilon\}}(i) \quad , \tag{5.102}$$
$$\psi_0^{ext}(i) = \psi_0(i) \cdot \mathbb{I}_{\{|i|\geq R_\varepsilon\}}(i) \quad . \tag{5.103}$$

Let

$$S_t^{I\_int} = \psi_0^{int} \quad , \quad S_t^{I\_ext} = \psi_0^{ext} \quad . \tag{5.104}$$

We conclude from (5.101):

$$\sup_{t\geq 0} \left\langle \mathfrak{n}, v_t^{S^{I\_ext}} \right\rangle \leq \varepsilon \quad . \tag{5.105}$$

Applying the subadditivity property[8] we get

$$\limsup_{t\to\infty} \left\langle \mathfrak{n}, v_t^{S^I} \right\rangle \leq \limsup_{t\to\infty} \left\langle \mathfrak{n}, v_t^{S^{I\_int}} \right\rangle + \sup_{t\geq 0} \left\langle \mathfrak{n}, v_t^{S^{I\_ext}} \right\rangle \tag{5.106}$$

We next verify

$$\limsup_{t\to\infty} \left\langle \mathfrak{n}, v_t^{S^{I\_int}} \right\rangle = 0 \quad . \tag{5.107}$$

Let $\check{Y}$ denote interacting Feller's branching diffusion with initial condition

$$\check{Y}_0 \equiv 1 \quad . \tag{5.108}$$

[8] $S^{I\_int} + S^{I\_ext} = S^I$

and shape parameter $c_y = \gamma_y = 1$. We know from Proposition 1 (p. 49) that the following holds true:

$$\mathbf{E}\left[\exp\left(-\sum_{i\in\mathbb{Z}^d}\check{Y}_t(i)\,\psi_0^{int}(i)\right)\right] = \mathbf{E}\left[\exp\left(-\left\langle v_t^{S^{I\_int}},\check{Y}_0\right\rangle\right)\right] \quad . \tag{5.109}$$

This gives us a relation between v_t^{S-int} and interacting Feller's branching diffusion at time t. It is a rather well-known fact that interacting Feller's branching diffusion is locally extinct in the recurrent case – e.g., this result can be found in [CFG96], Theorem 2. There, the ergodic theory of a wide class of interacting diffusions is discussed by means of a comparison argument.[9] Since $\left\{i : \psi_0^{int}(i) > 0\right\}$ is finite, the local extinction results yields:

$$\mathcal{L}\left\{\sum_{i\in\mathbb{Z}^d}\check{Y}_t(i)\,\psi_0^{int}(i)\right\} \quad \overset{t\to\infty}{\Longrightarrow} \delta_{\{0\}} \tag{5.110}$$

With (5.108) we get (5.107). Since $\varepsilon > 0$ is arbitrary, we finally get

$$\lim_{t\to\infty}\left\langle \mathfrak{n}, v_t^{S^I}\right\rangle = 0 \quad . \tag{5.111}$$

Proof of II.

By (5.100) there exists for arbitrary $\varepsilon > 0$ a time T_ε such that

$$\int_{T_\varepsilon}^{\infty}\sum_{i\in\mathbb{Z}^d} dS_t^{II}(i) < \varepsilon \quad . \tag{5.112}$$

We decompose S^{II} in the following way:

$$S^{II} = S^{II\_ante} + S^{II\_post} \tag{5.113}$$

where S^{II_ante} and S^{II_post} are given by

$$S_t^{II\_ante} = \begin{cases} S_t^{II} & \text{if } \; t < T_\varepsilon \\ \lim\limits_{s\to T_\varepsilon -} S_s^{II} & \text{if } \; t \geq T_\varepsilon \end{cases} \quad , \tag{5.114}$$

$$S_t^{II\_post} = \begin{cases} 0 & \text{if } \; t < T_\varepsilon \\ S_t^{II} - \lim\limits_{s\to T_\varepsilon -} S_s^{II} & \text{if } \; t \geq T_\varepsilon \end{cases} \quad . \tag{5.115}$$

By construction, we have

$$\sup_{t\geq 0}\left\langle \mathfrak{n}, v_t^{S^{II\_post}}\right\rangle \leq \varepsilon \quad . \tag{5.116}$$

[9] Namely, local extinction of interacting Feller's branching diffusion can be derived from extinction of interacting Fisher-Wright diffusion.

By subadditivity we have

$$\limsup_{t\to\infty}\left\langle \mathfrak{n}, v_t^{S^{II\_ante}}\right\rangle + \limsup_{t\to\infty}\left\langle \mathfrak{n}, v_t^{S^{II\_post}}\right\rangle$$
$$\geq \limsup_{t\to\infty}\left\langle \mathfrak{n}, v_t^{S^{II}}\right\rangle \quad . \tag{5.117}$$

Therefore it only remains to prove

$$\limsup_{t\to\infty}\left\langle \mathfrak{n}, v_t^{S^{II\_ante}}\right\rangle = 0 \quad . \tag{5.118}$$

Let $S_t^{III} := \lim_{s\to T_\varepsilon -} v_s^{S^{II\_ante}}$, $(t \geq 0)$. We can write

$$\left\langle \mathfrak{n}, v_{t+T_\varepsilon}^{S^{II\_ante}}\right\rangle = \left\langle \mathfrak{n}, v_t^{S^{III}}\right\rangle \quad . \tag{5.119}$$

By construction, we have $\langle \mathfrak{n}, S_0^{III}\rangle < \infty$. Since S^{III} is constant in time, we are in the situation of Step I. This yields

$$\lim_{t\to\infty}\left\langle \mathfrak{n}, v_t^{S^{III}}\right\rangle = 0 \quad . \tag{5.120}$$

Together with (5.119), this completes the proof.

■

5.4 Palm measure, Clustering and Kallenberg criterion

5.4.1 Introduction

The stochastic processes we discuss can be classified by their long-time behavior. In particular, there often is the following dichotomy: In the low dimensional case, a process may suffer local extinction, whereas its marginal distributions may converge to a stationary measure as time tends to infinity in the high-dimensional case.

This section is dedicated to show how the long time behavior can be characterized using arguments that are based on size-biasing. In particular, our discussion contains the following two aspects:

I. *Deriving persistence / local extinction from the Palm distribution:* We provide a proposition which we will need later to prove that the reactant persists. We do not use the branching property for these considerations.

II. *Kallenberg criterion:* The branching property and the infinite divisibility of the corresponding particle system allows a family decomposition. This leads to the notion of the canonical law. Applying size-biasing to the canonical law, one gets the canonical Palm distribution, which describes the shape of the family a typical particle belongs to. By means of the Kallenberg criterion, clustering/persistence can be derived by investigating the canonical Palm distribution.

5.4.2 Deriving persistence / local extinction from the Palm distribution

We have divided our considerations into two parts:

1. *Theory:* We introduce the notion of Palm measure. We provide a lemma giving the correspondance between long term properties of the Palm measure and long term properties of the process.
2. *Application:* We apply the general results of 1. to the quenched reactant process.

5.4.2.1 Theory

Definitions

Let $\mathbf{P} \in \mathcal{M}^1\left(\mathcal{M}\left(\mathbb{Z}^d\right)\right)$ be the probability law of a $\mathcal{M}\left(\mathbb{Z}^d\right)$-valued random measure $\mathcal{Y}$ (what we have in mind is a marginal of Feller's interacting branching diffusion). In the next few lines, we will define the intensity measure and the Palm distribution corresponding to $\mathbf{P}$ and give an interpretation thereof.

Definition 6 *The intensity measure $I \in \mathcal{M}\left(\mathbb{Z}^d\right)$ is given by*

$$I(B) = \int \sum_{i \in B} \mathcal{Y}(i)\, \mathbf{P}\,(d\mathcal{Y}) \tag{5.121}$$

for $B \subset \mathbb{Z}^d$.

The intensity measure informs about how much mass is expected to be found in a given set B.

Definition 7 *The family* $\{Palm^{(x)}\}_{x\in\mathbb{Z}^d}$ *is called* Palm distribution *associated with* $\mathbf{P}$ *if for all measurable* $g:\mathcal{M}(\mathbb{Z}^d)\to\mathbb{R}_+$ *we have*

$$\begin{aligned}&\int\limits_{\mathcal{M}(\mathbb{Z}^d)} g(\mathcal{Y})\ Palm^{(x)}(d\mathcal{Y})\, I(x)\\ &= \int\limits_{\mathcal{M}(\mathbb{Z}^d)} g(\mathcal{Y})\,\mathcal{Y}(\{x\})\ \mathbf{P}(d\mathcal{Y}) \quad .\end{aligned} \tag{5.122}$$

Interpretation: The Palm distribution at site x is a probability measure on $\mathcal{M}(\mathbb{Z}^d)$. It weights random configurations $\mathcal{M}(\mathbb{Z}^d)$ according to how much mass they contribute at a given site x.

Interpretation of $Palm^{(x)}$ in the particle picture

Considering a translation invariant setting, we can obtain the Palm distribution $Palm^{(0)}$ at site 0 as follows:

(a) Consider a (typical) realization $\mathcal{Y}\in\mathcal{M}(\mathbb{Z}^d)$.

(b) Consider a box $[-N,N]^d\subset\mathbb{Z}^d$.

(c) Randomly sample a particle ξ from the particles in $[-N,N]^d$ (each particle with probability $\frac{1}{\mathcal{Y}([-N,N]^d)}$)

(d) Let x be the site ξ is located at. Consider $\mathcal{Y}$ shifted by $-x$, denoted $\pi_{-x}\mathcal{Y}$, such that afterwards ξ is located at zero.

(e) The distribution of $\pi_{-x}\mathcal{Y}$ still depends on N. Letting $N\to\infty$, one obtains a weak limit, which is the Palm distribution at zero.

Summarizing we can say that the Palm distribution describes how the process looks like seen from a randomly sampled particle.

This provides some intuition, how local extincion could be characterized: If there is local extinction, a randomly sampled particle is typically surrounded by a huge cluster of particles.

Criteria for local extinction and persistence

The interpretation of the Palm law as "the process seen from a randomly sampled particle" motivates the following proposition, relating the properties of the Palm law $Palm^{(x)}$ to the dichotomy of extinction resp. persistence.

Proposition 2 *On a probability space $(\Omega, \mathfrak{A}, P)$ consider a $\mathcal{M}\left(\mathbb{Z}^d\right)$-valued stochastic process $(\mathcal{Y}_t)_{t\geq 0}$ with the property that it is translation invariant*

$$\mathcal{L}\left\{(\mathcal{Y}_t(\cdot + i))_{t\geq 0}\right\} = \mathcal{L}\left\{(\mathcal{Y}_t(\cdot))_{t\geq 0}\right\} \quad \forall i \in \mathbb{Z}^d \tag{5.123}$$

and preserves expectation

$$\exists I < \infty : I = \mathbf{E}\left[\mathcal{Y}_t(\{i\})\right] \quad \forall t \geq 0, i \in \mathbb{Z}^d \quad . \tag{5.124}$$

Let $\left(Palm_t^{(i)}\right)_{i\in\mathbb{Z}^d}$ denote the Palm distribution of $\mathcal{Y}_t$.

1. *If and only if*

$$\forall M > 0 : \lim_{t\to\infty} \int\limits_{\mathcal{M}(\mathbb{Z}^d)} \mathbb{1}_{\{\tilde{\mathcal{Y}}(\{0\})>M\}}\, Palm_t^{(0)}\left(d\tilde{\mathcal{Y}}\right) = 1 \quad , \tag{5.125}$$

   the process $(\mathcal{Y}_t)_{t\geq 0}$ suffers local extinction.

2. *If the Palm measures satisfy*

$$\lim_{M\to\infty} \limsup_{t\to\infty} \int\limits_{\mathcal{M}(\mathbb{Z}^d)} \mathbb{1}_{\{\mathcal{Y}(\{0\})>M\}}\, Palm_t^{(0)}(d\mathcal{Y}) = 0 \tag{5.126}$$

   the sequence $(\mathcal{L}\{\mathcal{Y}_t(\{0\})\})_{t\geq 0}$ is uniformly integrable, hence the process persists.

Proof.

1. Let $M > 0$. The definition of the Palm distribution (5.122) yields (with $g(\mathcal{Y}) = \mathbb{1}_{\{\mathcal{Y}(\{0\})>M\}}$) that we can conclude from (5.125):

$$\lim_{t\to\infty} \mathbf{E}\left[\mathbb{1}_{\{\mathcal{Y}_t(\{0\})>M\}}\mathcal{Y}_t(\{0\})\right] = I \quad . \tag{5.127}$$

   We conclude that for arbitrary $\varepsilon > 0$ we have

$$\liminf_{t\to\infty} \mathbf{E}\left[\mathbb{1}_{\{\mathcal{Y}_t(\{0\})>M\}}\mathcal{Y}_t(\{0\})\right] \geq I - \varepsilon^2 \quad . \tag{5.128}$$

   We get immediately

$$\limsup_{t\to\infty} P\left[\mathcal{Y}_t(\{0\}) \in [\varepsilon, M]\right] \leq \varepsilon \quad . \tag{5.129}$$

   This holds for arbitrary M. Since I is finite, we have

$$\lim_{M\to\infty} \sup_{t\in\mathbb{R}_+} P\left[\mathcal{Y}_t(\{0\}) > M\right] = 0 \quad . \tag{5.130}$$

We can conclude

$$\limsup_{t\to\infty} P\left[\mathcal{Y}_t(\{0\}) \geq \varepsilon\right] \leq \varepsilon \quad . \tag{5.131}$$

This yields

$$\mathcal{L}\{\mathcal{Y}_t(\{0\})\} \overset{t\to\infty}{\Longrightarrow} \delta_{\{0\}} \quad . \tag{5.132}$$

The backward direction can be proven as follows: The starting point is

$$\mathcal{L}\{\mathcal{Y}_t(\{0\})\} \overset{t\to\infty}{\Longrightarrow} \delta_{\{0\}} \quad . \tag{5.133}$$

This yields for arbitrary $\varepsilon > 0$:

$$\limsup_{t\to\infty} P\left[\mathcal{Y}_t(\{0\}) \in [\varepsilon, M]\right] \leq \varepsilon \quad . \tag{5.134}$$

We get:

$$\liminf_{t\to\infty} \mathbf{E}\left[\mathbb{1}_{\{\mathcal{Y}_t(\{0\})>M\}} \mathcal{Y}_t(\{0\})\right] \geq I - \varepsilon M - \varepsilon \quad . \tag{5.135}$$

Since we are free to choose ε, we get

$$\liminf_{t\to\infty} \mathbf{E}\left[\mathbb{1}_{\{\mathcal{Y}_t(\{0\})>M\}} \mathcal{Y}_t(\{0\})\right] = I \quad . \tag{5.136}$$

This is equivalent to (5.125), completing the proof.

2. With the definition of Palm distribution (5.122) and as a consequence of (5.126) we get

$$\begin{aligned} 0 &= \lim_{M\to\infty} \limsup_{t\to\infty} \int\limits_{\mathcal{M}(\mathbb{Z}^d)} \mathbb{1}_{\{\tilde{\mathcal{Y}}(\{0\})>M\}} \, Palm_t^{(0)}\left(d\tilde{\mathcal{Y}}\right) \cdot I \\ &= \lim_{M\to\infty} \limsup_{t\to\infty} \mathbf{E}\left[\mathbb{1}_{\{\mathcal{Y}_t(\{0\})>M\}} \mathcal{Y}_t(\{0\})\right] \end{aligned} \tag{5.137}$$

Since the right hand side describes uniform integrability, the limit of intensities and the intensity of the limit coincide. Hence we have persistence.

∎

5.4.2.2 Application

Intensity of the Palm distribution of interacting catalytic Feller diffusion for given catalyst

Let $Palm_T^{(0)}$ denote the Palm distribution of interacting catalytic Feller diffusion for given catalyst at time T.

In order to verify that the process persists, we have to know if (5.126) holds, namely.

$$\lim_{M\to\infty} \limsup_{t\to\infty} \int\limits_{\mathcal{M}(\mathbb{Z}^d)} \mathbb{1}_{\{\mathcal{X}(\{0\})>M\}} \, Palm_t^{(0)}(d\mathcal{X}) = 0 \quad . \tag{5.138}$$

By Markov's inequality, it suffices to prove that the intensity measure does not grow to infinity:

$$\limsup_{t\to\infty} \int\limits_{\mathcal{M}(\mathbb{Z}^d)} \mathcal{X}(\{0\}) \, Palm_t^{(0)}(d\mathcal{X}) < \infty \quad . \tag{5.139}$$

We calculate the intensity measure of $Palm_T^{(0)}$ using the definition of the Palm distribution (5.122):

$$\int_{\mathcal{M}(\mathbb{Z}^d)} \mathcal{X}(\{0\})\ Palm_T^{(0)}\ (d\mathcal{X}) = \frac{1}{\theta_x} \int_{\mathcal{M}(\mathbb{Z}^d)} X_T(\{0\}) \cdot X_T(\{0\}) \quad \mathbf{P}(dX_T)$$
$$= \frac{1}{\theta_x} \mathbf{E}^Y_{reac} \left[(X_T(0))^2 \right] \quad . \tag{5.140}$$

We will prove later(Lemma 4, page 69)

$$X_t(0) = \sum_{z\in\mathbb{Z}^d} a_{c_x t}(z,0)\, X_0(z)$$
$$+ \sum_{z\in\mathbb{Z}^d} \left(\int_0^t a_{c_x(t-s)}(z,0)\, \sqrt{2\gamma_x X_s(z)\, Y_s(z)} dB^x_s(z) \right) \quad . \tag{5.141}$$

This yields

$$\mathbf{E}^Y_{reac} \left[(X_t(0))^2 \right] = \mathbf{E}^Y_{reac} \left[\left(\sum_{z\in\mathbb{Z}^d} a_{c_x t}(z,0)\, X_0(z) \right)^2 \right]$$
$$+ 2\gamma_x \sum_{z\in\mathbb{Z}^d} \left(\int_0^t \left(a_{c_x(t-s)}(z,0)\right)^2 \mathbf{E}^Y_{reac}\left[X_s(z)\, Y_s(z)\right] ds \right) \quad . \tag{5.142}$$

Since - by assumption (1.17) - catalyst and reactant are and remain uncorrelated, we get

$$\mathbf{E}^Y_{reac} \left[(X_t(0))^2 \right] \quad = \quad \mathbf{E}^Y_{reac} \left[\left(\sum_{z\in\mathbb{Z}^d} a_{c_x t}(z,0)\, X_0(z) \right)^2 \right]$$
$$+ 2\gamma_x \theta_x \sum_{z\in\mathbb{Z}^d} \int_0^t \left(a_{c_x(t-s)}(z,0)\right)^2 Y_s(z)\, ds \quad . \tag{5.143}$$

Note that the first term is uniformly bounded as assumed in Section 1.2.4. We therefore can draw the following conclusion:

Conclusion 2 *To prove persistence of the catalyst process, it suffices to prove that*

$$\mathcal{L}_{cat} \left\{ \sum_{k\in\mathbb{Z}^d} \int_0^T (a_s(k,0))^2\, Y_{T-s}(k)\ ds \right\} . \tag{5.144}$$

is tight.

5.4.3 Kallenberg criterion

Introduction

So far, we have not taken into account that the processes we consider are infinitely divisible.

Considering the diffusion limit, infinite divisibility is an immediate consequence of the branching property; considering particle processes, we assume that the initial law is a Poisson point process.

In the sequel, we only consider the particle picture. To be more precise, we consider branching random walk in a catalytic medium, which is interacting Feller diffusion.[10] The only difference between the particle model considered here and the model considered by Greven, Klenke and Wakolbinger [GKW99] is that their catalyst is a particle process, wheras in our case only the reactant is a particle process, not the catalyst, which is a diffusion process. In the following, we reformulate the material presented in [GKW99] adjusting it to the situation discussed here.

Notation

We consider the following process:

- Catalyst: We consider the interacting Feller diffusion process given by (1.7).
- Reactant: The reactant process $\mathcal{X}$ - as a particle process - attains values in $\mathcal{N}\left(\mathbb{Z}^d\right)$. It describes the superposition of independent particles, with
  - migration with rate c_x
  - transition probabilities given by $a(\cdot,\cdot)$
  - branching rate γ_x, where each branching event is critical binary branching.

The reactant is initialized as follows: independently, we sample at each site $i \in \mathbb{Z}^d$ a Poisson number of particles, where - for simplicity - we assume that the intensity is constant[11]. This way, for given catalyst Y, the reactant $\mathcal{X}$ has an infinite divisible law at time zero. Since the reactant - given the catalyst - is a branching process, this infinite divisibility is preserved forever.

Canonical measure

We have just mentioned that the quenched law of the reactant $\mathcal{L}^Y_{reac}(\mathcal{X}_t)$ is infinitely divisible. Comparable to a Levy-Khinchine formula, the following decomposition is possible: For each catalyst realization Y there exists a σ-finite measure $Q^Y_t \in \mathcal{M}\left(\mathcal{N}\left(\mathbb{Z}^d\right)\right)$ such that for $\varphi \in \mathbb{R}_+^{\mathbb{Z}^d}$ with $card\{\varphi > 0\} < \infty$ we have

$$\mathbf{E}^Y_{reac}\left[\exp\left(-\langle\varphi, \mathcal{X}_t\rangle\right)\right] = \exp\left(-\int_{\mathcal{N}(\mathbb{Z}^d)} \left(1 - e^{-\langle\varphi,\chi\rangle}\right) Q^Y_t(d\chi)\right). \tag{5.145}$$

[10] Note that the long time behavior of particle system and diffusion limit coincide.

[11] In fact, we would have to choose X_0 as intensity of the Poisson field we start with. This would lead to a more complicated definition of the "backbone" of the backward tree. (In fact, one has to carry out a change of measure). We do not want to obscure the main ideas of the Kallenberg criterion; this is why we have decided to introduce this small simplification.

As mentioned in [GKW99], Q_t^Y describes the intensity of configurations of particles descending from the same ancestor.

Canonical Palm distribution

We apply a size-biasing procedure to the canonical measure. This way, we get the Palm distribution $\left\{\tilde{Q}_t^{Y,x}\right\}_{x\in\mathbb{Z}^d}$. Similar to (5.122), $\tilde{Q}_t^{Y,x}$ is defined by

$$\int\limits_{\mathcal{M}(\mathbb{Z}^d)} g(\chi)\,\tilde{Q}_t^{Y,x}(d\chi)\quad I(x) = \int\limits_{\mathcal{M}(\mathbb{Z}^d)} g(\chi)\,\chi(\{x\})\,Q_t^Y(d\chi) \tag{5.146}$$

for $g \in (\mathbb{R}_+)^{\mathcal{M}(\mathbb{Z}^d)}$ measurable.

Backward tree representation of the Canonical Palm distribution

It is possible to describe the canonical Palm distribution at x for given catalyst Y by means of a backward tree representation. For convenience, we choose $x = 0$.

Ingredients:

- Backbone: Let $(\xi_s)_{s\in[0,t]}$ denote a random walk with initial condition $\xi_s = 0$ and transition rates given by $c_x\bar{a}(\cdot,\cdot)$. This random walk serves as model of the ancestral path of a particle living at 0 at time t.
- Collection of trees: Let $\left\{(\mathcal{X}_s^{z,\tau})_{s\geq\tau}\right\}_{\tau\in\mathbb{R}_+,z\in\mathbb{Z}^d}$ be a collection of independent processes – independent of the backbone – with
  - initial conditions: $\mathcal{X}_\tau^{z,\tau} = z$
  - evolution mechanism: same as evolution of $\mathcal{X}$ (but all independent)
- Let $(\tau_i)_{i\in\mathbb{N}}$ denote the increasing sequence of jump times of a Poisson process with time-dependent rate $Y_s(\xi_{t-s})$ ([12])

The random population $\bar{\mathcal{X}}_t$ is defined as follows:

$$\bar{\mathcal{X}}_t := \delta_{\{0\}} + \sum_{i:\tau_i\leq t} \delta_{\left\{\mathcal{X}_t^{\xi_{t-s},\tau_i}\right\}} \quad . \tag{5.147}$$

In other words, $\bar{\mathcal{X}}_t$ contains all "relatives" of the considered particle at time t and site 0.We draw a picture for visualization:

[12] s denotes time, t is fixed

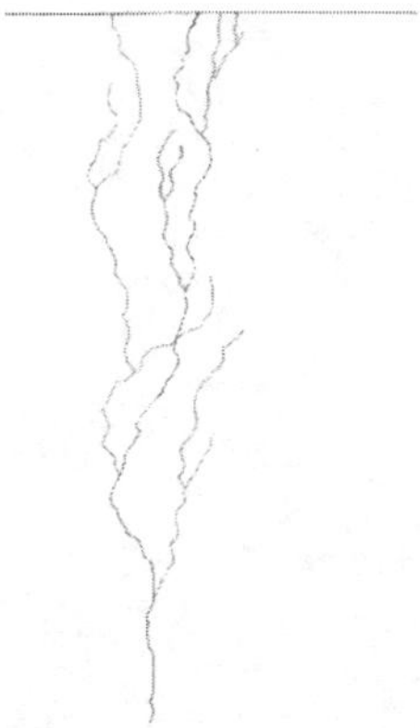

Particle with descendents

As mentioned in [GKW99], p. 23, $\mathcal{L}^Y\left\{\bar{\mathcal{X}}_t\right\}$ describes the canonical Palm distribution of the reactant (given the catalyst), namely

$$\tilde{Q}_t^{Y,0} = \mathcal{L}^Y\left\{\bar{\mathcal{X}}_t\right\} \quad .$$

Kallenberg criterion

It is possible to derive local extinction resp. persistence[13] considering the canonical Palm measure. This criterion goes back to Kallenberg [Kal77], compare also [GKW99], Proposition 2.5. For given catalyst, we can apply the following equivalence:

$$\begin{array}{c} \mathcal{X} \text{ is persistent as } t \to \infty \\ \Longleftrightarrow \\ \tilde{Q}_t^{Y,0} \text{ is stochastically bounded as } t \to \infty \text{ for a. a. } Y \\ \Longleftrightarrow \\ \limsup\limits_{T\to\infty} \int_0^T a_s\left(\xi_s, 0\right) Y_{T-s}\left(\xi_s\right) < \infty \quad \text{for a. a. } \left(\xi_s\right)_{s\in[0,t]}, Y. \end{array}$$

[13]Ther exists a weak limit law with full intensity

Chapter 6

Proof of Theorem 0

In this chapter we discuss and prove some results concerning the long time behavior of the process $(X_t, Y_t)_{t\geq 0}$ on the lattice. As there are different regimes of longtime behavior according to the dichotomy of recurrence and transience of the underlying random walk, we have to distinguish between the low-dimensional case $d \leq 2$, where the symmetrized migration matrix $\hat{a}$ is recurrent, and the high-dimensional case $d \geq 3$, where $\hat{a}$ is transient.

6.1 Low-dimensional case

Throughout this section, we assume $d \in \{1, 2\}$.

6.1.1 The reactant becomes locally flat in the long run

Now we prove the claim formulated by equation (3.1), namely for $\alpha, \beta \in \mathbb{Z}^d$

$$\mathcal{L}_{cat}\left\{\mathcal{L}^Y_{reac}\left\{X_t(\alpha) - X_t(\beta)\right\}\right\} \quad \overset{t\to\infty}{\Longrightarrow} \quad \delta_{\{\delta_0\}} \quad . \tag{6.1}$$

It suffices to prove

$$\mathcal{L}_{cat}\left\{\mathbf{E}^Y_{reac}\left[\left(X_t(\alpha) - X_t(\beta)\right)^2\right]\right\} \quad \overset{t\to\infty}{\Longrightarrow} \quad \delta_0 \quad . \tag{6.2}$$

We will make use of a special representation of X_t. This representation is provided by the following lemma:

Lemma 4 *Let $(X_t, Y_t)_{t\geq 0}$ be defined as (unique) strong solution of (1.6) and (1.7). Then $X_t(x)$ $\left(x \in \mathbb{Z}^d\right)$ has the following representation on the probability space the underlying Brownian motions are defined on:*

$$\begin{aligned} X_t(x) &= \sum_{z\in\mathbb{Z}^d} a_{c_x t}(z, x)\, X_0(z) \\ &+ \sum_{z\in\mathbb{Z}^d}\left(\int_0^t a_{c_x(t-s)}(z, x)\sqrt{2\gamma_x X_s(z)\, Y_s(z)}\, dB^x_s(z)\right) \quad . \end{aligned} \tag{6.3}$$

Proof. Fix $t > 0, x \in \mathbb{Z}^d$. Consider the process $\left(R_s^t\right)_{s\in[0,t]}$ given by

$$R_s^t := \sum_{z\in\mathbb{Z}^d} a_{c_x(t-s)}(z,x)\, X_s(z) \quad . \tag{6.4}$$

In the sequel we shall drop 't' since it is fixed and write R_s instead of R_s^t. This process is well-defined $\mathbf{P}$-almost surely, its second moments exist for all times:

$$\begin{aligned}
\mathbf{E}\left[(R_s)^2\right] &= \mathbf{E}\left[\left(\sum_{z\in\mathbb{Z}^d} a_{c_x(t-s)}(z,x)\, X_s(z)\right)^2\right] \\
&\le \sum_{z\in\mathbb{Z}^d} a_{c_x(t-s)}(z,x)\, \mathbf{E}\left[(X_s(z))^2\right] &(6.5)\\
&= \mathbf{E}\left[(X_s(0))^2\right] &(6.6)\\
&< \infty \quad ,(s\in[0,t])\,. &(6.7)
\end{aligned}$$

In (6.5) we have applied Jensen's inequality, (6.6) follows from translation invariance (compare Section 1.2.4) and (6.7) follows from the finiteness of second moments (compare Lemma 16, p. 257).

Note that we assume at the moment that t is fixed and treated as a constant parameter. $X_t(x)$ can be expressed in terms of R by

$$X_t(x) = R_t \quad . \tag{6.8}$$

In addition to that, it is clear that the initial condition of the process R is given by

$$R_0 = \sum_{z\in\mathbb{Z}^d} a_{c_x t}(z,x)\, X_0(z) \quad . \tag{6.9}$$

In order to write R_s as stochastic integral, we consider its (stochastic) differential with respect to s. We get:

$$\begin{aligned}
&dR_s \\
&= d\left\{\sum_{z\in\mathbb{Z}^d} a_{c_x(t-s)}(z,x)\, X_s(z)\right\} \quad . &(6.10)
\end{aligned}$$

Next we have to interchange the differential d with the summation $\sum_{z\in\mathbb{Z}^d}$. Since the sum $\sum_{z\in\mathbb{Z}^d}$ includes an infinite number of terms, it is not clear that this is allowed. We have to verify that the sum converges absolutely. But what does this mean in this case?

Consider the differential $d\left\{a_{c_x(t-s)}(z,x)\, X_s(z)\right\}$. By the product rule of stochastic differentiation (compare [Ok], exercise 4.3, or apply the time-dependent version of Ito's formula ([Ok], p. 48)) we get:

$$\begin{aligned}
& d\left\{a_{c_x(t-s)}(z,x)\,X_s(z)\right\} \\
&= a_{c_x(t-s)}(z,x)\,[dX_s(z)] + \left[\frac{\partial}{\partial s}a_{c_x(t-s)}(z,x)\right] X_s(z)\,ds \\
&= a_{c_x(t-s)}(z,x) \\
&\quad \cdot \left[\sum_{k\in\mathbb{Z}^d} c_x a(z,k)\,(X_s(k)-X_s(z))\,ds + \sqrt{2\gamma_x X_s(z)\,Y_s(z)}\,dB_s(z)\right] \\
&\quad - \sum_{k\in\mathbb{Z}^d} c_x a(z,k)\left[a_{c_x(t-s)}(k,x) - a_{c_x(t-s)}(z,x)\right] X_s(z)\,ds \qquad (6.11)
\end{aligned}$$

This family of stochastic differentials is absolutely summable if

- The sum of absolute values of the drift terms remains finite in L^2 :

$$\begin{aligned}
&\mathbf{E}\left[\left(\sum_{z\in\mathbb{Z}^d}\left|a_{c_x(t-s)}(z,x)\sum_{k\in\mathbb{Z}^d} c_x a(z,k)\,(X_s(k)-X_s(z))\right.\right.\right. \\
&\left.\left.\left. - \sum_{k\in\mathbb{Z}^d} c_x a(z,k)\left[a_{c_x(t-s)}(k,x) - a_{c_x(t-s)}(z,x)\right] X_s(z)\right|^2\right)\right] \\
&< \infty \qquad (6.12)
\end{aligned}$$

- The sum of increments of the increasing processes corresponding to the martingale terms remains finite in expectation:

$$\mathbf{E}\left[\sum_{z\in\mathbb{Z}^d}\left(a_{c_x(t-s)}(z,x)\sqrt{2\gamma_x X_s(z)\,Y_s(z)}\right)^2\right] < \infty \qquad (6.13)$$

The finiteness claimed in (6.12) follows from

$$\begin{aligned}
&\mathbf{E}\left[\left(\sum_{z\in\mathbb{Z}^d}\left|a_{c_x(t-s)}(z,x)\sum_{k\in\mathbb{Z}^d} c_x a(z,k)\,(X_s(k)-X_s(z))\right.\right.\right. \\
&\left.\left.\left. - \sum_{k\in\mathbb{Z}^d} c_x a(z,k)\left[a_{c_x(t-s)}(k,x) - a_{c_x(t-s)}(z,x)\right] X_s(z)\right|^2\right)\right] \\
&\le 2\mathbf{E}\left[\left(\sum_{z\in\mathbb{Z}^d}\sum_{k\in\mathbb{Z}^d} a_{c_x(t-s)}(z,x)\,c_x a(z,k)\,(X_s(k)+X_s(z))\right)^2\right] \\
&\quad + 2\mathbf{E}\left[\left(\sum_{z\in\mathbb{Z}^d}\sum_{k\in\mathbb{Z}^d} c_x a(z,k)\left[a_{c_x(t-s)}(k,x) + a_{c_x(t-s)}(z,x)\right] X_s(z)\right)^2\right] \qquad (6.14)
\end{aligned}$$

$$\leq 4\mathbf{E}\left[\left(\sum_{z\in\mathbb{Z}^d}\sum_{k\in\mathbb{Z}^d} a_{c_x(t-s)}(z,x)\, c_x a(z,k)\,(X_s(k))\right)^2\right]$$
$$+ 4\mathbf{E}\left[\left(\sum_{z\in\mathbb{Z}^d}\sum_{k\in\mathbb{Z}^d} a_{c_x(t-s)}(z,x)\, c_x a(z,k)\,(X_s(z))\right)^2\right]$$
$$+ 4\mathbf{E}\left[\left(\sum_{z\in\mathbb{Z}^d}\sum_{k\in\mathbb{Z}^d} c_x a(z,k)\, a_{c_x(t-s)}(k,x)\, X_s(z)\right)^2\right]$$
$$+ 4\mathbf{E}\left[\left(\sum_{z\in\mathbb{Z}^d}\sum_{k\in\mathbb{Z}^d} c_x a(z,k)\, a_{c_x(t-s)}(z,x)\, X_s(z)\right)^2\right] \quad . \tag{6.15}$$

Next, we apply Jensen's inequality:

$$\leq 4\mathbf{E}\left[\sum_{z\in\mathbb{Z}^d}\sum_{k\in\mathbb{Z}^d} a_{c_x(t-s)}(z,x)\, a(z,k)\,(c_x(X_s(k)))^2\right]$$
$$+ 4\mathbf{E}\left[\sum_{z\in\mathbb{Z}^d}\sum_{k\in\mathbb{Z}^d} a_{c_x(t-s)}(z,x)\, a(z,k)\,(c_x(X_s(z)))^2\right]$$
$$+ 4\mathbf{E}\left[\sum_{z\in\mathbb{Z}^d}\sum_{k\in\mathbb{Z}^d} a(z,k)\, a_{c_x(t-s)}(k,x)\,(c_x X_s(z))^2\right]$$
$$+ 4\mathbf{E}\left[\sum_{z\in\mathbb{Z}^d}\sum_{k\in\mathbb{Z}^d} a(z,k)\, a_{c_x(t-s)}(z,x)\,(c_x X_s(z))^2\right]$$
$$= 16 c_x^2 \mathbf{E}\left[(X_s(0))^2\right] < \infty \quad . \tag{6.16}$$

The finiteness claimed in (6.13) follows from:

$$\mathbf{E}\left[\sum_{z\in\mathbb{Z}^d}\left(a_{c_x(t-s)}(z,x)\sqrt{2\gamma_x X_s(z)\, Y_s(z)}\right)^2\right]$$
$$= 2\gamma_x \sum_{z\in\mathbb{Z}^d}\left(a_{c_x(t-s)}(z,x)\right)^2 \mathbf{E}\left[X_s(z)\, Y_s(z)\right]$$
$$\leq 2\gamma_x \sum_{z\in\mathbb{Z}^d}\left(a_{c_x(t-s)}(z,x)\right) \mathbf{E}\left[X_s(z)\, Y_s(z)\right]$$
$$= 2\gamma_x \mathbf{E}\left[X_s(0)\, Y_s(0)\right] = 2\gamma_x\theta_x\theta_y < \infty \quad . \tag{6.17}$$

Hence, applying (6.13) and (6.12) we get:

$$= \sum_{z\in\mathbb{Z}^d} d\left\{a_{c_x(t-s)}(z,x)\, X_s(z)\right\} \quad . \tag{6.18}$$

We get (with (6.10)):

$$
\begin{aligned}
& dR_s \\
& = \sum_{z\in\mathbb{Z}^d} a_{c_x(t-s)}(z,x)\left[dX_s(z)\right] + \sum_{z\in\mathbb{Z}^d}\left[\frac{\partial}{\partial s}a_{c_x(t-s)}(z,x)\right]X_s(z)\,ds \\
& = \sum_{z\in\mathbb{Z}^d} a_{c_x(t-s)}(z,x) \\
& \qquad \cdot\left[\sum_{k\in\mathbb{Z}^d} c_x a(z,k)\left(X_s(k)-X_s(z)\right)ds + \sqrt{2\gamma_x X_s(z)\,Y_s(z)}dB_s(z)\right] \\
& \quad -\sum_{z\in\mathbb{Z}^d}\sum_{k\in\mathbb{Z}^d} c_x a(z,k)\left[a_{c_x(t-s)}(k,x)-a_{c_x(t-s)}(z,x)\right]X_s(z)\,ds \\
& = \sum_{z\in\mathbb{Z}^d} a_{c_x(t-s)}(z,x)\sqrt{2\gamma_x X_s(z)\,Y_s(z)}dB_s(z) \\
& \quad + c_x\sum_{z\in\mathbb{Z}^d} a_{c_x(t-s)}(z,x)\sum_{k\in\mathbb{Z}^d} a(z,k)\left(X_s(k)-X_s(z)\right)ds \\
& \quad -\left[c_x\sum_{z\in\mathbb{Z}^d}\sum_{k\in\mathbb{Z}^d} a(z,k)\left(a_{c_x(t-s)}(k,x)\,X_s(z)-a_{c_x(t-s)}(z,x)\,X_s(z)\right)\right]ds \\
& = \sum_{z\in\mathbb{Z}^d} a_{c_x(t-s)}(z,x)\sqrt{2\gamma_x X_s(z)\,Y_s(z)}dB_s(z) \quad . \qquad (6.19)
\end{aligned}
$$

This yields:

$$
\begin{aligned}
X_t(x) &= \sum_{z\in\mathbb{Z}^d} a_{c_x(t-t)}(z,x)\,X_s(z) \\
&= R_t \\
&= R_0 + \int_0^t dR_s \\
&= \sum_{z\in\mathbb{Z}^d} a_{c_x t}(z,x)\,X_0(z) \\
&\quad + \sum_{z\in\mathbb{Z}^d}\int_0^t a_{c_x(t-s)}(z,x)\sqrt{2\gamma_x X_s(z)\,Y_s(z)}dB_s(z) \quad . \qquad (6.20)
\end{aligned}
$$

This is exactly what we have claimed. ■

With the representation just obtained, we can write:

$$
\begin{aligned}
X_t(\alpha) &= \sum_{z\in\mathbb{Z}^d} a_{c_x t}(z,\alpha)\,X_0(z) \\
&\quad + \sum_{z\in\mathbb{Z}^d}\left(\int_0^t a_{c_x(t-s)}(z,\alpha)\sqrt{2\gamma_x X_s(z)\,Y_s(z)}dB_s^x(z)\right) \quad , \qquad (6.21)
\end{aligned}
$$

$$X_t(\beta) = \sum_{z\in\mathbb{Z}^d} a_{c_x t}(z,\beta) X_0(z) + \sum_{z\in\mathbb{Z}^d} \left(\int_0^t a_{c_x(t-s)}(z,\beta) \sqrt{2\gamma_x X_s(z) Y_s(z)} dB_s^x(z) \right) \quad . \tag{6.22}$$

To prove (6.2) we have to carry out a two-step program, treating seperately the terms of the right hand sides of (6.21) and (6.22) respectively, calculating their differences.

1. We have to consider the difference

$$\Delta_t^{(1)} := \sum_{z\in\mathbb{Z}^d} a_{c_x t}(z,\alpha) X_0(z) - \sum_{z\in\mathbb{Z}^d} a_{c_x t}(z,\beta) X_0(z) \tag{6.23}$$

and verify

$$\mathcal{L}_{cat}\left\{ \mathbf{E}_{reac}^Y \left[\left(\Delta_t^{(1)} \right)^2 \right] \right\} \overset{t\to\infty}{\Longrightarrow} \delta_0 \quad . \tag{6.24}$$

2. Define

$$\Delta_t^{(2)} \; : \; = \sum_{z\in\mathbb{Z}^d} \left(\int_0^t a_{c_x(t-s)}(z,\alpha) \sqrt{2\gamma_x X_s(z) Y_s(z)} dB_s^x(z) \right) - \sum_{z\in\mathbb{Z}^d} \left(\int_0^t a_{c_x(t-s)}(z,\beta) \sqrt{2\gamma_x X_s(z) Y_s(z)} dB_s^x(z) \right) \quad . \tag{6.25}$$

We have to show that the laws of quenched variances of $\Delta_t^{(2)}$ converge weakly to the atomar law on $\mathbb{R}$ located at zero:

$$\mathcal{L}_{cat}\left\{ \mathbf{E}_{reac}^Y \left[\left(\Delta_t^{(2)} \right)^2 \right] \right\} \overset{t\to\infty}{\Longrightarrow} \delta_0 \quad . \tag{6.26}$$

ad 1. We have assumed that $\mathcal{L}\{(X_0, Y_0)\} \in S^{(\theta_x,\theta_y)}$. In particular this yields:

$$\lim_{t\to\infty} \mathbf{E}\left[\left(\sum_{z\in\mathbb{Z}^d} a_{c_x t}(z,i) X_0(z) - \theta_x \right)^2 \right] = 0 \quad \forall i \in \mathbb{Z}^d. \tag{6.27}$$

We obtain what we have claimed in (6.24) applying the defining equation (6.23) and the triangular inequality.

ad 2. Our goal is to prove (6.26), namely the weak convergence of the laws of quenched second moments to the atomar measure at zero. Quenched second moments

$$\left(\mathbf{E}_{reac}^Y \left[\left(\Delta_t^{(2)} \right)^2 \right] \right)_{t\geq 0}$$

are in fact random variables – they are catalyst-dependent. Our strategy is to prove that the Laplace transforms of these random variables converge to 1 as $t \to \infty$. We first have to express the random variables as functions of the process Y. For this purpose, we carry out the following calculation:

$$\begin{aligned} & \mathbf{E}^Y_{reac}\left[\left(\Delta^{(2)}_t\right)^2\right] \\ = \; & \mathbf{E}^Y_{reac}\left[\left(\sum_{z\in\mathbb{Z}^d}\left(\int_0^t \left(a_{c_x(t-s)}(z,\alpha) - a_{c_x(t-s)}(z,\beta)\right)\sqrt{2\gamma_x X_s(z) Y_s(z)} dB^x_s(z)\right)\right)^2\right] \end{aligned} .$$

Consider $(H_r)_{r\in[0,t]}$ defined by

$$H_r := \sum_{z\in\mathbb{Z}^d}\left(\int_0^r \left(a_{c_x(t-s)}(z,\alpha) - a_{c_x(t-s)}(z,\beta)\right)\sqrt{2\gamma_x X_s(z) Y_s(z)}\, dB^x_s(z)\right) . \tag{6.28}$$

This yields

$$\mathbf{E}^Y_{reac}\left[\left(\Delta^{(2)}_t\right)^2\right] = \mathbf{E}^Y_{reac}\left[(H_t)^2\right] . \tag{6.29}$$

By considerations similar to the proof of (6.13), the following line makes sense in L^2:

$$dH_r = \sum_{z\in\mathbb{Z}^d}\left(a_{c_x(t-r)}(z,\alpha) - a_{c_x(t-r)}(z,\beta)\right)\sqrt{2\gamma_x X_r(z) Y_r(z)}\, dB^x_r(z) . \tag{6.30}$$

Applying Itô's formula (the family of Brownian motions $\{B^x(z)\}_{z\in\mathbb{Z}^d}$ is a family of independent BM), we get:

$$\begin{aligned} dH_r^2 = {} & 2H_r \cdot \sum_{z\in\mathbb{Z}^d}\left(a_{c_x(t-r)}(z,\alpha) - a_{c_x(t-r)}(z,\beta)\right)\sqrt{2\gamma_x X_r(z) Y_r(z)} dB^x_r(z) \\ & + \frac{1}{2}\cdot 2\sum_{z\in\mathbb{Z}^d}\left(\left(a_{c_x(t-r)}(z,\alpha) - a_{c_x(t-r)}(z,\beta)\right)\right)^2 2\gamma_x X_r(z) Y_r(z)\, dr . \end{aligned} \tag{6.31}$$

In particular, we get:

$$\begin{aligned} & \mathbf{E}^Y_{reac}\left[\left(\Delta^{(2)}_t\right)^2\right] \\ = \; & \mathbf{E}^Y_{reac}\left[H_t^2\right] \\ = \; & \sum_{z\in\mathbb{Z}^d}\mathbf{E}^Y_{reac}\left[\int_0^t \left(\left(a_{c_x(t-s)}(z,\alpha) - a_{c_x(t-s)}(z,\beta)\right)\sqrt{2\gamma_x X_s(z) Y_s(z)}\right)^2 ds\right] \\ = \; & \sum_{z\in\mathbb{Z}^d}\int_0^t 2\gamma_x\left(a_{c_x(t-s)}(z,\alpha) - a_{c_x(t-s)}(z,\beta)\right)^2 \mathbf{E}^Y_{reac}\left[X_s(z)\right] Y_s(z)\, ds . \end{aligned} \tag{6.32}$$

Since $\forall z \in \mathbb{Z}^d : \mathbf{E}^Y_{reac}[X_s(z)] = \theta_x$ (this follows immediately from the evolution equation (1.6)) we get:

$$\begin{aligned} & \mathbf{E}^Y_{reac}\left[\left(\Delta^{(2)}_t\right)^2\right] \\ = & \sum_{z\in\mathbb{Z}^d}\int_0^t 2\gamma_x\left(a_{c_x(t-s)}(z,\alpha)-a_{c_x(t-s)}(z,\beta)\right)^2\theta_x Y_s(z)\,ds \\ = & 2\gamma_x\theta_x\int_0^t \left\langle \left(a_{c_x(t-s)}(\cdot,\alpha)-a_{c_x(t-s)}(\cdot,\beta)\right)^2, Y_s(\cdot)\right\rangle ds \quad . \end{aligned} \tag{6.33}$$

Now we have finished our preparations to calculate the Laplace transform of the l.h.s. of (6.33). The right hand side of this equation is a random occupation time – we have learned (compare p. 49) to characterize Laplace transforms of random occupation times. To abbreviate the right hand side of (6.33), we define:

$$\aleph(r) := 2\gamma_x\theta_x\int_0^r \left\langle \left(a_{c_x(t-s)}(\cdot,\alpha)-a_{c_x(t-s)}(\cdot,\beta)\right)^2, Y_s(\cdot)\right\rangle ds \quad (0\le r\le t)\,. \tag{6.34}$$

Then we have

$$\exp\left(-\mathbf{E}^Y_{reac}\left[\left(\Delta^{(2)}_t\right)^2\right]\right) = \exp\left(-\aleph(t)\right) \quad . \tag{6.35}$$

Let

$$g^{(\alpha,\beta)}_\rho(x) := 2\gamma_x\theta_x\left(a_{c_x\rho}(\alpha,x)-a_{c_x\rho}(\beta,x)\right)^2, \quad \left(x\in\mathbb{Z}^d, \rho\in\mathbb{R}_+\right). \tag{6.36}$$

Then $\aleph$ can be described as a random occupation time by

$$\aleph(r) = \int_0^r \left\langle g^{(\alpha,\beta)}_{t-s}, Y_s\right\rangle ds \quad . \tag{6.37}$$

We have learned how to handle laws of random occupation times in terms of Laplace transforms (compare Proposition 1, p. 49). In particular, we know that the following equation of duality holds:

$$\exp\left(-\int_0^r \left\langle g^{(\alpha,\beta)}_{t-s}, Y_s\right\rangle ds\right) = \exp(-\langle Y_0, h^r_r\rangle \tag{6.38}$$

where the family of dual processes $\left\{(h^r_s)_{s\in[0,r]}\right\}_{r\in[0,t]}$ is defined as follows

$$\begin{aligned} h^r_0 &\equiv 0 \quad , \qquad (6.39) \\ \frac{\partial}{\partial s}h^r_s(i) &= \frac{c_y}{2}\sum_{j\in\mathbb{Z}^d} a(j,i)\left[h_s(j)-h_s(i)\right] \\ &\quad -\left(h^r_s(i)\right)^2 + g^{(\alpha,\beta)}_{r-(t-s)}(i) \quad (s\in(0,r])\,. \qquad (6.40) \end{aligned}$$

In particular, we get

$$\exp\left(-\int_0^t \left\langle g_{t-s}^{(\alpha,\beta)}, Y_s\right\rangle ds\right) = \exp(-\left\langle Y_0, h_t^t\right\rangle \tag{6.41}$$

with

$$\begin{aligned} h_0^t &\equiv 0 \ , & (6.42)\\ \frac{\partial}{\partial s} h_s^t(i) &= \frac{c_y}{2}\sum_{j\in\mathbb{Z}^d} a(j,i)\left[h_s^t(j) - h_s^t(i)\right] - \left(h_s^t(i)\right)^2 + g_s^{(\alpha,\beta)}(i) \quad . & (6.43)\end{aligned}$$

Since h^t does not depend on t, we drop the upper index t and define $(h_t)_{t\geq 0}$ by

$$\begin{aligned} h_0 &\equiv 0 \ , & (6.44)\\ \frac{\partial}{\partial s} h_s(i) &= \frac{c_y}{2}\sum_{j\in\mathbb{Z}^d} a(j,i)\left[h_s(j) - h_s(i)\right] - \left(h_s(i)\right)^2 + g_s^{(\alpha,\beta)}(i) \quad . & (6.45)\end{aligned}$$

Then we get for arbitrary t

$$\exp\left(-\mathbf{E}_{reac}^Y\left[\left(\Delta_t^{(2)}\right)^2\right]\right) = \exp\left(-\int_0^t \left\langle g_{t-s}^{(\alpha,\beta)}, Y_s\right\rangle ds\right) = \exp(-\left\langle h_t, Y_0\right\rangle) \quad . \tag{6.46}$$

Thus – taking $\mathbf{E}_{cat}$ on both sides of (6.46) – we have obtained a representation for the Laplace transform

$$\mathbf{E}_{cat}\left[\exp\left(-\lambda\mathbf{E}_{reac}^Y\left[\left(\Delta_t^{(2)}\right)^2\right]\right)\right] \quad (\lambda > 0) \tag{6.47}$$

at $\lambda = 1$.

Since Laplace transforms are convex and decreasing, we only have to prove

$$\lim_{t\to\infty} \mathbf{E}_{cat}\left[\exp\left(-\mathbf{E}_{reac}^Y\left[\left(\Delta_t^{(2)}\right)^2\right]\right)\right] = 1 \tag{6.48}$$

to obtain

$$\mathcal{L}\left\{\mathbf{E}_{reac}^Y\left[\left(\Delta_t^{(2)}\right)^2\right]\right\} \overset{t\to\infty}{\Longrightarrow} \delta_{\{0\}} \quad . \tag{6.49}$$

We prove (6.48) as follows. We consider the expectation $\mathbf{E}_{cat}$ of the right hand side of (6.46) and look for a lower bound:

$$\begin{aligned} &\liminf_{t\to\infty} \mathbf{E}_{cat}\left[\exp\left(-\left\langle Y_0, h_t\right\rangle\right)\right] \\ \geq\ &\liminf_{t\to\infty} \mathbf{E}_{cat}\left[1 - \left\langle Y_0, h_t\right\rangle\right] \\ =\ &\liminf_{t\to\infty} \left(1 - \left\langle \theta_y \mathfrak{n}, h_t\right\rangle\right) \quad . \end{aligned} \tag{6.50}$$

Hence we are left with showing

$$\lim_{t\to\infty} \left\langle \mathfrak{n}, h_t\right\rangle = 0 \quad . \tag{6.51}$$

By the "finite input vanishes" property (5.64) it is sufficient for (6.51) that we have

$$\int_0^\infty \left\langle \mathfrak{n}, g_s^{(\alpha,\beta)} \right\rangle ds < \infty \quad . \tag{6.52}$$

To prove this, we need to know how to calculate $\sum_{z\in\mathbb{Z}^d} \int_0^t a_{c_x s}(\beta, z)\, a_{c_x s}(\alpha, z)\, ds$:

How to calculate $\sum_{z\in\mathbb{Z}^d} \int_0^t a_{c_x s}(\beta, z)\, a_{c_x s}(\alpha, z)\, ds$

Let $(\xi_t^\alpha)_{t\geq 0}$, $\left(\xi_t^\beta\right)_{t\geq 0}$ be two independent random walks, defined on a common probability space, with

$$P\left[\xi_t^\alpha = z\right] = a_{c_x t}(\alpha, z) \quad , \tag{6.53}$$

$$P\left[\xi_t^\beta = z\right] = a_{c_x t}(\beta, z) \quad , \tag{6.54}$$

Considering $\left(\xi_t^\alpha - \xi_t^\beta\right)_{t\geq 0}$, which is a random walk with jump rate $2c_x$ and jump kernel $\hat{a}$, we get:

$$P\left[\xi_t^\alpha - \xi_t^\beta = z\right] = \hat{a}_{2c_x t}(\alpha - \beta, z) \quad . \tag{6.55}$$

In particular, this yields:

$$\begin{aligned} &\sum_{z\in\mathbb{Z}^d} \int_0^t a_{c_x s}(\beta, z)\, a_{c_x s}(\alpha, z)\, ds \\ &= \int_0^t P\left[\xi_s^\alpha = \xi_s^\beta\right] ds \\ &= \int_0^t \hat{a}_{2c_x s}(\alpha, \beta)\, ds \quad . \end{aligned} \tag{6.56}$$

We get:

$$\begin{aligned} &\int_0^\infty \left\langle \mathfrak{n}, g_s^{(\alpha,\beta)} \right\rangle ds \\ &= \lim_{t\to\infty} \sum_{z\in\mathbb{Z}^d} \int_0^t \left(a_{c_x s}(\alpha, z) - a_{c_x s}(\beta, z)\right)^2 ds \\ &= \lim_{t\to\infty} \sum_{z\in\mathbb{Z}^d} \int_0^t \Bigg(\left(a_{c_x s}(\alpha, z)\right)^2 + \left(a_{c_x s}(\beta, z)\right)^2 \\ &\quad -2 a_{c_x s}(\beta, z)\, a_{c_x s}(\alpha, z) \Bigg) ds \quad . \end{aligned} \tag{6.57}$$

Applying (6.56), we get

$$= 2 \lim_{t\to\infty} \int_0^t \left(\hat{a}_{2c_x s}(0,0) - \hat{a}_{2c_x s}(\alpha,\beta) \right) ds$$

$$= 2 \lim_{t\to\infty} \int_0^t \left(\hat{a}_{2c_x s}(0,0) - \hat{a}_{2c_x s}(0,\beta-\alpha) \right) ds \quad . \tag{6.58}$$

The fact that the right hand side is finite is a pure random walk property and will be proven in the appendix (page 243).

Relation (6.52) allows to apply the "finite input vanishes" property (5.64): We get

$$\lim_{t\to\infty} \langle \mathfrak{n}, h_t \rangle = 0 \quad . \tag{6.59}$$

Together with (6.50) and (6.46) this yields the claimed relation (6.26), namely

$$\mathcal{L}_{cat} \left\{ \mathbf{E}^Y_{reac} \left[\left(\Delta^{(2)}_t \right)^2 \right] \right\} \quad \overset{t\to\infty}{\Longrightarrow} \quad \delta_{\{0\}} \quad . \tag{6.60}$$

This concludes the proof of (6.1).

6.1.2 Non-extinction of the reactant

We can give a proof that the reactant is persistent under the following condition:

Condition 1 *There exists* $K > 0$ *with*

$$a_{c_x t}(\cdot, 0) \leq K \cdot a_{c_y t}(\cdot, 0) \quad . \tag{6.61}$$

Remark 14 *The following is necessary for (6.61) to be true:*

- *There is no drift:*

$$\sum_{i \in \mathbb{Z}^d} i a(0, i) = 0 \quad . \tag{6.62}$$

- *The mobility of the catalyst is at least as large as the mobility of the reactant:*

$$c_x \leq c_y \quad . \tag{6.63}$$

- *In particular, (6.61) holds if* $a(\cdot, \cdot)$ *is the symmetric next-neighbor random walk transition kernel.*

Theorem 0 claims that the catalyst is not locally extinct in the long run if both $d \leq 2$ and (6.61) hold. Conclusion 2 (p. 64) provides us with a criterion for persistence. We have to prove that

$$\left\{ \mathcal{L}_{cat} \left\{ \int_0^t \sum_{k \in \mathbb{Z}^d} (a_s(k, 0))^2 Y_{t-s}(k)\, ds \right\} \right\}_{t \geq 0} \tag{6.64}$$

is tight.

We can formulate this claim in terms of Laplace transforms:

Lemma 5 *Assume that we have*

$$\lim_{\lambda \to 0} \liminf_{t \to \infty} \mathbf{E} \left[\exp \left(-\lambda \int_0^t \sum_{k \in \mathbb{Z}^d} (a_s(k, 0))^2 Y_{t-s}(k)\, ds \right) \right] = 1 \quad . \tag{6.65}$$

Then there is tightness of (6.64).

Proof. We assume for the moment that (6.64) is not tight. Then there exists $\beta > 0$ such that for all $K > 0$ there exists an increasing sequence $(\vartheta_n^K)_{n \in \mathbb{N}}$ with $\lim_{n \to \infty} \vartheta_n^K = \infty$ and

$$P \left[\int_0^{\vartheta_n^K} \sum_{\ell \in \mathbb{Z}^d} (a_s(\ell, 0))^2 Y_{\vartheta_n^K - s}(\ell)\, ds > K \right] \geq \beta \quad \forall n \in \mathbb{N}. \tag{6.66}$$

We can draw the conclusion

$$\begin{aligned}
&\mathbf{E}\left[\exp\left(-\lambda\int_0^{\vartheta_n^K}\sum_{\ell\in\mathbb{Z}^d}(a_s(\ell,0))^2\,Y_{\vartheta_n-s}(\ell)\,ds\right)\right]\\
&= \mathbf{E}\left[\exp\left(-\lambda\int_0^{\vartheta_n^K}\sum_{\ell\in\mathbb{Z}^d}(a_s(\ell,0))^2\,Y_{\vartheta_n^K-s}(\ell)\,ds\right)\mathbb{1}_{\left\{\int_0^{\vartheta_n^K}\sum_{\ell\in\mathbb{Z}^d}(a_s(\ell,0))^2Y_{\vartheta_n^K-s}(\ell)\,ds>K\right\}}\right]\\
&\quad+\mathbf{E}\left[\exp\left(-\lambda\int_0^{\vartheta_n^K}\sum_{\ell\in\mathbb{Z}^d}(a_s(\ell,0))^2\,Y_{\vartheta_n^K-s}(\ell)\,ds\right)\mathbb{1}_{\left\{\int_0^{\vartheta_n^K}\sum_{\ell\in\mathbb{Z}^d}(a_s(\ell,0))^2Y_{\vartheta_n^K-s}(\ell)\,ds\le K\right\}}\right]\\
&\le \exp(-\lambda K)\,P\left[\int_0^{\vartheta_n^K}\sum_{\ell\in\mathbb{Z}^d}(a_s(\ell,0))^2\,Y_{\vartheta_n^K-s}(\ell)\,ds>K\right]\\
&\quad+P\left[\int_0^{\vartheta_n^K}\sum_{\ell\in\mathbb{Z}^d}(a_s(\ell,0))^2\,Y_{\vartheta_n^K-s}(\ell)\,ds\le K\right]\\
&\le \exp(-\lambda K)+(1-\beta)\quad.
\end{aligned}\tag{6.67}$$

Since for all $K>0$ we have

$$\begin{aligned}
&\liminf_{t\to\infty}\mathbf{E}\left[\exp\left(-\lambda\int_0^t\sum_{k\in\mathbb{Z}^d}(a_s(k,0))^2\,Y_{t-s}(k)\,ds\right)\right]\\
&\le \mathbf{E}\left[\exp\left(-\lambda\int_0^{\vartheta_n^K}\sum_{\ell\in\mathbb{Z}^d}(a_s(\ell,0))^2\,Y_{\vartheta_n-s}(\ell)\,ds\right)\right]\quad,
\end{aligned}\tag{6.68}$$

we can conclude (compare (6.67)) that for arbitrary $\lambda>0$ we have

$$\liminf_{t\to\infty}\mathbf{E}\left[\exp\left(-\lambda\int_0^t\sum_{k\in\mathbb{Z}^d}(a_s(k,0))^2\,Y_{t-s}(k)\,ds\right)\right]\le 1-\beta\quad.\tag{6.69}$$

In particular, this yields

$$\limsup_{\lambda\to0+}\liminf_{t\to\infty}\mathbf{E}\left[\exp\left(-\lambda\int_0^t\sum_{k\in\mathbb{Z}^d}(a_s(k,0))^2\,Y_{t-s}(k)\,ds\right)\right]\le 1-\beta\quad.\tag{6.70}$$

This is a contradiction to (6.65). ∎

Using Proposition 1, p. 49, we characterize the Laplace transform

$$\mathbf{E}\left[\exp\left(-\lambda\int_0^t\sum_{\ell\in\mathbb{Z}^d}(a_s(\ell,0))^2\,Y_{t-s}(\ell)\,ds\right)\right]\tag{6.71}$$

by its dual process

$$u_0^\lambda \equiv 0 \tag{6.72}$$

$$\begin{aligned}\frac{\partial}{\partial t}u_t^\lambda(i) &= c_y \sum_{j\in\mathbb{Z}^d} a(j,i)\left(u_t^\lambda(j) - u_t^\lambda(i)\right) \\ &\quad - \gamma_y \left(u_t^\lambda(i)\right)^2 \\ &\quad + \lambda \left(a_{c_x t}(j,0)\right)^2 \quad \left(i \in \mathbb{Z}^d\right) \quad .\end{aligned} \tag{6.73}$$

The following duality relation holds:

$$\mathbf{E}\left[\exp\left(-\lambda \int_0^t \sum_{k\in\mathbb{Z}^d} (a_s(k,0))^2\, Y_{t-s}(k)\, ds\right)\right] = \mathbf{E}\left[\exp\left(-\left\langle Y_0, u_t^\lambda\right\rangle\right)\right] \quad . \tag{6.74}$$

Hence, we have to prove

$$\lim_{\lambda\to 0+} \liminf_{t\to\infty} \mathbf{E}\left[\exp\left(-\left\langle Y_0, u_t^\lambda\right\rangle\right)\right] = 1 \quad . \tag{6.75}$$

With Jensen's inequality and $\mathbf{E}Y_0 = \theta_y \mathfrak{n}$ we get[1]

$$\begin{aligned}\mathbf{E}\left[\exp\left(-\left\langle Y_0, u_t^\lambda\right\rangle\right)\right] &\geq \exp \mathbf{E}\left[\left(-\left\langle Y_0, u_t^\lambda\right\rangle\right)\right] \\ &\geq \exp\left(-\left\langle \theta_y \mathfrak{n}, u_t^\lambda\right\rangle\right) \quad .\end{aligned} \tag{6.76}$$

Hence, it remains to prove

$$\lim_{\lambda\to 0+} \limsup_{t\to\infty} \left\langle \mathfrak{n}, u_t^\lambda\right\rangle = 0 \quad . \tag{6.77}$$

As we have assumed in (6.61) there exists $K < \infty$ with

$$a_{c_x t}(\cdot,0) \leq K \cdot a_{c_y t}(\cdot,0) \quad . \tag{6.78}$$

This bound enables us to construct $\hat{u}^\lambda$ with $\hat{u}^\lambda > u^\lambda$. This way, we obtain a supersolution of equation (6.73) with initial condition (6.72). The function $\hat{u}^\lambda$ is given by

$$\hat{u}_0^\lambda \equiv K\sqrt{\frac{\lambda}{\gamma_y}}\delta_0 \tag{6.79}$$

$$\begin{aligned}\frac{\partial}{\partial t}\hat{u}_t^\lambda(i) &= c_y \sum_{j\in\mathbb{Z}^d} a(j,i)\left[\hat{u}_t^\lambda(j) - \hat{u}_t^\lambda(i)\right] \\ &\quad + \lambda\left(K \cdot a_{c_y t}(\cdot,0)\right)^2 \quad \left(i\in\mathbb{Z}^d\right) \quad .\end{aligned} \tag{6.80}$$

In fact

$$\hat{u}^\lambda > u^\lambda$$

[1] Note that u^λ does not depend from Y.

follows from the monotonicity property provided by Lemma 3, p. 51. The equations (6.79), (6.80) can be solved explicitly:

$$\hat{u}_t^\lambda(i) = K\sqrt{\frac{\lambda}{\gamma_y}}a_{c_y t}(i,0) \tag{6.81}$$

We can conclude from

$$\left\langle \mathfrak{n}, \hat{u}_t^\lambda \right\rangle = K\sqrt{\frac{\lambda}{\gamma_y}} \overset{\lambda\to 0+}{\longrightarrow} 0 \tag{6.82}$$

that the reactant does not suffer local extinction.

Remark 15 *We can even obtain more precise results for $c_x = c_y$. In this case, the equation corresponding to (6.81) is*

$$\hat{u}_t^\lambda(i) = \sqrt{\frac{\lambda}{\gamma_y}}a_{c_y t}(i,0) \quad . \tag{6.83}$$

This supersolution of (6.73) exceeds the solution u_t^λ only as a consequence of its initial condition, not as a consequence of its evolution. Hence it is tempting to introduce processes $\left(q_t^\lambda\right)_{t\geq 0}$ $(\lambda \geq 0)$ as upper bound of the differences $\left(\hat{u}_t^\lambda - u_t^\lambda\right)_{t\geq 0}$. We prove that the following is a good choice: Let

$$\frac{\partial}{\partial t}q_t^\lambda(i) = c_y \sum_{j\in\mathbb{Z}^d} a(j,i)\left(q_t^\lambda(j) - q_t^\lambda(i)\right) - \gamma_y\left(q_t^\lambda(i)\right)^2 \quad \left(i\in\mathbb{Z}^d\right) \tag{6.84}$$

$$q_0^\lambda \equiv \sqrt{\frac{\lambda}{\gamma_y}}\delta_0 \quad . \tag{6.85}$$

We have the following properties of $\left(q_t^\lambda\right)_{t\geq 0}$ at our disposal:

a) We can use the "finite input vanishes" property (5.64), from which we can conclude

$$\lim_{t\to\infty}\left\langle \mathfrak{n}, q_t^\lambda\right\rangle = 0 \quad . \tag{6.86}$$

b) In addition to that, by (5.62) there is subadditivity :

$$u_t^\lambda + q_t^\lambda \geq \hat{u}_t^\lambda \quad , \tag{6.87}$$

hence

$$\hat{u}_t^\lambda - u_t^\lambda \leq q_t^\lambda \quad .$$

We can now conclude:

$$\lim_{t\to\infty}\left\langle \mathfrak{n}, \hat{u}_t^\lambda - u_t^\lambda\right\rangle = 0 \quad . \tag{6.88}$$

Hence we know that in this special case the distribution of the expected family size converges to a $\frac{1}{2}$-stable distribution. This result implies non-extinction of the reactant.

6.2 High-dimensional case

Now and henceforth we assume $d \geq 3$. We have to prove the following:

a) Assume
$$\mathcal{L}\{(X_0, Y_0)\} \in S^{(\theta_x, \theta_y)} \quad . \tag{6.89}$$
Let $(\varsigma_n) \in \mathbb{R}_+^{\mathbb{N}}$ denote an increasing sequence with $\lim\limits_{n\to\infty} \varsigma_n = \infty$.
Then any accumulation point (w.r.t. the weak topology) of $(\mathcal{L}\{(X_{\varsigma_n}, Y_{\varsigma_n})\})$ is a probability measure on $\mathcal{M}(\mathbb{Z}^d) \times \mathcal{M}(\mathbb{Z}^d)$ with intensity (θ_x, θ_y).

b) For given $\theta_x, \theta_y > 0$ there exists a stationary process with law
$$\nu_{(\theta_x,\theta_y)} \in \mathcal{M}^1\left(\mathcal{C}\left(\mathbb{R}_+, \mathcal{M}\left(\mathbb{Z}^d\right) \times \mathcal{M}\left(\mathbb{Z}^d\right)\right)\right) \tag{6.90}$$
such that for any initial condition
$$\mathcal{L}\{(X_0, Y_0)\} \in S^{(\theta_x, \theta_y)} \tag{6.91}$$
we have convergence of the laws of paths, namely
$$\mathcal{L}\left\{(X_{t+s}, Y_{t+s})_{s\geq 0}\right\} \overset{t\to\infty}{\Longrightarrow} \nu_{(\theta_x,\theta_y)} \quad . \tag{6.92}$$

Proof of a)

We have to verify uniform integrability of the family of random variables $\{X_t(i), Y_t(i)\}_{t\geq 0}$ for every $i \in \mathbb{Z}^d$. It suffices to show
$$\sup_{t\in\mathbb{R}_+} \mathbf{E}\left[(X_t(i))^2\right] < \infty \tag{6.93}$$
and
$$\sup_{t\in\mathbb{R}_+} \mathbf{E}\left[(Y_t(i))^2\right] < \infty \quad . \tag{6.94}$$
This follows immediately from the moment lemma proven in Appendix C (Lemma 16, p. 257).

Proof of b)

The proof consists of three parts. We first give a survey what will be proved in each of the three parts.

I. *Convergence of finite dimensional distributions*
Let $n \in \mathbb{N}$, $0 \leq t_1 < t_2 < ... < t_n < \infty$. Let
$$\mathcal{T} := (t_1, ..., t_n) \in \mathbb{R}_+^n \quad . \tag{6.95}$$
We have to show that for any choice of $n, \mathcal{T}$ there exists a unique measure
$$\mu^{\mathcal{T}}_{(\theta_x,\theta_y)} \in \mathcal{M}^1\left(\left(\mathcal{M}\left(\mathbb{Z}^d\right) \times \mathcal{M}\left(\mathbb{Z}^d\right)\right)^{\otimes n}\right) \tag{6.96}$$

with

$$\mathcal{L}\left\{(X_{s+t}, Y_{s+t})_{t\in\mathcal{T}}\right\} \overset{s\to\infty}{\Longrightarrow} \mu^{\mathcal{T}}_{(\theta_x,\theta_y)} \quad . \tag{6.97}$$

In particular, we claim that there exists $\mu^{(0)}_{(\theta_x,\theta_y)} \in \mathcal{M}^1\left(\mathcal{M}\left(\mathbb{Z}^d\right) \times \mathcal{M}\left(\mathbb{Z}^d\right)\right)$ with

$$\mathcal{L}\left\{(X_s, Y_s)\right\} \overset{s\to\infty}{\Longrightarrow} \mu^{(0)}_{(\theta_x,\theta_y)} \quad . \tag{6.98}$$

II. *Pathwise convergence*
We define $\nu_{(\theta_x,\theta_y)}$ as the law of the process given by (1.6), (1.7) and initial condition $\mu^{(0)}_{(\theta_x,\theta_y)}$.
We know from I. that $\mathcal{L}\left\{(X_{\varsigma+t}, Y_{\varsigma+t})_{t\geq 0}\right\}$ converges to $\nu_{(\theta_x,\theta_y)}$ as $\varsigma \to \infty$ in terms of finite dimensional distributions. Let $(\varsigma_i)_{i\in\mathbb{N}} \in \mathbb{R}^{\mathbb{N}}_+$ be a strictly increasing sequence with

$$\lim_{i\to\infty} \varsigma_i = \infty \quad . \tag{6.99}$$

We have to show that $\mathcal{L}\left\{(X_{\varsigma_i+t}, Y_{\varsigma_i+t})_{t\geq 0}\right\}$ converges to $\nu_{(\theta_x,\theta_y)}$ in terms of weak convergence on $\mathcal{M}^1\left(\mathcal{C}\left(\mathbb{R}_+, \mathcal{M}\left(\mathbb{Z}^d\right) \times \mathcal{M}\left(\mathbb{Z}^d\right)\right)\right)$. We know from (6.97) that the only candidate for being an accumulation point of

$$\left(\mathcal{L}\left\{(X_{\varsigma_i+t}, Y_{\varsigma_i+t})_{t\geq 0}\right\}\right)_{i\in\mathbb{N}} \tag{6.100}$$

is $\nu_{(\theta_x,\theta_y)}$. It remains to verify the existence of accumulation points of the set

$$\mathcal{L}^{(\varsigma_i)} := \left\{\mathcal{L}\left\{(X_{\varsigma+t}, Y_{\varsigma+t})_{t\geq 0}\right\}, \varsigma \in \{\varsigma_1, \varsigma_2, \ldots\}\right\} \quad .$$

As there exist accumulation points if $\mathcal{L}^{(\varsigma_i)}$ is relatively compact with respect to the underlying topology induced by weak convergence on the set of probability measures on the pathspace $\mathcal{M}^1\left(\mathcal{C}\left(\mathbb{R}_+, \mathcal{M}\left(\mathbb{Z}^d\right) \times \mathcal{M}\left(\mathbb{Z}^d\right)\right)\right)$, and as relative compactness of sets containing probability laws on Polish spaces is equivalent to the tightness property of these laws, the following claim is sufficient for (6.92).

Claim 1 *We claim:*

$$\mathcal{L}^{(\varsigma_i)} \textit{ is tight.} \tag{6.101}$$

III. *Stationarity:*
Consider the measure $\mu^{(0)}_{(\theta_x,\theta_y)}$ introduced in (6.98). It is uniquely characterized as weak limit point of $\mathcal{L}\left\{(X_s, Y_s)\right\}$ as $s \to \infty$. We claim:

Claim 2 *The measure $\mu^{(0)}_{(\theta_x,\theta_y)}$ is stationary with respect to the evolution semigroup corresponding to the evolution equations (1.6) and (1.7).*

Proof of I. - Convergence of finite dimensional distributions

Thoughout the proof, we will change our point of view: Instead of using time $t = 0$ as time of initialization, we rather start the processes at times $T < 0$ and compare their distributions at times $-t_1, -t_2, ..., -t_n$ letting $T \to -\infty$.
Let $(\mathfrak{X}, \mathfrak{Y})$ be a random variable with values in $\mathcal{M}(\mathbb{Z}^d) \times \mathcal{M}(\mathbb{Z}^d)$ and $\mathcal{L}\{(\mathfrak{X}, \mathfrak{Y})\} \in S^{(\theta_x, \theta_y)}$.

This random variable serves as initialization of processes $\left(X_t^T, Y_t^T\right)_{t \in [T,0]}$. Let these processes solve the evolution equations (1.6) and (1.7).

Namely, we can write

$$\left(X_T^T, Y_T^T\right) = (\mathfrak{X}, \mathfrak{Y}) \quad , \tag{6.102}$$

and for $t \in (T, \infty)$ let the evolution be given by the system of stochastic differential equations

$$\begin{aligned} dX_t^T(i) &= c_x \sum_{j \in \mathbb{Z}^d} a(j,i) \left[X_t^T(j) - X_t^T(i)\right] dt \\ &\quad + \sqrt{2\gamma_x X_t^T(i) Y_t^T(i)}\, dB_t^x(i) \quad , \end{aligned} \tag{6.103}$$

$$\begin{aligned} dY_t^T(i) &= c_y \sum_{j \in \mathbb{Z}^d} a(j,i) \left[Y_t^T(j) - Y_t^T(i)\right] dt \\ &\quad + \sqrt{2\gamma_y Y_t^T(i)}\, dB_t^y(i) \quad \left(i \in \mathbb{Z}^d\right). \end{aligned} \tag{6.104}$$

Let $\tilde{\mathcal{T}} := \{-t_n, ..., -t_1\}$. We have to prove that there exists

$$\mu_{(\theta_x,\theta_y)}^{\tilde{\mathcal{T}}} \in \mathcal{M}^1\left(\left(\mathcal{M}\left(\mathbb{Z}^d\right) \times \mathcal{M}\left(\mathbb{Z}^d\right)\right)^{\otimes n}\right) \tag{6.105}$$

with

$$\mathcal{L}\left\{\left(X_\tau^T, Y_\tau^T\right)_{\tau \in \tilde{\mathcal{T}}}\right\} \overset{T \to -\infty}{\Longrightarrow} \mu_{(\theta_x,\theta_y)}^{\tilde{\mathcal{T}}} \quad . \tag{6.106}$$

Let $B \subset\subset \mathbb{Z}^d$. Consider a finite set of test functions $(f_k, g_k)_{k \in \{1,...,n\}}$ with

$$f_k, g_k \in \mathbb{R}_+^{\mathbb{Z}^d} \quad (k \in \{1, ..., n\}) \tag{6.107}$$

and

$$\{f_k > 0\} \subset B, \quad \{g_k > 0\} \subset B \quad . \tag{6.108}$$

Consider the family

$$\left\{\mathcal{L}\left\{\sum_{k=1}^n \left(\langle f_k, X_{-t_k}^T\rangle + \langle g_k, Y_{-t_k}^T\rangle\right)\right\}\right\}_{T<0} \quad . \tag{6.109}$$

We have to verify that we have weak convergence of (6.109) as $T \to -\infty$ for any collection of test functions $(f_k), (g_k)$ satisfying (6.108). Since we consider a family of non-negative random variables, and since expectations are uniformly bounded, namely

$$\sup_{T<0} \mathbf{E}\left[\sum_{k=1}^n \left(\langle f_k, X_{-t_k}^T\rangle + \langle g_k, Y_{-t_k}^T\rangle\right)\right] = \sum_{k=1}^n \left(\langle f_k, \theta_x\rangle + \langle g_k, \theta_y\rangle\right) < \infty \quad , \tag{6.110}$$

the family $\left\{\mathcal{L}\left\{\sum_{k=1}^{n}\left(\langle f_k, X_{-t_k}^T\rangle + \langle g_k, Y_{-t_k}^T\rangle\right)\right\}\right\}_{T<0}$ is a tight family of probability measures. Then the limit of the Laplace transforms

$$\mathbf{E}\left[\exp\left(-\lambda\sum_{k=1}^{n}\left(\langle f_k, X_{-t_k}^T\rangle + \langle g_k, Y_{-t_k}^T\rangle\right)\right)\right] \quad ,(\lambda > 0) \tag{6.111}$$

– if it exists – is again a Laplace transform of a probability measure on $\mathbb{R}_+$.[2] Since this limit Laplace transform characterizes the limit probability law, it only remains to prove that the limit of (6.111) exists as $T \to -\infty$ for all choices of $(f_k), (g_k)$.

At this point, we can make use of the fact that the catalyst converges weakly on the path space: There exists $\nu_{\theta_y} \in \mathcal{M}^1\left(\mathcal{C}\left(\mathbb{R}_+, \mathcal{M}\left(\mathbb{Z}^d\right)\right)\right)$ with

$$\mathcal{L}\left\{\left(Y_{t+s}\right)_{s\geq 0}\right\} \underset{t\to\infty}{\Longrightarrow} \nu_{\theta_y} \quad . \tag{6.115}$$

This result can be found in [CG94] and [Sh92], but only in terms of convergence with respect to finite dimensional distributions. The tightness argument needed to obtain (6.115) looks similar to the proof of tightness we give for the catalyst-reactant system later on (compare II.)[3]. As law of a stationary process, ν_{θ_y} can be extended canonically to

$$\nu_{\theta_y}^{\pm} \in \mathcal{M}^1\left(\mathcal{C}\left(\mathbb{R}, \mathcal{M}\left(\mathbb{Z}^d\right)\right)\right), \tag{6.116}$$

($\pm$ denotes that the paths are indexed by $\mathbb{R}$) describing the stationary process living on $(-\infty, +\infty)$. [This is an immediate consequence of the Kolmogoroff extension theorem. By time shift, it is possible to construct consistent families of marginal distributions even for negative times. Then the process can be constructed as a projective limit.] In terms of our setup to start processes at times $T < 0$, letting $T \to -\infty$, we get the following reformulation of (6.115):

$$\mathcal{L}\left\{\left(Y_{t\vee T}^T\right)_{t\in\mathbb{R}}\right\} \underset{T\to-\infty}{\Longrightarrow} \nu_{\theta_y}^{\pm} \quad . \tag{6.117}$$

---

[2]According to [Fe], XIII.1, Theorem 2 it is sufficient for a pointwise limit $(\varphi(\lambda))_{\lambda\geq 0}$ of Laplace transforms $(\varphi_n(\lambda))_{\lambda\geq 0}$ $(n \in \mathbb{N})$ to be a Laplace transform that

$$\lim_{\lambda\to 0+} \varphi(\lambda) = 1. \tag{6.112}$$

Consider a tight family $(\Xi_n)_{n\in\mathbb{N}}$ of $\mathbb{R}_+$-valued random variables with Laplace transforms φ_n. Property (6.112) follows from tightness as the following calculation shows:

$$\begin{aligned}
&\lim_{\lambda\to 0+}\lim_{n\to\infty} \mathbf{E}\left[\exp\left(-\lambda\Xi_n\right)\right] \\
&= \lim_{\lambda\to 0+}\lim_{n\to\infty}\left(\mathbf{E}\left[\exp\left(-\lambda\Xi_n\right)\mathbf{I}_{\left\{\Xi_n > \frac{1}{\sqrt{\lambda}}\right\}}\right] + \mathbf{E}\left[\exp\left(-\lambda\Xi_n\right)\mathbf{I}_{\left\{\Xi_n \leq \frac{1}{\sqrt{\lambda}}\right\}}\right]\right) \\
&\geq \lim_{\lambda\to 0+}\inf_{n\to\infty}\left(\mathbf{E}\left[\exp\left(-\frac{\lambda}{\sqrt{\lambda}}\right)\mathbf{I}_{\left\{\Xi_n \leq \frac{1}{\sqrt{\lambda}}\right\}}\right]\right) \\
&\geq \lim_{\lambda\to 0+}\left(1-\sqrt{\lambda}\right)\inf_{n\to\infty}\left(\mathbf{P}\left[\Xi_n \leq \frac{1}{\sqrt{\lambda}}\right]\right) \\
&= 1 \quad .
\end{aligned} \tag{6.113}$$

In the last step, we have applied the tightness property. We finally get:

$$\lim_{\lambda\to 0+}\lim_{n\to\infty} \mathbf{E}\left[\exp\left(-\lambda\Xi_n\right)\right] = 1 \quad . \tag{6.114}$$

[3]The only thing one would have to do is to replace the catalyst by a constant.

Remark 16 *In (6.117), we had to write "$t \vee T$" to obtain a sequence of processes mapping $\mathbb{R}$ onto the path-space. By this construction, the limiting objects on the l.h.s. of (6.117) and the limit object on the r.h.s. of (6.117) are of the same type. We want to simplify notation in the forthcoming paragraphs:*

We regard processes (with values in some set E) given by an initial condition at time T and a certain evolution mechanism not as processes with paths in $\mathcal{C}([T,\infty), E)$, but in $\mathcal{C}(\mathbb{R}, E)$ by the following convention:
Consider $(Z_t)_{t\in[T,\infty)} \in \mathcal{C}([T,\infty), E)$. Then, let Z_t be continued canonically by

$$Z_t := Z_T \text{ for } t < T \quad . \tag{6.118}$$

According to the Skorohod representation (compare page 40) there exists a probability space with probability measure denoted[4] by $\mathbb{P}$, on which we can define the following two objects:

- a family of processes with state space $\mathcal{M}(\mathbb{Z}^d) \times \mathcal{M}(\mathbb{Z}^d)$

$$\left\{ \left(\hat{X}_t^T, \hat{Y}_t^T \right)_{t\in\mathbb{R}} \right\}_{T<0} \tag{6.119}$$

- a limit catalyst

$$\left(\hat{Y}_t^{-\infty} \right)_{t\in\mathbb{R}} \tag{6.120}$$

  with

$$\mathbb{L}\left\{ \left(\hat{Y}_t^{-\infty} \right)_{t\in\mathbb{R}} \right\} = \nu_{\theta_y}^{\pm} \quad . \tag{6.121}$$

These objects shall satisfy

- *The laws of the processes coincide*

$$\forall T < 0 : \mathbb{L}\left\{ \left(\hat{X}_t^T, \hat{Y}_t^T \right)_{t\in\mathbb{R}} \right\} = \mathcal{L}\left\{ \left(X_t^T, Y_t^T \right)_{t\in\mathbb{R}} \right\} \tag{6.122}$$

- *Application of the Skorohod representation to the catalyst*

$$\forall B \subset \mathbb{Z}^d \text{ bounded, } \forall \vartheta > 0 :$$
$$\lim_{T\to-\infty} \sup_{\substack{r\in[-\vartheta,0] \\ i\in B}} d\left(\hat{Y}_r^T(i), \hat{Y}_r^{-\infty}(i) \right) = 0 \quad \mathbb{P}\text{-a.s.} \tag{6.123}$$

4

a) We denote the law of a random variable X with respect to $\mathbb{P}$ by $\mathbb{L}$.

b) The catalyst-reactant decomposition is denoted as follows:

$$\mathbb{P}[\cdot] = \int \mathbb{P}_{reac}^{\hat{Y}}[\cdot] \quad \mathbb{P}_{cat}\left(d\hat{Y}\right)$$

c) Expectation is denoted by $\mathbb{E}$, quenched expectation by $\mathbb{E}_{reac}^{\hat{Y}}$.

We have claimed that (6.111) converges for all choices of $\lambda, (f_k), (g_k)$. It is possible to absorbe λ in (f_k) and (g_k). This way, we can reformulate the claimed convergence of (6.111) as follows:

Claim 3 *The limit*

$$\lim_{T\to-\infty} \mathbf{E}\left[\exp\left(-\sum_{k=1}^{n}\left(\left\langle f_k, \hat{X}^T_{-t_k}\right\rangle + \left\langle g_k, \hat{Y}^T_{-t_k}\right\rangle\right)\right)\right] \tag{6.124}$$

exists.

Next, we make use of the catalyst-reactant setting (compare the discussion in the "Methods and Tools" Section carried out in 5.2). We consider the Laplace transforms describing the finite dimensional distributions of the reactant $\hat{X}^T$ for given catalyst paths $\left\{\hat{Y}^T\right\}_{T<0}$, namely

$$\mathbb{E}^{\hat{Y}}_{reac}\left[\exp\left(-\sum_{k=1}^{n}\left\langle f_k, \hat{X}^T_{-t_k}\right\rangle\right)\right] \quad . \tag{6.125}$$

Note that this suffices since by (6.123) the catalyst paths converge as $T \to -\infty$ (in the sense of compact convergence, as described by (6.123)) $\mathbb{P}$-a. s.
To calculate the Laplace transforms, we can make use of the following dual representation[5] (compare Proposition 1, p. 49).

$$\mathbb{E}^{\hat{Y}}_{reac}\left[\exp\left(-\sum_{k=1}^{n}\left\langle f_k, \hat{X}^T_{-t_k}\right\rangle\right)\right] = \mathbb{E}^{\hat{Y}}_{reac}\left[\exp\left(-\left\langle u^T_{-T}, \mathfrak{X}\right\rangle\right)\right] \tag{6.126}$$

where $\left(u^T_t(i)\right)_{i\in\mathbb{Z}^d}$ $(T \le t \le 0)$ is given by

$$u^T_0 \equiv 0 \quad , \tag{6.127}$$

$$\frac{\partial}{\partial t}u^T_t(i) = \sum_{j\in\mathbb{Z}^d} c_x a(j,i)\left[u^T_t(j) - u^T_t(i)\right] - \gamma_x \hat{Y}^T_{-t}(i)\left(u^T_t(i)\right)^2 \quad \text{for } t \notin \{t_1,...,t_n\} \quad . \tag{6.128}$$

$$u^T_t(i) = \lim_{s\to t-} u^T_t(i) + f_k(i) \quad \text{for } t = t_k, (k \in \{1,...,n\}) \quad . \tag{6.129}$$

We have to prove that the right hand side of (6.126) converges in L^1 as $T \to -\infty$. We have to compare $\left(u^T_t\right)$ with $\left(u^{-\infty}_t\right)$, which we define next. We have in mind that $\left(u^{-\infty}_t\right)$ shall correspond to the limit case $T \to -\infty$, in particular with limit catalyst $\hat{Y}^{-\infty}$. Therefore the following definition intuitively makes sense:

$$u^{-\infty}_0 \equiv 0 \quad , \tag{6.130}$$

$$\frac{\partial}{\partial t}u^{-\infty}_t(i) = \sum_{j\in\mathbb{Z}^d} c_x a(j,i)\left[u^{-\infty}_t(j) - u^{-\infty}_t(i)\right] - \gamma_x \hat{Y}^{-\infty}_{-t}(i)\left(u^{-\infty}_t(i)\right)^2 \quad \text{for } t \notin \{t_1,...,t_n\} \quad . \tag{6.131}$$

$$u^{-\infty}_t(i) = \lim_{s\to t-} u^{-\infty}_t(i) + f_k(i) \quad \text{for } t = t_k, (k \in \{1,...,n\}) \quad . \tag{6.132}$$

[5] it can be applied for P_{cat}-almost all catalysts

We claim that – taking (6.126) into account the limit of (6.111) is related to $\left(u_t^{-\infty}\right)_{t\geq 0}$ in the following way:

$$\begin{aligned}\lim_{T\to-\infty} \mathbb{E}_{cat}\Big|&\mathbb{E}_{reac}^{\hat{Y}}\left[\exp\left(-\left\langle u_{-T}^{T},\mathfrak{X}\right\rangle\right)\right]\\ &-\lim_{\tilde{T}\to-\infty}\mathbb{E}_{reac}^{\hat{Y}}\left[\exp\left(-\left\langle u_{-\tilde{T}}^{-\infty},\theta_x\mathfrak{n}\right\rangle\right)\right]\Big|=0\end{aligned} \tag{6.133}$$

Before proving (6.133) we still have to prove that the limit occurring in the second line is well-defined: By (6.131) we get:

$$\frac{\partial}{\partial t}\left\langle u_{-\tilde{T}}^{-\infty},\mathfrak{n}\right\rangle=-\sum_{i\in\mathbb{Z}^d}\gamma_x\hat{Y}_{-t}^{-\infty}(i)\left(u_t^{-\infty}(i)\right)^2<0\quad\text{for } t\notin\{t_1,...,t_n\}. \tag{6.134}$$

Therefore, $\left\langle u_t^{-\infty},\mathfrak{n}\right\rangle$ is decreasing for $t\in(t_n,\infty)$. Since it is non-negative and finite (by (6.108)), it converges as $t\to+\infty$.

The proof of (6.133) consists of two steps:

1. We have to verify

$$\lim_{T\to-\infty}\mathbb{E}_{cat}\left|\mathbb{E}_{reac}^{\hat{Y}}\left[\exp\left(-\left\langle u_{-T}^{T},\mathfrak{X}\right\rangle\right)\right]-\mathbb{E}_{reac}^{\hat{Y}}\left[\exp\left(-\left\langle u_{-T}^{T},\theta_x\mathfrak{n}\right\rangle\right)\right]\right|=0\quad. \tag{6.135}$$

2. We have to prove

$$\lim_{T\to-\infty}\mathbb{E}_{cat}\left|\mathbb{E}_{reac}^{\hat{Y}}\left[\exp\left(-\left\langle u_{-T}^{T},\theta_x\mathfrak{n}\right\rangle\right)\right]-\mathbb{E}_{reac}^{\hat{Y}}\left[\exp\left(-\left\langle u_{-\tilde{T}}^{-\infty},\theta_x\mathfrak{n}\right\rangle\right)\right]\right|=0\quad. \tag{6.136}$$

Proof of 1.: With dominated convergence, it suffices to prove

$$\begin{aligned}&\lim_{T\to-\infty}\mathbb{E}_{reac}^{\hat{Y}}\left[\exp\left(-\left\langle u_{-T}^{T},\mathfrak{X}\right\rangle\right)\right]\\ =&\lim_{T\to-\infty}\mathbb{E}_{reac}^{\hat{Y}}\left[\exp\left(-\left\langle u_{-T}^{T},\theta_x\mathfrak{n}\right\rangle\right)\right]\quad\mathbb{P}_{cat}-a.s.\end{aligned} \tag{6.137}$$

To obtain this, we prove

$$\lim_{T\to-\infty}\mathbb{E}_{reac}^{\hat{Y}}\left[\left(\left\langle u_{-T}^{T},\mathfrak{X}\right\rangle-\left\langle u_{-T}^{T},\theta_x\mathfrak{n}\right\rangle\right)^2\right]=0\quad\mathbb{P}_{cat}\text{-a. s.}\quad. \tag{6.138}$$

Since for arbitrary $T<0$

$$\mathbb{E}_{reac}^{\hat{Y}}\left[\left\langle u_{-T}^{T},\mathfrak{X}\right\rangle\right]=\mathbb{E}_{reac}^{\hat{Y}}\left[\left\langle u_{-T}^{T},\theta_x\mathfrak{n}\right\rangle\right]\quad\mathbb{P}_{cat}-a.s. \tag{6.139}$$

holds, we have to prove

$$\lim_{T\to-\infty}var_{reac}^{\hat{Y}}\left[\left\langle u_{-T}^{T},\mathfrak{X}\right\rangle\right]=0\quad\mathbb{P}_{cat}-a.s. \tag{6.140}$$

We know from the definition of u^T (compare (6.128)) that for arbitrary $i\in\mathbb{Z}^d$

$$u_{-T}^{T}(i)\leq\sum_{j\in\mathbb{Z}^d}\sum_{k=1}^{n}f_k(j)\,a_{-T-t_k}(j,i) \tag{6.141}$$

holds. To continue, note that we have assumed (compare (1.21)) that $\mathfrak{X}(i)$ and $\mathfrak{X}(j)$ have non-negative correlation for any $i, j \in \mathbb{Z}^d$. This yields by (6.141):

$$\begin{aligned} & var^{\hat{Y}}_{reac}\left[\langle u^T_{-T}, \mathfrak{X}\rangle\right] \\ \leq \; & var^{\hat{Y}}_{reac}\left[\sum_{j\in\mathbb{Z}^d}\sum_{i\in\mathbb{Z}^d}\sum_{k=1}^{n} f_k(j)\, a_{-T-t_k}(j,i)\mathfrak{X}(i)\right] \quad \mathbb{P}_{cat}-a.s. \end{aligned} \tag{6.142}$$

Let

$$\bar{f} := \max\{f_k(j) : j \in B, k \in \{1, ..., n\}\} \quad . \tag{6.143}$$

We get (with the property of non-negative correlation mentioned above):

$$\begin{aligned} & var^{\hat{Y}}_{reac}\left[\sum_{j\in\mathbb{Z}^d}\sum_{i\in\mathbb{Z}^d}\sum_{k=1}^{n} f_k(j)\, a_{-T-t_k}(j,i)\mathfrak{X}(i)\right] \\ \leq \; & var^{\hat{Y}}_{reac}\left[\sum_{j\in B}\sum_{i\in\mathbb{Z}^d}\sum_{k=1}^{n} \bar{f}\cdot a_{-T-t_k}(j,i)\mathfrak{X}(i)\right] \\ = \; & \bar{f}^2\cdot var^{\hat{Y}}_{reac}\left[\sum_{j\in B}\sum_{i\in\mathbb{Z}^d}\sum_{k=1}^{n} a_{-T-t_k}(j,i)\mathfrak{X}(i)\right] \\ = \; & \bar{f}^2\cdot var^{\hat{Y}}_{reac}\left[\sum_{j\in B}\sum_{k=1}^{n}\sum_{i\in\mathbb{Z}^d} a_{-T-t_k}(j,i)\mathfrak{X}(i)\right] \\ \leq \; & \bar{f}^2\mathbf{E}^{\hat{Y}}_{reac}\left[n\,|B|\sum_{j\in B}\sum_{k=1}^{n}\left(\sum_{i\in\mathbb{Z}^d} a_{-T-t_k}(j,i)\left(\mathfrak{X}(i)-\theta_x\right)\right)^2\right] \end{aligned} \tag{6.144}$$

Since catalyst and reactant are independent at initialization, we have (applying that the expectation is linear)

$$= n\,|B|\,\bar{f}^2\cdot\sum_{j\in B}\mathbf{E}\left[\sum_{k=1}^{n}\left(\sum_{i\in\mathbb{Z}^d} a_{-T-t_k}(j,i)\left(\mathfrak{X}(i)-\theta_x\right)\right)^2\right]$$

Together with

$$\lim_{t\to\infty} var\left[\sum_{j\in\mathbb{Z}^d} a_t(j,i)\mathfrak{X}(i)\right] = 0 \tag{6.145}$$

– (this was an assumption on the initial condition, compare (1.38) and (1.39)) – we get (6.138).

Proof of 2.: We have to prove that the following holds:

$$\lim_{T\to-\infty}\mathbb{E}_{cat}\left|\mathbb{E}^{\hat{Y}}_{reac}\left[\exp\left(-\langle u^T_{-T}, \theta_x\mathfrak{n}\rangle\right)\right]-\mathbb{E}^{\hat{Y}}_{reac}\left[\exp\left(-\langle u^{-\infty}_{-T}, \theta_x\mathfrak{n}\rangle\right)\right]\right| = 0 \quad . \tag{6.146}$$

We consider

$$\mathbb{E}_{reac}^{\hat{Y}}\left[\left|\exp\left(-\left\langle u_{-T}^{T},\theta_x\mathfrak{n}\right\rangle\right)-\exp\left(-\left\langle u_{-T}^{-\infty},\theta_x\mathfrak{n}\right\rangle\right)\right|\right] \quad . \tag{6.147}$$

It is sufficient to prove

$$\lim_{T\to-\infty}\mathbf{E}\left[\left\langle\left|u_{-T}^{T}-u_{-T}^{-\infty}\right|,\mathfrak{n}\right\rangle\wedge 1\right]=0 \quad . \tag{6.148}$$

For that purpose, we have to compare $\left(u_t^T\right)_{t\geq 0}$ and $\left(u_t^{-\infty}\right)_{t\geq 0}$ which satisfy:

$$\begin{aligned}\frac{\partial}{\partial t}u_t^T(i) &= \sum_{j\in\mathbb{Z}^d}c_x a(j,i)\left[u_t^T(j)-u_t^T(i)\right]\\ &\quad -\gamma_x\hat{Y}_{-t}^T(i)\left(u_t^T(i)\right)^2 \quad \text{for } t\notin\{t_1,...,t_n\}. &(6.149)\\ u_t^T(i) &= \lim_{s\to t-}u_t^T(i)+f_k(i) \quad \text{for } t=t_k,(k\in\{1,...,n\}) &(6.150)\end{aligned}$$

and

$$\begin{aligned}\frac{\partial}{\partial t}u_t^{-\infty}(i) &= \sum_{j\in\mathbb{Z}^d}c_x a(j,i)\left[u_t^{-\infty}(j)-u_t^{-\infty}(i)\right]\\ &\quad -\gamma_x\hat{Y}_{-t}^{-\infty}(i)\left(u_t^{-\infty}(i)\right)^2 \quad \text{for } t\notin\{t_1,...,t_n\} \quad , &(6.151)\\ u_t^{-\infty}(i) &= \lim_{s\to t-}u_t^{-\infty}(i)+f_k(i) \quad \text{for } t=t_k,(k\in\{1,...,n\}) \quad . &(6.152)\end{aligned}$$

Note that by (6.150) and (6.152) the jumps coincide for both, $\left(u_t^T\right)_{t\geq 0}$ and $\left(u_t^{-\infty}\right)_{t\geq 0}$. Introducing a notation for the difference

$$D_t^T(i):=u_t^T(i)-u_t^{-\infty}(i) \quad \left(i\in\mathbb{Z}^d\right) \tag{6.153}$$

and using the identity

$$ab-a'b'=\frac{1}{2}\left[\left(a+a'\right)\left(b-b'\right)+\left(a-a'\right)\left(b+b'\right)\right] \quad , \tag{6.154}$$

we get[6]

$$\begin{aligned}\frac{\partial}{\partial t}D_t^T(i) &= \sum_{j\in\mathbb{Z}^d}c_x a(j,i)\left[D_t^T(j)-D_t^T(i)\right]\\ &\quad -\gamma_x\left[\hat{Y}_{-t}^T(i)\left(u_t^T(i)\right)^2-\hat{Y}_{-t}^{-\infty}(i)\left(u_t^{-\infty}(i)\right)^2\right]\\ &= \sum_{j\in\mathbb{Z}^d}c_x a(j,i)\left[D_t^T(j)-D_t^T(i)\right]\\ &\quad -\frac{1}{2}\gamma_x\left(\hat{Y}_{-t}^T(i)+\hat{Y}_{-t}^{-\infty}(i)\right)D_t^T(i)\left(u_t^T(i)+u_t^{-\infty}(i)\right)\\ &\quad +\frac{1}{2}\gamma_x\left(\hat{Y}_{-t}^{-\infty}(i)-\hat{Y}_{-t}^T(i)\right)\left(\left(u_t^T(i)\right)^2+\left(u_t^{-\infty}(i)\right)^2\right) &(6.155)\end{aligned}$$

[6]At jump-times, the right sided differential is considered.

We want to prove

$$\lim_{T\to-\infty} \mathbf{E}\left[\left\langle \left|D^T_{-T}\right|, \mathfrak{n}\right\rangle \wedge 1\right] = 0 \quad . \tag{6.156}$$

To approximate $x \longmapsto |x|$, which is not differentiable at zero, we introduce the following sequence of approximation functions. Let $(g_n) \in \left(\mathbb{R}_+^{\mathbb{R}_+}\right)^{\mathbb{N}}$ be defined by

$$g_n(x) := \frac{1}{n} \ln \cosh(nx) \quad (n \in \mathbb{N})\,. \tag{6.157}$$

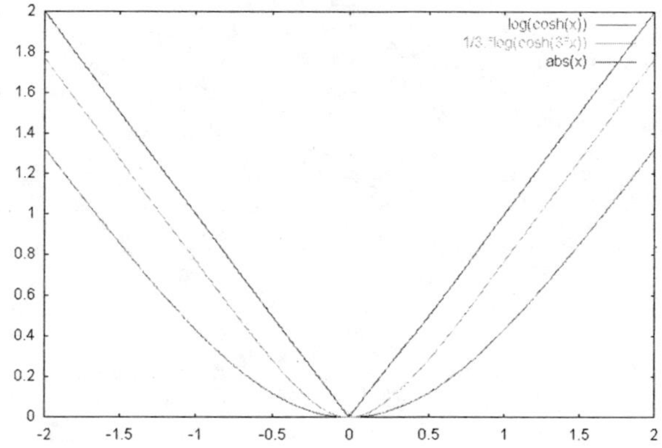

Approximation of $|\bullet|$ by $1/n \ln \cosh(n\bullet)$

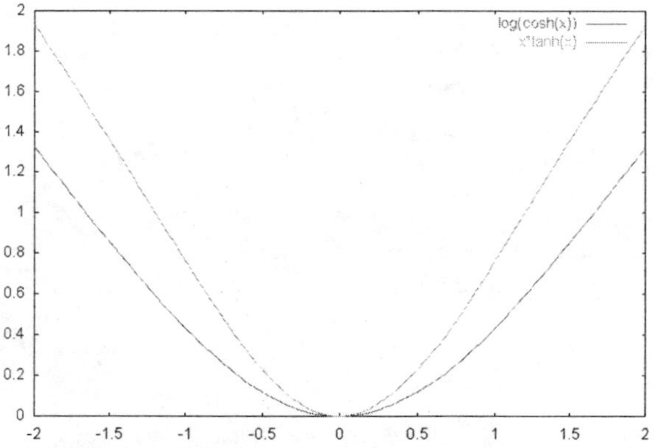

Approximation by $1/n \ln \cosh(nx)$ and $x \tanh(nx)$

We need the following properties of $(g_n)_{n\in\mathbb{N}}$:

1. $g_n'(x) = \tanh(nx)$, $n \in \mathbb{N}$

2. $g_n(x) \leq x g_n'(x) \leq |x| \leq g_n(x) + \frac{\ln 2}{n}$, $n \in \mathbb{N}$.

Proof of 1. This is obtained immediately by differentiation.

Proof of 2. This follows from

$$\begin{aligned} -\sinh(|nx|) &\leq 0 \\ &\Longleftrightarrow \\ \cosh(|nx|) &\leq e^{|nx|} \\ &\Longleftrightarrow \\ \frac{1}{n}\ln\cosh(|nx|) &\leq |x| \quad . \end{aligned}$$

and

$$\begin{aligned} e^{|nx|} &\leq e^{|nx|} + e^{-|nx|} \\ &\Longleftrightarrow \\ e^{|nx|} &\leq 2\cosh(|nx|) \\ &\Longleftrightarrow \\ x &\leq \frac{1}{n}(\ln 2 + \ln\cosh(|nx|)) \quad . \end{aligned} \tag{6.158}$$

The property $x g_n'(x) \leq |x|$ immediately follows from $|\tanh(nx)| < 1$. There remains to prove

$$g_n(x) \leq x g_n'(x) \quad (x \in \mathbb{R}). \tag{6.159}$$

For $x = 0$, we have equality. With symmetry, we only have to discuss the case $x > 0$. We obtain (6.159) by differentiation:

$$\begin{aligned} \frac{\partial}{\partial x}\left(x g_n'(x) - g_n(x)\right) &= \frac{\partial}{\partial x}\left(x\tanh(nx) - \frac{1}{n}\ln\cosh(nx)\right) \\ &= \left[\tanh nx + nx - nx\tanh^2 nx\right] - \tanh nx \\ &= nx\left(1 - \tanh^2 nx\right) \\ &> 0 \text{ for } x > 0 \quad . \end{aligned} \tag{6.160}$$

Applying the chain-rule to (6.155), we get:

$$\begin{aligned} \frac{\partial}{\partial t} g_n\left(D_t^T(i)\right) = &\sum_{j\in\mathbb{Z}^d} c_x a(j,i)\left[g_n'\left(D_t^T(i)\right) D_t^T(j) - g_n'\left(D_t^T(i)\right) D_t^T(i)\right] \\ &-\frac{1}{2}\gamma_x\left(\hat{Y}_{-t}^T(i) + \hat{Y}_{-t}^{-\infty}(i)\right) g_n'\left(D_t^T(i)\right) D_t^T(i)\left(u_t^T(i) + u_t^{-\infty}(i)\right) \\ &+g_n'\left(D_t^T(i)\right)\frac{1}{2}\gamma_x\left(\hat{Y}_{-t}^{-\infty}(i) - \hat{Y}_{-t}^T(i)\right)\left(\left(u_t^T(i)\right)^2 + \left(u_t^{-\infty}(i)\right)^2\right) \end{aligned}$$

Since $g_n'\left(D_t^T(i)\right) D_t^T(i) > 0$,we get an upper bound by leaving out the second line:

$$\begin{aligned}
&\leq \sum_{j\in\mathbb{Z}^d} c_x a(j,i)\left[g_n'\left(D_t^T(i)\right) D_t^T(j) - g_n'\left(D_t^T(i)\right) D_t^T(i)\right] \\
&\quad + \frac{1}{2}\gamma_x \left|\hat{Y}_{-t}^{-\infty}(i) - \hat{Y}_{-t}^T(i)\right| \left(\left(u_t^T(i)\right)^2 + \left(u_t^{-\infty}(i)\right)^2\right) \\
&\leq \sum_{j\in\mathbb{Z}^d} c_x a(j,i)\left[\left|D_t^T(j)\right| - g_n\left(D_t^T(i)\right)\right] \\
&\quad + \frac{1}{2}\gamma_x \left|\hat{Y}_{-t}^{-\infty}(i) - \hat{Y}_{-t}^T(i)\right| \left(\left(u_t^T(i)\right)^2 + \left(u_t^{-\infty}(i)\right)^2\right) \quad . \qquad (6.161)
\end{aligned}$$

Define $\tilde{n} : \mathbb{N} \times \mathbb{Z}^d \to \mathbb{R}_+$ by

$$\tilde{n}(n,i) = \exp\left(\left(|i|+1\right)n\right) \quad . \qquad (6.162)$$

We get:

$$\begin{aligned}
&\frac{\partial}{\partial t} g_{\tilde{n}(n,i)}\left(D_t^T(i)\right) \\
&\leq \sum_{j\in\mathbb{Z}^d} c_x a(j,i)\left[\left|D_t^T(j)\right| - g_{\tilde{n}(n,i)}\left(D_t^T(i)\right)\right] \\
&\quad + \frac{1}{2}\gamma_x \left|\hat{Y}_{-t}^{-\infty}(i) - \hat{Y}_{-t}^T(i)\right| \left(\left(u_t^T(i)\right)^2 + \left(u_t^{-\infty}(i)\right)^2\right) \\
&\leq \sum_{j\in\mathbb{Z}^d} c_x a(j,i)\left[g_{\tilde{n}(n,j)}\left(D_t^T(j)\right) - g_{\tilde{n}(n,i)}\left(D_t^T(i)\right)\right] \\
&\quad + \sum_{j\in\mathbb{Z}^d} c_x a(j,i) \frac{\ln 2}{\exp\left(\left(|j|+1\right)n\right)} \\
&\quad + \frac{1}{2}\gamma_x \left|\hat{Y}_{-t}^{-\infty}(i) - \hat{Y}_{-t}^T(i)\right| \left(\left(u_t^T(i)\right)^2 + \left(u_t^{-\infty}(i)\right)^2\right) \quad . \qquad (6.163)
\end{aligned}$$

This yields:

$$\begin{aligned}
&\frac{\partial}{\partial t}\left(\sum_{i\in\mathbb{Z}^d} g_{\tilde{n}(n,i)}\left(D_t^T(i)\right)\right) \\
&\leq \sum_{i\in\mathbb{Z}^d}\sum_{j\in\mathbb{Z}^d} c_x a(j,i) \frac{\ln 2}{\exp\left(\left(|j|+1\right)n\right)} \\
&\quad + \sum_{i\in\mathbb{Z}^d} \frac{1}{2}\gamma_x \left|\hat{Y}_{-t}^{-\infty}(i) - \hat{Y}_{-t}^T(i)\right| \left(\left(u_t^T(i)\right)^2 + \left(u_t^{-\infty}(i)\right)^2\right) \\
&\leq \sum_{j\in\mathbb{Z}^d} c_x \frac{\ln 2}{\exp\left(\left(|j|+1\right)n\right)} \\
&\quad + \sum_{i\in\mathbb{Z}^d} \frac{1}{2}\gamma_x \left|\hat{Y}_{-t}^{-\infty}(i) - \hat{Y}_{-t}^T(i)\right| \left(\left(u_t^T(i)\right)^2 + \left(u_t^{-\infty}(i)\right)^2\right) \quad . \qquad (6.164)
\end{aligned}$$

By integration, we get (recall that we consider $T < 0$)

$$\begin{aligned}
&\sum_{i\in\mathbb{Z}^d} g_{\tilde{n}(n,i)}\left(D_{-T}^{T}(i)\right)\\
\leq\ & |T|\cdot\sum_{j\in\mathbb{Z}^d} c_x\frac{\ln 2}{\exp\left(\left(|j|+1\right)n\right)}\\
&+\int_0^{|T|}\sum_{i\in\mathbb{Z}^d}\frac{1}{2}\gamma_x\left|\hat{Y}_{-s}^{-\infty}(i)-\hat{Y}_{-s}^{T}(i)\right|\left(\left(u_s^T(i)\right)^2+\left(u_s^{-\infty}(i)\right)^2\right)ds \quad . \qquad (6.165)
\end{aligned}$$

We evidently have constructed $\tilde{n}$ in such a way that we have

- $$\lim_{n\to\infty}\sum_{i\in\mathbb{Z}^d} g_{\tilde{n}(n,i)}\left(D_{-T}^{T}(i)\right)=\left\langle\mathfrak{n},\left|D_{-T}^{T}\right|\right\rangle \quad , \qquad (6.166)$$

- $$\lim_{n\to\infty}|T|\sum_{j\in\mathbb{Z}^d} c_x\frac{\ln 2}{\exp\left(\left(|j|+1\right)n\right)}=0 \quad . \qquad (6.167)$$

This is the reason why the following is sufficient for (6.148) to be true:

$$\lim_{T\to-\infty}\mathbb{E}_{cat}\left[\int_0^{-T}\sum_{i\in\mathbb{Z}^d}\left|\hat{Y}_{-t}^{-\infty}(i)-\hat{Y}_{-t}^{T}(i)\right|\left(\left(u_t^T(i)\right)^2+\left(u_t^{-\infty}(i)\right)^2\right)dt\wedge 1\right]=0 \quad . \qquad (6.168)$$

We only increase the values of $u_t^T(i)$ and $u_t^{-\infty}(i)$ if we neglect the negative summands of the dual evolution equations (6.149) and (6.151). This motivates the following definition. For $t\geq 0$ and $\mathfrak{f}=(f_k)_{k\in\{1,\ldots,n\}}$, let

$$\tilde{S}_t\mathfrak{f}(\cdot):=\sum_{j\in\mathbb{Z}^d}\sum_{k=1}^{n} f_k(j)\,a_{t-t_k}(\cdot,j)\mathbb{1}_{\{t>t_k\}} \quad . \qquad (6.169)$$

In particular, we clearly have

$$\left(u_t^T(i)\right)^2+\left(u_t^{-\infty}(i)\right)^2\leq 2\left(\tilde{S}_t\mathfrak{f}(i)\right)^2 \quad . \qquad (6.170)$$

As next step, we want to prove

$$\int_0^{\infty}\sum_{i\in\mathbb{Z}^d}\left(\tilde{S}_t\mathfrak{f}(i)\right)^2 dt<\infty \quad . \qquad (6.171)$$

This follows from

$$\int_0^\infty \sum_{i\in\mathbb{Z}^d} \left(\tilde{S}_t\mathfrak{f}(i)\right)^2 dt$$

$$= \int_0^\infty \sum_{i\in\mathbb{Z}^d} \left(\sum_{j\in\mathbb{Z}^d} \sum_{k=1}^n f_k(j)\, a_{t-t_k}(i,j) \mathbb{I}_{\{t>t_k\}} \right)^2 dt$$

$$= \int_0^\infty \sum_{i\in\mathbb{Z}^d} \sum_{j_1\in\mathbb{Z}^d} \sum_{j_2\in\mathbb{Z}^d} \sum_{k_1=1}^n \sum_{k_2=1}^n f_{k_1}(j_1)\, f_{k_2}(j_2)$$

$$\cdot\, a_{t-t_{k_1}}(i,j_1) a_{t-t_{k_2}}(i,j_2) \mathbb{I}_{\{t>t_{k_1}\vee t_{k_2}\}} dt$$

$$= \int_0^\infty \sum_{j_1\in\mathbb{Z}^d} \sum_{j_2\in\mathbb{Z}^d} \sum_{k_1=1}^n \sum_{k_2=1}^n f_{k_1}(j_1)\, f_{k_2}(j_2)\, \hat{a}_{2t-t_{k_1}-t_{k_2}}(j_2,j_1) \mathbb{I}_{\{t>t_{k_1}\vee t_{k_2}\}} dt \quad . \tag{6.172}$$

In the last step, we have applied (6.56) – that's where the symmetrized random walk kernel comes from.

By (6.108), the sum (6.172) contains only finitely many non-vanishing terms. Since the transient case is considered, this yields (6.171), namely

$$\int_0^\infty \sum_{i\in\mathbb{Z}^d} \left(\tilde{S}_t\mathfrak{f}(i)\right)^2 dt < \infty \quad . \tag{6.173}$$

We can conclude that for given $\varepsilon > 0$ there exists a bounded set $A \subset \mathbb{R}_+ \times \mathbb{Z}^d$ s. t.

$$\int_0^\infty \sum_{i\in\mathbb{Z}^d} \left(\tilde{S}_t\mathfrak{f}(i)\right)^2 \mathbb{I}_{\{(t,i)\notin A\}}(t,i) \le \varepsilon \quad . \tag{6.174}$$

Using the upper bound provided by (6.170) and splitting up summation and integration with respect to A, we get a chain of estimates

$$\begin{aligned}
&\limsup_{T\to-\infty}\mathbb{E}_{cat}\left[\int_0^{-T}\sum_{i\in\mathbb{Z}^d}\left|\hat{Y}_{-t}^{-\infty}(i)-\hat{Y}_{-t}^{T}(i)\right|\left(\left(u_t^T(i)\right)^2+\left(u_t^{-\infty}(i)\right)^2\right)dt\wedge 1\right]\\
\leq\;&\limsup_{T\to-\infty}2\mathbb{E}_{cat}\left[\int_0^{-T}\sum_{i\in\mathbb{Z}^d}\left|\hat{Y}_{-t}^{-\infty}(i)-\hat{Y}_{-t}^{T}(i)\right|\left(\tilde{S}_t\mathfrak{f}(i)\right)^2 dt\wedge 1\right]\\
\leq\;&2\mathbb{E}_{cat}\left[\int_0^{\infty}\sum_{i\in\mathbb{Z}^d}\left|\hat{Y}_{-t}^{-\infty}(i)-\hat{Y}_{-t}^{T}(i)\right|\left(\tilde{S}_t\mathfrak{f}(i)\right)^2\mathbb{1}_{\{(t,i)\notin A\}}(t,i)\,dt\wedge 1\right]\\
&+2\limsup_{T\to-\infty}\mathbb{E}_{cat}\left[\int_0^{\infty}\sum_{i\in\mathbb{Z}^d}\left|\hat{Y}_{-t}^{-\infty}(i)-\hat{Y}_{-t}^{T}(i)\right|\left(\tilde{S}_t\mathfrak{f}(i)\right)^2\mathbb{1}_{\{(t,i)\in A\}}(t,i)\,dt\wedge 1\right]\\
\leq\;&\limsup_{T\to-\infty}2\mathbb{E}_{cat}\left[\int_0^{\infty}\sum_{i\in\mathbb{Z}^d}\left|\hat{Y}_{-t}^{-\infty}(i)-\hat{Y}_{-t}^{T}(i)\right|\left(\tilde{S}_t\mathfrak{f}(i)\right)^2\mathbb{1}_{\{(t,i)\in A\}}(t,i)\,dt\wedge 1\right]\\
&+2\theta_y\varepsilon
\end{aligned}\tag{6.175}$$

As A is a bounded set, we can conclude from (6.123) that the first summand of (6.175) converges to zero. Since $\varepsilon>0$ was arbitrarily chosen and $\tilde{S}_t\mathfrak{f}$ does not depend from T, we finally get:

$$\lim_{T\to-\infty}\mathbb{E}_{cat}\left[\left\langle\left|u_{-T}^T-u_{-T}^{-\infty}\right|,\mathfrak{n}\right\rangle\right]=0 \quad . \tag{6.176}$$

This completes the proof of 2., i. e. (6.146).

Proof of II. - Pathwise convergence

We have proved step I of our original three step program. For step II it remains to show that there is convergence not only in terms of finite dimensional distributions, but weakly on the pathspace. We have to prove that for any $\vartheta<0$ the family

$$\left\{\mathcal{L}\left\{\left(X_{t\vee T}^T,Y_{t\vee T}^T\right)_{t\in[\vartheta,0]}\right\}\right\}_{T<0} \tag{6.177}$$

is a tight family. We need to prove the following:

- Tightness of initial conditions (at $t=\vartheta$)
- The Kolmogorov criterion (see [EK86], Proposition 3.6.3) holds for all component processes

The laws of initial conditions (in this case: laws of the processes $\left(X^T,Y^T\right)$ at time ϑ) are a tight family, since for $i\in\mathbb{Z}^d$:

$$\mathbf{E}\left[\left(X_{\vartheta\vee T}^T(i)\right)\right]=\theta_x<\infty,\quad \mathbf{E}\left[\left(Y_{\vartheta\vee T}^T(i)\right)\right]=\theta_y<\infty \tag{6.178}$$

holds by assumption on the initial state (compare 1.2.4, p. 15)

We have to show that the Kolmogorov criterion is fulfilled, namely:

Claim 4 *There exist $C < \infty$, $\tilde{p} > 2$ such that for $t, u \in [0, |\vartheta|]$ (the considered processes are defined on $[-|\vartheta|, 0]$) we have*

$$\limsup_{T \to -\infty} \mathbf{E}\left[\left|X^T_{\vartheta+t}(0) - X^T_{\vartheta+u}(0)\right|^{\tilde{p}}\right] \leq C\,|t-u|^{\frac{\tilde{p}}{2}} \tag{6.179}$$

and

$$\limsup_{T \to -\infty} \mathbf{E}\left[\left|Y^T_{\vartheta+t}(0) - Y^T_{\vartheta+u}(0)\right|^{\tilde{p}}\right] \leq C\,|t-u|^{\frac{\tilde{p}}{2}} \quad . \tag{6.180}$$

Without loss of generality, assume $t > u$.

We only provide a proof of (6.179) – (6.180) can be verified by similar - but simpler - calculations.

By writing in integral form the stochastic differential equation (1.6) defining the reactant process we get

$$\begin{aligned} & X^T_{\vartheta+t}(0) - X^T_{\vartheta+u}(0) \\ = & \int\limits_{\vartheta+u}^{\vartheta+t} \sum_{j \in \mathbb{Z}^d} c_x a(0,j)\left[X^T_\varrho(j) - X^T_\varrho(0)\right] d\varrho \\ & + \int\limits_{\vartheta+u}^{\vartheta+t} \sqrt{2\gamma_x X^T_\varrho(0)\, Y^T_\varrho(0)} dB^x_\varrho(0) \quad . \end{aligned} \tag{6.181}$$

We consider $\tilde{p}$-th moments, namely

$$\begin{aligned} & \mathbf{E}\left[\left|X^T_{\vartheta+t}(0) - X^T_{\vartheta+u}(0)\right|^{\tilde{p}}\right] \\ = & \mathbf{E}\Bigg[\Bigg(\int\limits_{\vartheta+u}^{\vartheta+t} \sum_{j \in \mathbb{Z}^d} c_x a(0,j)\left[X^T_\varrho(j) - X^T_\varrho(0)\right] d\varrho \\ & + \int\limits_{\vartheta+u}^{\vartheta+t} \sqrt{2\gamma_x X^T_\varrho(0)\, Y^T_\varrho(0)} dB^x_\varrho(0)\Bigg)^{\tilde{p}}\Bigg] \quad . \end{aligned} \tag{6.182}$$

By Lemma 4 (p. 69) the following inequality holds

$$\exists C_{\tilde{p}} > 0 : (\alpha+\beta)^{\tilde{p}} \leq (\tilde{p}-1)\,\alpha^{\tilde{p}} + C_{\tilde{p}}\beta^{\tilde{p}} \quad \forall \alpha, \beta \geq 0 \quad . \tag{6.183}$$

Applying this inequality to (6.182) we get:

$$\begin{aligned} & \mathbf{E}\left[\left|X^T_{\vartheta+t}(0) - X^T_{\vartheta+u}(0)\right|^{\tilde{p}}\right] \\ \leq & C_{\tilde{p}}\mathbf{E}\left[\left|\int\limits_{\vartheta+u}^{\vartheta+t} \sum_{j \in \mathbb{Z}^d} c_x a(0,j)\left[X^T_\varrho(j) - X^T_\varrho(0)\right] d\varrho\right|^{\tilde{p}}\right] \\ & + (\tilde{p}-1)\,\mathbf{E}\left[\left|\int\limits_{\vartheta+u}^{\vartheta+t} \sqrt{2\gamma_x X^T_\varrho(0)\, Y^T_\varrho(0)} dB^x_\varrho(0)\right|^{\tilde{p}}\right] \quad . \end{aligned} \tag{6.184}$$

As next step, we normalize the integral $\int_{\vartheta+u}^{\vartheta+t}$ in order to apply Jensen's inequaltity afterwards:

$$= C_{\tilde{p}}\mathbf{E}\left[(t-u)^{\tilde{p}}\left|\frac{1}{t-u}\int_{\vartheta+u}^{\vartheta+t}\sum_{j\in\mathbb{Z}^d}c_x a(0,j)\left[X_\varrho^T(j)-X_\varrho^T(0)\right]d\varrho\right|^{\tilde{p}}\right]$$

$$+(\tilde{p}-1)\,\mathbf{E}\left[\left|\int_{\vartheta+u}^{\vartheta+t}\sqrt{2\gamma_x X_\varrho^T(0)\,Y_\varrho^T(0)}dB_\varrho^x(0)\right|^{\tilde{p}}\right] \quad . \tag{6.185}$$

We can apply the following inequality (which is nothing else but the well-known Jensen inequality)

$$\forall f\in\mathbb{R}_+^{\mathbb{Z}^d}:$$

$$\left(\frac{1}{t-u}\int_u^t\sum_{j\in\mathbb{Z}^d}a(0,j)\left(f(j)\right)d\varrho\right)^{\tilde{p}}\le\frac{1}{t-u}\int_u^t\sum_{j\in\mathbb{T}_N^d}a(0,j)\left(f(j)\right)^{\tilde{p}}d\varrho \tag{6.186}$$

and obtain (in continuation of (6.185)):

$$\le C_{\tilde{p}}c_x^{\tilde{p}}\mathbf{E}\left[\frac{(t-u)^{\tilde{p}}}{t-u}\int_{\vartheta+u}^{\vartheta+t}\sum_{j\in\mathbb{Z}^d}a(0,j)\left|X_\varrho^T(j)-X_\varrho^T(0)\right|^{\tilde{p}}d\varrho\right]$$

$$+(\tilde{p}-1)\,\mathbf{E}\left[\left|\int_{\vartheta+u}^{\vartheta+t}\sqrt{2\gamma_x X_\varrho^T(0)\,Y_\varrho^T(0)}dB_\varrho^x(0)\right|^{\tilde{p}}\right] \quad . \tag{6.187}$$

The process $M(t)$ given by

$$M(t):=\int_{\vartheta+u}^{\vartheta+t}\sqrt{2\gamma_x X_\varrho^T(0)\,Y_\varrho^T(0)}dB_\varrho^x(0) \tag{6.188}$$

is a local martingale[7]. This is an immediate consequence of the fact that its increasing process

$$\langle M\rangle(t)=\int_{\vartheta+u}^{\vartheta+t}2\gamma_x X_\varrho^T(0)\,Y_\varrho^T(0)\,d\varrho \quad . \tag{6.189}$$

has finite expectation as a consequence of the finiteness of second moments proven in the appendix, where we prove (Lemma 16, p. 257) that there exists $p>2$ such that

$$\sup_{\varrho\in[T,0]}\mathbf{E}\left[\left(X_\varrho^T(0)\right)^p+\left(Y_\varrho^T(0)\right)^p\right]<\infty \quad . \tag{6.190}$$

[7]There exists an increasing sequence of stopping times (T_n) with $\lim_{n\to\infty}T_n=\infty$, s. t. $(M(t\wedge T_n))_{t\ge0}$ is a martingale (compare [KS], I.5.15)

The fact that M is a local martingale enables us to apply the Burkholder-Davis-Gundy inequality (compare [KS], III.3.28). Namely, we have

$$\forall \tilde{p} > 2 \quad \exists c_{\tilde{p}} < \infty : \sup_{0 \le r \le t-u} \mathbf{E}\left[|M_r|^{\tilde{p}}\right] \le c_{\tilde{p}} \mathbf{E}\left[\left(\langle M \rangle_{t-u}\right)^{\frac{\tilde{p}}{2}}\right] \quad .$$

We get (in continuation of (6.187))

$$\le \quad c_x^{\tilde{p}} \mathbf{E}\left[C_{\tilde{p}} (t-u)^{\tilde{p}-1} \int_{\vartheta+u}^{\vartheta+t} \sum_{j \in \mathbb{Z}^d} a(0,j) \left|X_\varrho^T(j) - X_\varrho^T(0)\right|^{\tilde{p}} d\varrho \right] \tag{6.191}$$

$$+ (\tilde{p}-1) c_{\tilde{p}} \mathbf{E}\left[\left| \int_{\vartheta+u}^{\vartheta+t} 2\gamma_x X_\varrho^T(0) Y_\varrho^T(0) d\varrho \right|^{\frac{\tilde{p}}{2}} \right] \quad . \tag{6.192}$$

We insert $\frac{1}{t-u}$ just before $\int_{\vartheta+u}^{\vartheta+t}$, compensation it by the factor $(t-u)^{\frac{\tilde{p}}{2}}$ outside of the expectation. This prepares (6.192) for another application of Jensen's inequality:

$$\le \quad c_x^{\tilde{p}} C_{\tilde{p}} (t-u)^{\tilde{p}-1} \int_{\vartheta+u}^{\vartheta+t} 2\mathbf{E}\left[\left(X_\varrho^T(0)\right)^{\tilde{p}}\right] d\varrho$$

$$+ (2\gamma_x)^{\frac{\tilde{p}}{2}} (\tilde{p}-1) c_{\tilde{p}} (t-u)^{\frac{\tilde{p}}{2}} \mathbf{E}\left[\left(\frac{1}{t-u} \int_{\vartheta+u}^{\vartheta+t} X_\varrho^T(0) Y_\varrho^T(0) d\varrho \right)^{\frac{\tilde{p}}{2}} \right] \quad . \tag{6.193}$$

We apply Jensen's inequality[8]:

$$\le \quad c_x^{\tilde{p}} C_{\tilde{p}} (t-u)^{\tilde{p}-1} \int_{\vartheta+u}^{\vartheta+t} 2\mathbf{E}\left[\left(X_\varrho^T(0)\right)^{\tilde{p}}\right] d\varrho$$

$$+ (2\gamma_x)^{\frac{\tilde{p}}{2}} (\tilde{p}-1) c_{\tilde{p}} (t-u)^{\frac{\tilde{p}}{2}} \mathbf{E}\left[\frac{1}{t-u} \int_{\vartheta+u}^{\vartheta+t} \left(X_\varrho^T(0) Y_\varrho^T(0) \right)^{\frac{\tilde{p}}{2}} d\varrho \right] \quad . \tag{6.194}$$

With

$$\left|X_\varrho^T(0) Y_\varrho^T(0)\right|^{\frac{\tilde{p}}{2}} \le \left(X_\varrho^T(0)\right)^{\tilde{p}} + \left(Y_\varrho^T(0)\right)^{\tilde{p}} \tag{6.195}$$

8

$$\left(\frac{1}{t-u} \int_{\vartheta+u}^{\vartheta+t} X_\varrho^T(0) Y_\varrho^T(0) d\varrho \right)^{\frac{\tilde{p}}{2}}$$

$$\le \frac{1}{t-u} \int_{\vartheta+u}^{\vartheta+t} \left(X_\varrho^T(0) Y_\varrho^T(0) d\varrho \right)^{\frac{\tilde{p}}{2}}$$

and

$$\int_{\vartheta+u}^{\vartheta+t} 2\mathbf{E}\left[\left|X_{\varrho}^{T}(0)\right|^{\tilde{p}}\right] d\varrho$$
$$\leq 2(t-u) \sup_{\varrho\in[T,0]} \mathbf{E}\left[\left|X_{\varrho}^{T}(0)\right|^{\tilde{p}}\right] \tag{6.196}$$

we get:

$$\begin{aligned}
&\leq 2c_x^{\tilde{p}} C_{\tilde{p}} (t-u)^{\tilde{p}} \sup_{\varrho\in[T,0]} \mathbf{E}\left[\left|X_{\varrho}^{T}(0)\right|^{\tilde{p}}\right] \\
&\quad + (2\gamma_x)^{\frac{\tilde{p}}{2}} (\tilde{p}-1)^{\tilde{p}+1} (t-u)^{\frac{\tilde{p}}{2}} \frac{1}{t-u} \int_{\vartheta+u}^{\vartheta+t} \mathbf{E}\left[\left(\left(X_{\varrho}^{T}(0)\right)^{\tilde{p}} + \left(Y_{\varrho}^{T}(0)\right)^{\tilde{p}}\right) d\varrho\right] \\
&\leq 2(t-u)^{\frac{\tilde{p}}{2}} \\
&\quad \cdot \Bigg(c_x^{\tilde{p}} C_{\tilde{p}} (t-u)^{\frac{\tilde{p}}{2}} \sup_{\varrho\in[T,0]} \mathbf{E}\left[\left|X_{\varrho}^{T}(0)\right|^{\tilde{p}}\right] \\
&\quad + (2\gamma_x)^{\frac{\tilde{p}}{2}} (\tilde{p}-1)^{\tilde{p}+1} \sup_{\varrho\in[T,0]} \mathbf{E}\left[\left(X_{\varrho}^{T}(0)\right)^{\tilde{p}} + \left(Y_{\varrho}^{T}(0)\right)^{\tilde{p}}\right] \Bigg) \quad .
\end{aligned} \tag{6.197}$$

We prove in the appendix (Lemma 16, p. 257) that there exists $\tilde{p} > 2$ such that

$$\sup_{\varrho\in[T,0]} \mathbf{E}\left[\left(X_{\varrho}^{T}(0)\right)^{\tilde{p}} + \left(Y_{\varrho}^{T}(0)\right)^{\tilde{p}}\right]$$

is finite. Together with (6.197), this completes the proof of (6.179).

Proof of III. - Stationarity

We want to prove that $\mu^{(0)}_{(\theta_x,\theta_y)}$ is stationary. Recall that $\mu^{(0)}_{(\theta_x,\theta_y)}$ was defined by

$$\mathcal{L}\left\{\left(X_0^T, Y_0^T\right)\right\} \overset{T\to-\infty}{\Longrightarrow} \mu^{(0)}_{(\theta_x,\theta_y)} \quad . \tag{6.198}$$

Clearly, $\mu^{(0)}_{(\theta_x,\theta_y)}$ is stationary iff the process $\tilde{\nu}_{(\theta_x,\theta_y)}$ defined by its initial law $\mu^{(0)}_{(\theta_x,\theta_y)}$ and the evolution equations (1.6), (1.7) is stationary[9]. We can describe $\tilde{\nu}_{(\theta_x,\theta_y)}$ as weak limit point:

$$\mathcal{L}\left\{\left(X_t^T, Y_t^T\right)_{t\in\mathbb{R}}\right\} \overset{T\to-\infty}{\Longrightarrow} \tilde{\nu}_{(\theta_x,\theta_y)} \quad . \tag{6.200}$$

By construction, $\tilde{\nu}_{(\theta_x,\theta_y)}$ follows the evolution mechanism given by (1.6) and (1.7) and has marginal distribution $\mu^{(0)}_{(\theta_x,\theta_y)}$ at time zero (by (6.198)). To prove that $\tilde{\nu}_{(\theta_x,\theta_y)}$ is the law

[9] A process $(\Xi_t)_{t\geq 0}$ is called stationary if

$$\mathcal{L}\left\{(\Xi_t)_{t\geq 0}\right\} = \mathcal{L}\left\{(\Xi_{t+s})_{t\geq 0}\right\} \tag{6.199}$$

for any $s > 0$.

of a stationary process, it suffices to ensure that it is invariant under any finite shift of time. In other words, the distribution of paths is stationary with respect to any semigroup describing time-shift. It suffices to consider left-shifts, which are defined as follows:

Definition 8 *Let E be a Polish space. The family of* left-shifts $(\mu L_s)_{s\geq 0}$ of a probability measure μ *on* $\mathcal{C}(\mathbb{R}, \mathcal{M}(E))$ *is defined as follows. Consider a continuous process* $(Z_t)_{t\in\mathbb{R}}$ *with* $\mu = \mathcal{L}\{Z\}$. *Then* μL_s $(s \geq 0)$ *is given by*

$$\mu L_s = \mathcal{L}\left\{(Z_{t+s})_{t\geq 0}\right\} \quad . \tag{6.201}$$

We claim:

$$\forall s \geq 0 : \tilde{\nu}_{(\theta_x,\theta_y)} = \tilde{\nu}_{(\theta_x,\theta_y)} L_s \quad . \tag{6.202}$$

Clearly, there is a Markov semigroup (L_s) acting on

$$\boldsymbol{C} := \mathcal{C}_b\left(\mathcal{C}\left(\mathbb{R}, \mathcal{M}\left(\mathbb{Z}^d\right) \times \mathcal{M}\left(\mathbb{Z}^d\right)\right), \mathbb{R}_+\right),$$

which describes left-shift. The domain $\boldsymbol{C}$ of this semigroup has the property that it is invariant under the semigroup:

$$\forall s \geq 0, f \in \boldsymbol{C} : L_s f \in \boldsymbol{C} \quad . \tag{6.203}$$

Consider a process with law

$$\Upsilon \in \mathcal{M}_1\left(\mathcal{C}\left(\mathbb{R}, \mathcal{M}\left(\mathbb{Z}^d\right) \times \mathcal{M}\left(\mathbb{Z}^d\right)\right)\right)$$

given by its initial law

$$\mathfrak{Z} \in \mathcal{M}_1\left(\mathcal{M}\left(\mathbb{Z}^d\right) \times \mathcal{M}\left(\mathbb{Z}^d\right)\right)$$

with $\mathfrak{Z} \in S^{(\theta_x,\theta_y)}$. By (6.200), we know that for $f \in \mathcal{C}_b\left(\mathcal{C}\left(\mathbb{R}, \mathcal{M}\left(\mathbb{Z}^d\right) \times \mathcal{M}\left(\mathbb{Z}^d\right)\right), \mathbb{R}_+\right)$ we have

$$\left\langle \tilde{\nu}_{(\theta_x,\theta_y)}, f\right\rangle = \lim_{t\to\infty} \langle \Upsilon, L_t f\rangle \quad . \tag{6.204}$$

We have to prove that it is allowed to choose $\tilde{\nu}_{(\theta_x,\theta_y)}$ as initial law, namely

$$\tilde{\nu}_{(\theta_x,\theta_y)} \in S^{(\theta_x,\theta_y)} \quad . \tag{6.205}$$

The following points have to be checked (compare 1.2.4)

1. homogeneity in space
2. finite pth moments for a $p > 2$
3. catalyst and reactant not correlated
4. non-negative auto-correlation

1. By the homogeneity of Υ and the homogeneity of the evolution equations (1.6) and (1.7) we immediately get that $\tilde{\nu}_{(\theta_x,\theta_y)}$ is homogeneous, that means invariant under any shift of space.

2. By Lemma 16 we know that there exists $p^* > 2$ such that p^*th moments remain bounded uniformly for all times. This implies uniform integrability of p-th moments with $p < p^*$. Hence we obtain that the p-th moments of the limit law $\tilde{\nu}_{(\theta_x,\theta_y)}$ are finite.

3., 4. With uniform integrability of p-th moments for a $p > 2$ we also have uniform integrability of correlation coefficients.

Now we can write

$$\begin{aligned} \left\langle \tilde{\nu}_{(\theta_x,\theta_y)}, L_s f \right\rangle &= \lim_{t\to\infty} \langle \Upsilon, L_{s+t} f \rangle \\ &= \lim_{t\to\infty} \langle \Upsilon, L_t f \rangle \\ &= \left\langle \tilde{\nu}_{(\theta_x,\theta_y)}, f \right\rangle \quad . \end{aligned} \tag{6.206}$$

This completes the proof of stationarity of $\tilde{\nu}_{(\theta_x,\theta_y)}$ under the semigroup of left-shifts L.

□

Chapter 7

Proof of Theorem 1

We want to prove the rescaling results formulated as "Finite System Scheme" in Chapter 3 (compare p. 28). These investigations center on a comparison of the considered process on large tori and on the lattice respectively. As we have claims regarding both the macroscopic and the microscopic aspect, this Chapter is devided in two parts:

At first, we analyse how empirical averages fluctuate after a rescaling of time if the system size grows to infinity.

At second, we discuss the limiting behavior from a microscopic point of view, comparing large finite systems and infinite systems locally.

7.1 Proof of the first part of Theorem 1 – Fluctuations of averages

7.1.1 Preliminaries

We consider the processes of empirical averages. We use the following notation:

$$\bar{X}_t^N = \frac{1}{\left|\mathbb{T}_N^d\right|} \sum_{i \in \mathbb{T}_N^d} X_t^N(i) \text{ and } \bar{Y}_t^N = \frac{1}{\left|\mathbb{T}_N^d\right|} \sum_{i \in \mathbb{T}_N^d} Y_t^N(i) \quad . \tag{7.1}$$

To study the fluctuations of averages, we have to rescale time. The proper rescaling factor β_N turns out to be the system size $\left|\mathbb{T}_N^d\right|$:

$$\beta_N := \left|\mathbb{T}_N^d\right| \quad . \tag{7.2}$$

We denote the sequences of the time-rescaled averages by $\left(\hat{X}^N\right)$ and $\left(\hat{Y}^N\right)$:

$$\hat{X}_t^N := \bar{X}_{\beta_N t}^N, \quad \hat{Y}_t^N := \bar{Y}_{\beta_N t}^N \quad . \tag{7.3}$$

By the evolution equations (1.6) and (1.7) we get:

$$d\hat{X}_t^N = \frac{1}{\left|\mathbb{T}_N^d\right|} \sum_{i \in \mathbb{T}_N^d} \sqrt{2\gamma_x X_{\beta_N t}^N(i)\, Y_{\beta_N t}^N(i)}\, dB_{(\beta_N t)}^x(i) \tag{7.4}$$

and

$$d\hat{Y}_t^N = \frac{1}{|\mathbb{T}_N^d|} \sum_{i \in \mathbb{T}_N^d} \sqrt{2\gamma_y Y_{\beta_N t}^N(i)}\, dB_{(\beta_N t)}^y(i) \quad . \tag{7.5}$$

This yields that both processes, $\hat{X}^N$ and $\hat{Y}^N$ are martingales.

For given planar Wiener process $(B_t^x, B_t^y)_{t\geq 0}$ consider the processes $\left(\tilde{X}_t, \tilde{Y}_t\right)_{t\geq 0}$ given as strong solutions of

$$\tilde{X}_0 = \theta_x \quad , \tag{7.6}$$

$$\tilde{Y}_0 = \theta_y \quad , \tag{7.7}$$

$$d\tilde{X}_t = \sqrt{2\gamma_x \tilde{X}_t \tilde{Y}_t}\, dB_t^x \quad , \tag{7.8}$$

$$d\tilde{Y}_t = \sqrt{2\gamma_y \tilde{Y}_t}\, dB_t^y \quad . \tag{7.9}$$

Then the claim of the first part of Theorem 1 can be formulated as follows:

$$\mathcal{L}\left\{\left(\hat{X}_t^N, \hat{Y}_t^N\right)_{t\geq 0}\right\} \overset{N\to\infty}{\Longrightarrow} \mathcal{L}\left\{\left(\tilde{X}_t, \tilde{Y}_t\right)_{t\geq 0}\right\} \quad . \tag{7.10}$$

7.1.2 Tightness

The first step of our proof is to verify that there exist accumulation points of

$$\left\{\mathcal{L}\left\{\left(\hat{X}^N, \hat{Y}^N\right)\right\}\right\}_{N\in\mathbb{N}} \quad . \tag{7.11}$$

In other words, we have to prove relative compactness of the sequence (7.11). Since the paths $\left(\hat{X}^N, \hat{Y}^N\right)$ are random variables with values in a Polish space, Prohorov's theorem (compare e.g. [KS], Theorem II.4.7) is applicable, and relative compactness is equivalent to tightness. To prove tightness, we need to show

1. tightness of the laws of initial states;

2. uniform regularity of sample paths.

The *first part* of the proof of tightness is easy: Since $\hat{X}_0^N$ and $\hat{Y}_0^N$ are non-negative random variables with finite expectation, the family of their laws is tight.

For the *second part* we apply the Kolmogorov criterion of tightness of processes with values in a Polish space and continuous paths (see [EK86], Proposition 3.6.3). We have to verify that there exist $C < \infty$, $\tilde{p} > 2$ such that for all $u, t \in \mathbb{R}_+$ with $u + 1 > t > u > 0$ we have

$$\limsup_{N\to\infty} \mathbf{E}\left[\left|\hat{X}_t^N - \hat{X}_u^N\right|^{\tilde{p}}\right] \leq C\,|t-u|^{\frac{\tilde{p}}{2}} \tag{7.12}$$

and

$$\limsup_{N\to\infty} \mathbf{E}\left[\left|\hat{Y}_t^N - \hat{Y}_u^N\right|^{\tilde{p}}\right] \leq C\,|t-u|^{\frac{\tilde{p}}{2}} \quad . \tag{7.13}$$

We will only provide a proof of (7.12). The proof of (7.13) is skipped – it is similar, but easier.

We know by (7.4) that $\hat{X}^N$ is a martingale. By Lemma 14 (Appendix B, page 247), it is even a L^p-martingale[1].

Due to Burkholder [Bu] we can apply the following martingale inequality for $L^{\tilde{p}}$-martingales M with increasing process $\langle M \rangle$

$$\sup_{r\in[0,R]} \mathbf{E}\left[|M_r|^{\tilde{p}}\right] \leq (\tilde{p}-1)^{\tilde{p}}\,\mathbf{E}\left[(\langle M\rangle_R)^{\frac{\tilde{p}}{2}}\right] \quad . \tag{7.14}$$

We apply this inequality to the martingale process $\left(\hat{X}_t^N - \hat{X}_u^N\right)_{t\geq 0}$ (u fixed). Using (7.4) to identify the increasing process, we get

$$\begin{aligned}
&\mathbf{E}\left[\left|\hat{X}_t^N - \hat{X}_u^N\right|^{\tilde{p}}\right] \\
&\leq (\tilde{p}-1)^{\tilde{p}}\,\mathbf{E}\left[\left(\int_u^t 2\gamma_x \frac{1}{\left|\mathbb{T}_N^d\right|}\sum_{i\in\mathbb{T}_N^d} X_{\beta_N r}^N(i)\,Y_{\beta_N r}^N(i)\,dr\right)^{\frac{\tilde{p}}{2}}\right] \\
&\leq (\tilde{p}-1)^{\tilde{p}}\,\mathbf{E}\left[\left(\frac{1}{t-u}\int_u^t \frac{1}{\left|\mathbb{T}_N^d\right|}\sum_{i\in\mathbb{T}_N^d} (t-u)\,2\gamma_x X_{\beta_N r}^N(i)\,Y_{\beta_N r}^N(i)\,dr\right)^{\frac{\tilde{p}}{2}}\right]
\end{aligned} \tag{7.15}$$

Next, we apply Jensen's inequality with respect to the product measure of the uniform distributions on $[u,t]$ and $\mathbb{T}_N^d$. This yields

$$\leq (\tilde{p}-1)^{\tilde{p}}\,\mathbf{E}\left[\frac{1}{t-u}\int_u^t \frac{1}{\left|\mathbb{T}_N^d\right|}\sum_{i\in\mathbb{T}_N^d}\left((t-u)\,2\gamma_x X_{\beta_N r}^N(i)\,Y_{\beta_N r}^N(i)\right)^{\frac{\tilde{p}}{2}} dr\right] \quad . \tag{7.16}$$

We combine the powers of $(t-u)$. With[2]

$$\left(X_{\beta_N r}^N(i)\,Y_{\beta_N r}^N(i)\right)^{\frac{\tilde{p}}{2}} \leq \left(X_{\beta_N r}^N(i)\right)^{\tilde{p}} + \left(Y_{\beta_N r}^N(i)\right)^{\tilde{p}} \tag{7.17}$$

we finally get (in continuation of (7.16)):

$$\begin{aligned}
&\leq (2\gamma_x)^{\frac{\tilde{p}}{2}}(\tilde{p}-1)^{\tilde{p}}(t-u)^{\frac{\tilde{p}}{2}-1}\int_u^t \mathbf{E}\left[\frac{1}{\left|\mathbb{T}_N^d\right|}\sum_{i\in\mathbb{T}_N^d}\left(\left(X_{\beta_N r}^N(i)\right)^{\tilde{p}} + \left(Y_{\beta_N r}^N(i)\right)^{\tilde{p}}\right)\right] dr \\
&\leq (2\gamma_x)^{\frac{\tilde{p}}{2}}(\tilde{p}-1)^{\tilde{p}}(t-u)^{\frac{\tilde{p}}{2}} \sup_{r\in[0,\beta_N t]}\left(\mathbf{E}\left[\left(X_r^N(0)\right)^{\tilde{p}}\right] + \mathbf{E}\left[\left(Y_r^N(0)\right)^{\tilde{p}}\right]\right) \quad .
\end{aligned} \tag{7.18}$$

By Lemma 14 (Appendix B, page 247) we know

$$\sup_{N\in\mathbb{N}}\sup_{r\in[0,\beta_N t]}\left(\mathbf{E}\left[\left(X_r^N(0)\right)^{\tilde{p}}\right] + \mathbf{E}\left[\left(Y_r^N(0)\right)^{\tilde{p}}\right]\right) < \infty \quad , \tag{7.19}$$

if $\tilde{p} > 2$ is chosen sufficiently close to 2. Applying this to (7.18), we see that the claim (7.12) is true. This completes the proof of tightness.

[1] By assumption, pth moments exist at the beginning.

[2] $0 \leq (x-y)^2 = x^2 + y^2 - 2xy$.

7.1.3 Convergence of finite dimensional distributions

We have just proven that the set of accumulation points of the sequence

$$\left(\mathcal{L}\left\{\hat{X}^N, \hat{Y}^N\right\}\right)_{N\in\mathbb{N}} \tag{7.20}$$

is not empty. The next step is to give a characterization of these accumulation points in order to obtain that there exists only one accumulation point, which is then automatically the limit point of the sequence. By proving that the considered sequence of stochastic processes converges in finite dimensional marginal distributions we give such a characterization of the limit process. The proof of the latter makes extensive use of the fact that we have a catalyst-reactant setting: convergence of catalysts and convergence of reactant processes conditioned on the catalyst processes[3] can be proved seperately. In particular, we will need that it is no loss of generality to assume that there is not only weak convergence of the laws of the catalysts, but also almost sure convergence of the catalyst paths. This is possible by defining the processes on a common probability space; we explain this in the next paragraph.

Skorohod representation of converging catalysts

We know (compare (7.5) resp. the infinite divisibility of interacting Feller diffusion) that the sequence of processes $\left(\hat{Y}^N\right)$ is in law equal to a sequence of branching diffusions all having the same branching rate. In addition to that, we have assumed by (3.19) that their initial conditions converge.

$$\lim_{N\to\infty} \hat{Y}_0^N = \theta_y \quad \mathbf{P}\text{-a.s.} \tag{7.21}$$

This is the reason why the laws of paths converge weakly

$$\mathcal{L}\left\{\hat{Y}^N\right\} \overset{N\to\infty}{\Longrightarrow} \mathcal{L}\left\{\hat{Y}^\infty\right\} \qquad (N\in\mathbb{N}) \quad . \tag{7.22}$$

We want this convergence not only to hold in law, but absolutely surely pathwise. Again, the idea is to apply the Skorohod representation (compare 5.1.1)[4]: there exist versions of the processes $\left(\hat{X}_t^N, \hat{Y}_t^N\right)_{t\geq 0}$ $(N\in\mathbb{N})$ on a common probability space such that the catalyst paths converge. We work on such a probability space with probability measure P_{cat} (instead of $\mathbf{P}$), assuming

Assumption: For $T>0$:

$$\lim_{N\to\infty} \sup_{t\in[0,T]} \left|\hat{Y}_t^N - \hat{Y}_t^\infty\right| = 0 \quad P_{cat}\text{-a.s.} \tag{7.23}$$

Quenched approach

By means of the Skorohod representation, we have constructed a probability measure $P_{cat} \in \mathcal{M}^1\left(\mathcal{C}\left(\mathbb{R}_+, \bigotimes_{N\in\mathbb{N}}\left(\mathcal{M}\left(\mathbb{T}_N^d\right)\right)\right)\right)$ with

$$P_{cat}\left[\left(Y^N\right)\in\cdot\right] = \mathbf{P}\left[\left(Y^N\right)\in\cdot\right] \quad \forall N\in\mathbb{N} \quad . \tag{7.24}$$

[3] To be more precise: convergence of the laws of conditioned laws of reactants for given catalysts.

[4] Since the space of paths is Polish (compare [EK86], III.5.6), this is possible.

By (7.23) it is possible to define a $\mathcal{C}(\mathbb{R}_+,\mathbb{R}_+)$-valued random variable $\hat{Y}^\infty$ being the P_{cat}-a.s. limit point of $\left(\hat{Y}^N\right)_{N\in\mathbb{N}}$. Note that $\hat{Y}^\infty$ is well-defined P_{cat}-a.s..

Next we construct the law of the whole catalyst-reactant process by defining its conditioned laws, namely the joint law of the family $\left\{\left(X^N,Y^N\right)\right\}_{N\in\mathbb{N}}$ for given $\left\{\left(Y^N\right)\right\}_{N\in\mathbb{N}}$.

In the sequel, fix a realisation $\left\{\left(Y^N\right)\right\}_{N\in\mathbb{N}}$ of converging catalyst processes.

Let $\left(\Omega^{II},\mathfrak{A}^{II},P^{II}\right)$ be a probability space. On this probability space, we define an independent family of Brownian motions $\left(B^x(i)\right)_{i\in\mathbb{Z}^d}$.

Given $\left\{\left(Y^N\right)\right\}_{N\in\mathbb{N}}$, we define for each $N\in\mathbb{N}$ the random path $X^N\in\mathcal{C}\left(\mathbb{R}_+,\mathcal{M}\left(\mathbb{T}_N^d\right)\right)$ as unique strong solution of the system of stochastic differential equations :

$$\begin{aligned} dX_t^N(i) &= c_x\sum_{j\in\mathbb{T}_N^d} a(i,j)\left[X_t^N(j)-X_t^N(i)\right]dt \\ &+ \sqrt{2\gamma_x X_t^N(i)\,Y_t^N(i)}\,dB_t^x(i) \qquad \left(i\in\mathbb{T}_N^d\right). \end{aligned} \tag{7.25}$$

For given $Y=\left(Y^1,Y^2,\ldots\right)$, let $\mathcal{L}_{reac}^Y$ be defined by

$$\mathcal{L}_{reac}^Y := \mathcal{L}^{II}\left\{\left(X^N,Y^N\right)_{N\in\mathbb{N}}\right\} \quad . \tag{7.26}$$

Let P_{reac}^Y denote the corresponding probability operator.

Next, we need the following lemma:

Lemma 6 *The mapping*

$$f:\begin{cases} \mathop{\times}\limits_{i=1}^{\infty}\mathcal{M}\left(\mathbb{T}_i^d\right) \longrightarrow \mathcal{M}^1\left(\mathop{\times}\limits_{i=1}^{\infty}\left(\mathcal{M}\left(\mathbb{T}_i^d\right)\times\mathcal{M}\left(\mathbb{T}_i^d\right)\right)\right) \\ \qquad Y \longmapsto \mathcal{L}_{reac}^Y \end{cases} \tag{7.27}$$

is Borel-measurable.

Proof.

Notation 4 *We introduce the following sigmaalgebras:*

$\mathcal{B}$:	*Borel-Sigmaalgebra on* $\mathbb{R}_+$
$\mathcal{B}_1$:	*Borel-Sigmaalgebra on* $\mathop{\times}\limits_{i=1}^{\infty}\mathcal{M}\left(\mathbb{T}_i^d\right)$
$\mathcal{B}_2$:	*Borel-Sigmaalgebra on* $\mathcal{M}^1\left(\mathop{\times}\limits_{i=1}^{\infty}\left(\mathcal{M}\left(\mathbb{T}_i^d\right)\times\mathcal{M}\left(\mathbb{T}_i^d\right)\right)\right)$
$\mathcal{B}_3$:	*Borel-Sigmaalgebra on* $\mathcal{M}^1\left(\mathop{\times}\limits_{i=1}^{\infty}\mathcal{M}\left(\mathbb{T}_i^d\right)\right)$

Denote by $\pi:\mathcal{M}^1\left(\mathop{\times}\limits_{i=1}^{\infty}\left(\mathcal{M}\left(\mathbb{T}_i^d\right)\times\mathcal{M}\left(\mathbb{T}_i^d\right)\right)\right)\to\mathcal{M}^1\left(\mathop{\times}\limits_{i=1}^{\infty}\mathcal{M}\left(\mathbb{T}_i^d\right)\right)$ the projection on the second marginal; in particular we have

$$\pi\circ\mathcal{L}_{reac}^Y\left\{\left(X^N,Y^N\right)_{N\in\mathbb{N}}\right\}=\mathcal{L}_{reac}^Y\left\{\left(Y^N\right)_{N\in\mathbb{N}}\right\} \quad . \tag{7.28}$$

With the definition of $\mathcal{L}_{reac}^Y\left\{\left(Y^N\right)_{N\in\mathbb{N}}\right\}$ we get

$$\mathcal{L}_{reac}^Y\left\{\left(Y^N\right)_{N\in\mathbb{N}}\right\}=\delta_{\left\{\left(Y^N\right)_{N\in\mathbb{N}}\right\}} \quad . \tag{7.29}$$

Let

$$\psi : \begin{cases} \bigtimes_{i=1}^{\infty} \mathcal{M}\left(\mathbb{T}_i^d\right) & \longrightarrow & \mathcal{M}^1\left(\bigtimes_{i=1}^{\infty}\left(\mathcal{M}\left(\mathbb{T}_i^d\right)\right)\right) \\ Y & \longmapsto & \delta_{\{Y\}} \end{cases} \quad . \tag{7.30}$$

We have $\psi = \pi \circ f$. To verify that f is measurable, it suffices to prove:

1. ψ is measurable.
2. π preserves measurability

ad 1. By the definition of the Prohorov metric, ψ is continuous. Hence it is measurable.

ad 2. Consider $A \subset \mathcal{M}^1\left(\bigtimes_{i=1}^{\infty}\left(\mathcal{M}\left(\mathbb{T}_i^d\right) \times \mathcal{M}\left(\mathbb{T}_i^d\right)\right)\right)$. A is $\mathcal{B}_2$-measurable if and only if for any choice of

$$J \subset\subset \mathbb{N} \quad , \tag{7.31}$$

$$\left(\left(\alpha_j\right)_{j\in J}, \left(\beta_j\right)_{j\in J}\right) \in \mathbb{R}_+^{|J|} \times \mathbb{R}_+^{|J|} \quad , \tag{7.32}$$

there is $\mathcal{B}$-measurability of the set

$$\bigcup_{a\in A}\left\{\int \prod_{j\in J} \exp\left(-\alpha_j \cdot X(j) - \beta_j \cdot Y(j)\right)\, a\left(d(X,Y)\right)\right\} \quad . \tag{7.33}$$

Then we have also $\mathcal{B}$-measurability of

$$\bigcup_{a\in A}\left\{\int \prod_{j\in J} \exp\left(-\beta_j \cdot Y(j)\right)\, \pi(a)\left(d(Y)\right)\right\} \quad . \tag{7.34}$$

for arbitrary $J \subset\subset \mathbb{N}$, $\beta_j \in \mathbb{R}_+^j$. This is sufficient for $\pi(A)$ to be $\mathcal{B}_3$-measurable. Therefore we obtain that the image

$$\pi(A) \subset \mathcal{M}^1\left(\bigtimes_{i=1}^{\infty} \mathcal{M}\left(\mathbb{T}_i^d\right)\right) \tag{7.35}$$

of a $\mathcal{B}_2$-measurable set

$$A \subset \mathcal{M}^1\left(\bigtimes_{i=1}^{\infty}\left(\mathcal{M}\left(\mathbb{T}_i^d\right) \times \mathcal{M}\left(\mathbb{T}_i^d\right)\right)\right)$$

is $\mathcal{B}_3$-measurable. This completes the proof of the lemma.

■

With the lemma proven above, the following definition of P as a mixture of quenched laws is well-defined:

$$P[\cdot] := \int P_{reac}^{Y}[\cdot]\, P_{cat}(dY) \quad . \tag{7.36}$$

The important point is that $Y \longmapsto P^Y_{reac}$ is measurable, therefore we have integrability of P^Y_{reac}.

As a survey of the notation we have introduced, remember that always P_{cat} denotes the joint law of the catalyst processes, wheras P^Y_{reac} denotes the law of the catalysts and reactants given all the catalyst processes $\{Y^N\}_{N\in\mathbb{N}}$.

By this construction, we have constructed a probability measure

$$P \in \mathcal{M}^1\left(\mathcal{C}\left(\mathbb{R}_+, \bigotimes_{N\in\mathbb{N}}\left(\mathcal{M}\left(\mathbb{T}^d_N\right)\times\mathcal{M}\left(\mathbb{T}^d_N\right)\right)\right)\right)$$

with

$$P\left[\left(X^N, Y^N\right)\in\cdot\right] = \mathbf{P}\left[\left(X^N, Y^N\right)\in\cdot\right] \quad \forall N\in\mathbb{N} \tag{7.37}$$

and

$$\lim_{N\to\infty}\sup_{t\in[0,T]}\left|\hat{Y}^N_t - \hat{Y}^\infty_t\right| = 0 \quad P-a.s. \quad .$$

It is proved in Section 5.2, p. 42that we have to show the following to obtain that $\mathcal{L}\left\{\left(\hat{X}^N_t, \hat{Y}^N_t\right)_{t\geq 0}\right\}$ converges to $\mathcal{L}\left\{\left(\hat{X}^\infty_t, \hat{Y}^\infty_t\right)_{t\geq 0}\right\}$ as $N\to\infty$.

Claim 5

1. *The laws $\mathcal{L}_{cat}\left\{\mathcal{L}^Y_{reac}\left\{\hat{X}^N\right\}\right\}$ of $\mathcal{M}^1\left(\mathcal{C}\left(\mathbb{R}_+,\mathbb{R}_+\right)\right)$-valued random variables $\mathcal{L}^Y_{reac}\left\{\hat{X}^N\right\}$ converge weakly as $N\to\infty$.*

2. *The limit can be described as law of quenched laws – denoted $\mathcal{L}_{cat}\left\{\mathcal{L}^Y_{reac}\left\{\hat{X}^\infty\right\}\right\}$ – of a process $\hat{X}^\infty$ given as strong solution of the SDE*

$$\hat{X}^\infty_0 = \theta_x \tag{7.38}$$

$$d\hat{X}^\infty_t = \sqrt{2\gamma_x\hat{X}^\infty_t\hat{Y}^\infty_t}dB_t \quad . \tag{7.39}$$

for fixed $\hat{Y}^\infty$. (Note that by (7.23) $\hat{Y}^\infty$ is well-defined P_{cat}-a.s.)

We prove this claim below. The prove contains two steps, namely

1. Step 1: Cramér-Wold device – convergence of Laplace transforms
2. Step 2: Duality

Step 1: Cramér-Wold device – Convergence of Laplace transforms

Let $(t_k)_{k\in\{1,\dots,n\}}$ be a sequence of observation times with

$$0 < t_1 < t_2 < \ldots < t_n < T < \infty \quad . \tag{7.40}$$

We want to prove that

$$\mathcal{L}\left\{\mathcal{L}^Y_{reac}\left\{\left(\hat{X}^N_t\right)_{t\in\{t_1,\dots,t_n\}}\right\}\right\} \overset{N\to\infty}{\Longrightarrow} \mathcal{L}\left\{\mathcal{L}^Y_{reac}\left\{\left(\hat{X}^\infty_t\right)_{t\in\{t_1,\dots,t_n\}}\right\}\right\} \tag{7.41}$$

holds. Please note that by (7.23) (Skorohod representation) there is P_{cat}-a.s. convergence of catalysts. We have to consider the Laplace transforms characterizing

$$\mathcal{L}^{Y}_{reac}\left\{\left(\hat{X}^{N}_{t}\right)_{t\in\{t_1,\dots,t_n\}}\right\}$$

and

$$\mathcal{L}^{Y}_{reac}\left\{\left(\hat{X}^{\infty}_{t}\right)_{t\in\{t_1,\dots,t_n\}}\right\}$$

for given catalyst processes Y. We will call these conditioned Laplace transforms Y-Laplace transforms.

To obtain (7.41) we have to show that a sequence of differences of Y-Laplace transforms converges P_{cat}-stochastically to zero:

Let $(a_k)_{k\in\{1,\dots,n\}} \in \mathbb{R}^n_+$, $\varepsilon > 0$. We claim:

$$\lim_{N\to\infty} P_{cat}\left\{\left|\mathbf{E}^{Y}_{reac}\left[\exp\left(-\sum_{k=1}^{n} a_k \hat{X}^{N}_{t_k}\right)\right] - \mathbf{E}^{Y}_{reac}\left[\exp\left(-\sum_{k=1}^{n} a_k \hat{X}^{\infty}_{t_k}\right)\right]\right| > \varepsilon\right\} = 0 \quad . \tag{7.42}$$

Step 2: Duality

It is possible to characterize the Y-Laplace transforms of finite dimensional distributions by a dual approach. We can refer to our results formulated in Chapter 5, in particular Proposition 1 on page 49. Making use of these results, we get:

$$\mathbf{E}^{Y}_{reac}\left[\exp\left(-\sum_{k=1}^{n} a_k \hat{X}^{N}_{t_k}\right)\right] = \mathbf{E}^{Y}_{reac}\left[\exp\left(-\left\langle w^{N}_{\beta_N T}, X^{N}_{0}\right\rangle\right)\right] \tag{7.43}$$

where $\left(w^{N}_{t}\right)_{t\geq 0} \in \left(\left(\mathbb{R}_+\right)^{\mathbb{Z}^d}\right)^{\mathbb{R}_+} = \left(\mathcal{M}\left(\mathbb{Z}^d\right)\right)^{\mathbb{R}_+}$ is given uniquely by the following equations.

initial condition:

$$w^{N}_{0} \equiv 0 \quad . \tag{7.44}$$

evolution: if $\beta_N T - t \notin \{\beta_N t_1, \dots, \beta_N t_n\}$:

$$\frac{\partial}{\partial t} w^{N}_{t}(i) = c_x \sum_{j\in\mathbb{T}^d_N} a(j,i)\left[w^{N}_{t}(j) - w^{N}_{t}(i)\right] - \gamma_x Y^{N}_{\beta_N T - t}(i)\left(w^{N}_{t}(i)\right)^2 \quad \left(i \in \mathbb{T}^d_N\right). \tag{7.45}$$

jumps: if $\beta_N T - t \in \{\beta_N t_1, \dots, \beta_N t_n\}$:

$$w^{N}_{t} = \lim_{s\to t-} w^{N}_{s} + \frac{a_k}{\beta_N}\mathfrak{n} \quad . \tag{7.46}$$

As it is our objective to prove weak convergence by comparing Y-Laplace transforms, we describe the finite dimensional distributions of the claimed limit process (given by (7.38) and (7.39)) in a similar way. The equation corresponding to (7.43) is

$$\mathbf{E}^{Y}_{reac}\left[\exp\left(-\sum_{k=1}^{n} a_k \hat{X}^{\infty}_{t_k}\right)\right] = \mathbf{E}^{Y}_{reac}\left[\exp\left(-q_T \theta_x\right)\right] \quad , \tag{7.47}$$

where $(q_t)_{t\geq 0}$ is required to satisfy

- *initial condition:*

$$q_0 = 0 \tag{7.48}$$

- *jumps:* if $T - t \in \{t_1, ..., t_n\}$:

$$q_t = \lim_{\varsigma \to t-} q_\varsigma + a_k \tag{7.49}$$

- *evolution:* if $T - t \notin \{t_1, ..., t_n\}$:

$$\frac{\partial}{\partial t} q_t = -\gamma_x \hat{Y}^{\infty}_{T-t} q_t^2 \quad . \tag{7.50}$$

To prove our claim (7.42), we have to verify

$$\lim_{N\to\infty} \mathbf{E}\left[\left| \mathbf{E}^Y_{reac}\left[\exp\left(-\left\langle w^N_{\beta_N T}, X^N_0 \right\rangle \right) \right] - \mathbf{E}^Y_{reac}\left[\exp\left(-q_T \theta_x \right) \right] \right| \right] = 0 \quad . \tag{7.51}$$

Clearly, it is sufficient for (7.51) to prove that[5]

$$\lim_{N\to\infty} \mathbf{E}\left[\left| \left\langle w^N_{\beta_N T}, X^N_0 \right\rangle - q_T \theta_x \right| \wedge 1 \right] = 0 \quad . \tag{7.52}$$

The proof of (7.52) consists of two parts:

1.

$$\lim_{N\to\infty} \mathbf{E}\left[\left| \left\langle w^N_{\beta_N T}, X^N_0 \right\rangle - \left\langle w^N_{\beta_N T}, \theta_x \mathfrak{n} \right\rangle \right| \right] = 0 \quad . \tag{7.53}$$

2.

$$\lim_{N\to\infty} \mathbf{E}\left[\left| q_T - \left\langle w^N_{\beta_N T}, \mathfrak{n} \right\rangle \right| \wedge 1 \right] = 0 \quad . \tag{7.54}$$

ad 1. We prove this by controlling first and second moments. The calculation carried out in the following is similar to some earlier considerations – compare Section 6.2, p.90. Our starting point is

$$\mathbf{E}\left[\left\langle w^N_{\beta_N T}, X^N_0 \right\rangle - \left\langle w^N_{\beta_N T}, \theta_x \mathfrak{n} \right\rangle \right] = 0 \quad . \tag{7.55}$$

This follows from the fact that the catalyst evolution and the initial condition of the reactant are independent.

We prove next that the variances vanish in the limit $N \to \infty$:

$$\lim_{N\to\infty} var\left(\left\langle w^N_{\beta_N T}, X^N_0 \right\rangle \right) = 0 \quad . \tag{7.56}$$

We know from the evolution equations (7.44), (7.45) and (7.46) that there is a catalyst-independent upper bound of $w^N_{\beta_N T}$, namely:

$$w^N_{\beta_N T}(i) \leq \frac{1}{\beta_N} \sum_{k=1}^{n} a_k \quad \left(i \in \mathbb{T}^d_N \right) . \tag{7.57}$$

[5] by $|\exp(-x) - \exp(-y)| \leq |x - y| \wedge 1$ for $x, y \geq 0$.

Note that this equation corresponds to (6.141). Since by assumption, the initial condition is positively auto-correlated in the sense of (1.17), we get:

$$\begin{aligned} var\left(\left\langle w^N_{\beta_N T}, X^N_0\right\rangle\right) &\leq var\left(\frac{1}{\beta_N}\sum_{i\in\mathbb{T}^d_N}\sum_{k=1}^{n} a_k X^N_0(i)\right) \\ &\leq var\left(\left(\sum_{k=1}^{n} a_k\right)\bar{X}^N_0\right) \end{aligned} \tag{7.58}$$

The latter converges to zero as $N\to\infty$, since, by assumption (3.19)

$$\lim_{N\to\infty} var\left(\bar{X}^N_0\right) = 0 \quad . \tag{7.59}$$

This completes the proof of 1.

ad 2. We have to prove

$$\lim_{N\to\infty} \mathbf{E}\left[\left|q_T - \left\langle w^N_{\beta_N T}, \mathfrak{n}\right\rangle\right| \wedge 1\right] = 0 \quad . \tag{7.60}$$

Let

$$w^N_{\beta_N T}\left(\mathbb{T}^d_N\right) := \left\langle w^N_{\beta_N T}, \mathfrak{n}\right\rangle \quad . \tag{7.61}$$

We want to prove:

$$\lim_{N\to\infty} \mathbf{E}\left[\left|q_T - w^N_{\beta_N T}\left(\mathbb{T}^d_N\right)\right| \wedge 1\right] = 0 \quad . \tag{7.62}$$

For non-jump times we have

$$\frac{\partial}{\partial t} w^N_{\beta_N t}\left(\mathbb{T}^d_N\right) = -\gamma_x \beta_N \sum_{i\in\mathbb{T}^d_N} Y^N_{\beta_N(T-t)}(i)\left(w^N_{\beta_N t}(i)\right)^2 \quad . \tag{7.63}$$

By definition, the jumps of $(q_t)_{t\geq 0}$ and $\left(w^N_{\beta_N t}\left(\mathbb{T}^d_N\right)\right)_{t\geq 0}$ coincide. Therefore, $\left(w^N_{\beta_N t}\left(\mathbb{T}^d_N\right) - q_t\right)_{t\geq 0}$ is continuous and piecewise differentiable. We use $\frac{\partial}{\partial t}$ as notation for right-sided derivatives. We get the following telescopic expansion:

$$\begin{aligned} &\frac{\partial}{\partial t}\left(w^N_{\beta_N t}\left(\mathbb{T}^d_N\right) - q_t\right) \\ &= -\gamma_x \sum_{i\in\mathbb{T}^d_N} \beta_N Y^N_{\beta_N(T-t)}(i)\left(w^N_{\beta_N t}(i)\right)^2 \\ &\quad + \gamma_x \hat{Y}^\infty_{T-t} q_t^2 \\ &= \gamma_x \sum_{i\in\mathbb{T}^d_N} \beta_N Y^N_{\beta_N(T-t)}(i)\left(\left(\frac{w^N_{\beta_N t}\left(\mathbb{T}^d_N\right)}{\beta_N}\right)^2 - \left(w^N_{\beta_N t}(i)\right)^2\right) \\ &\quad - \gamma_x \hat{Y}^N_{(T-t)}\left(\left(w^N_{\beta_N t}\left(\mathbb{T}^d_N\right)\right)^2 - q_t^2\right) \\ &\quad + \gamma_x\left(\hat{Y}^\infty_{T-t} - \hat{Y}^N_{T-t}\right) q_t^2 \quad . \end{aligned} \tag{7.64}$$

We get[6]:

$$
\begin{aligned}
&\frac{\partial}{\partial t}\left|w^N_{\beta_N t}\left(\mathbb{T}^d_N\right)-q_t\right| \\
&\leq -\gamma_x \hat{Y}^N_{(T-t)}\left|\left(w^N_{\beta_N t}\left(\mathbb{T}^d_N\right)\right)^2-q_t^2\right| \\
&\quad+\gamma_x \sum_{i\in\mathbb{T}^d_N} \beta_N Y^N_{\beta_N (T-t)}(i)\left|\left(\frac{w^N_{\beta_N t}\left(\mathbb{T}^d_N\right)}{\beta_N}\right)^2-\left(w^N_{\beta_N t}(i)\right)^2\right| \\
&\quad+\gamma_x\left|\hat{Y}^\infty_{T-t}-\hat{Y}^N_{T-t}\right| q_t^2 \quad .
\end{aligned} \tag{7.65}
$$

In particular, this yields:

$$
\begin{aligned}
&\left|w^N_{\beta_N T}\left(\mathbb{T}^d_N\right)-q_T\right|\wedge 1 \\
&\leq \int_0^T \sum_{i\in\mathbb{T}^d_N} \beta_N Y^N_{\beta_N (T-t)}(i)\left|\left(\frac{w^N_{\beta_N t}\left(\mathbb{T}^d_N\right)}{\beta_N}\right)^2-\left(w^N_{\beta_N t}(i)\right)^2\right| dt \\
&\quad+\int_0^T \gamma_x\left|\hat{Y}^\infty_{T-t}-\hat{Y}^N_{T-t}\right| q_t^2\, dt \quad .
\end{aligned} \tag{7.66}
$$

Since now the l.h.s is bounded above by 1, we are free to introduce a cut-off term on the r.h.s. as well. In particular, we can replace $\int_0^T \gamma_x\left|\hat{Y}^\infty_{T-t}-\hat{Y}^N_{T-t}\right| q_t^2\, dt$ by $\int_0^T \gamma_x\left|\hat{Y}^\infty_{T-t}-\hat{Y}^N_{T-t}\right| q_t^2\, dt \wedge 1$ on the right hand side. The purpose of this operation is to guarantee integrability. In fact, we can write:

$$
\begin{aligned}
&\mathbf{E}\left[\left|w^N_{\beta_N T}\left(\mathbb{T}^d_N\right)-q_T\right|\wedge 1\right] \\
&\leq \mathbf{E}\left[\int_0^T \sum_{i\in\mathbb{T}^d_N} \beta_N Y^N_{\beta_N (T-t)}(i)\left|\left(\frac{w^N_{\beta_N t}\left(\mathbb{T}^d_N\right)}{\beta_N}\right)^2-\left(w^N_{\beta_N t}(i)\right)^2\right| dt\right] \\
&\quad+\mathbf{E}\left[\int_0^T \gamma_x\left|\hat{Y}^\infty_{T-t}-\hat{Y}^N_{T-t}\right| q_t^2\, dt\wedge 1\right] \quad .
\end{aligned} \tag{7.67}
$$

With

$$
w^N_0\left(\mathbb{T}^d_N\right)=q_0=0 \tag{7.68}
$$

and (by (7.23) and considerations similar to (6.175) (p. 98))

$$
\lim_{N\to\infty}\int_0^T \gamma_x\left|\hat{Y}^\infty_{T-t}-\hat{Y}^N_{T-t}\right| q_t^2 dt = 0 \quad P_{cat}\text{-a.s.} \tag{7.69}
$$

[6]This is similar to the calculations carried out on page 92. The key idea is to approximate $|\cdot|$ by the sequence $\left(\frac{1}{n}\ln\cosh(n\cdot)\right)_{n\in\mathbb{N}}$.

we get:

$$\lim_{N\to\infty} \mathbf{E}\left[\int_0^T \gamma_x \left|\hat{Y}^{\infty}_{T-t} - \hat{Y}^N_{T-t}\right| q_t^2 dt \wedge 1\right] = 0 \tag{7.70}$$

Therefore, we get:

$$\begin{aligned}&\limsup_{N\to\infty} \mathbf{E}\left[\left|w^N_{\beta_N T}\left(\mathbb{T}^d_N\right) - q_T\right| \wedge 1\right] \\ &\leq \limsup_{N\to\infty} \mathbf{E}\left[\int_0^T \gamma_x \sum_{i\in\mathbb{T}^d_N} \beta_N Y^N_{\beta_N(T-t)}(i) \left|\left(\frac{w^N_{\beta_N t}\left(\mathbb{T}^d_N\right)}{\beta_N}\right)^2 - \left(w^N_{\beta_N t}(i)\right)^2\right| dt\right] \quad .\end{aligned} \tag{7.71}$$

It only remains to prove:

Claim 6 *The following holds:*

$$\limsup_{N\to\infty} \mathbf{E}\left[\int_0^T \sum_{i\in\mathbb{T}^d_N} \beta_N Y^N_{\beta_N(T-t)}(i) \left|\left(\frac{w^N_{\beta_N t}\left(\mathbb{T}^d_N\right)}{\beta_N}\right)^2 - \left(w^N_{\beta_N t}(i)\right)^2\right| dt\right] = 0 \tag{7.72}$$

Below, we give a lemma which provides a representation of the terms $\frac{w^N_{\beta_N t}(\mathbb{T}^d_N)}{\beta_N}$ and $w^N_{\beta_N t}(i)$ by Feynman-Kac like formulas. But first, we define the ingredients we need – as we give a Feynman-Kac representation, we need Markov processes; in our case continuous time random walks.

Ingredients: Consider a probability space $(\mathbb{O}, \mathfrak{A}, \mathbb{P})$ (as notation we need the expectation $\mathbb{E}$, and the law $\mathbb{L}$). With respect to this probability space, we consider for each $N \in \mathbb{N}$ a pair of continuous time random walks denoted by $\left(\xi^{i,N}_r\right)_{r\geq 0}$ and $\left(\bar{\xi}^N_r\right)_{r\geq 0}$ both attaining values in $\mathbb{T}^d_N$. Suppose that the initial condition is given by

$$\mathbb{L}\left\{\left(\xi^{i,N}_0, \bar{\xi}_0\right)\right\} = \delta_{\{i\}} \otimes \frac{\mathfrak{n}}{\beta_N} \quad . \tag{7.73}$$

Furthermore, suppose that the evolution mechanism of the random walks has the following properties:

$$\begin{aligned}\mathbb{P}\left[\xi^{i,N}_t = y \,\middle|\, \xi^{i,N}_s = x\right] &= a^N_{c_x(t-s)}(x,y) \quad ; \left(t > s > 0; x, y \in \mathbb{T}^d_N\right), \quad (7.74)\\ \mathbb{P}\left[\bar{\xi}_t = y \,\middle|\, \bar{\xi}_s = x\right] &= a^N_{c_x(t-s)}(x,y) \quad ; \left(t > s > 0; x, y \in \mathbb{T}^d_N\right). \quad (7.75)\end{aligned}$$

Note that we are still free to specify interdependencies between both random walks. In fact, we will introduce coupling assumptions later in order to prove (7.72). The specification of coupling assumptions is not necessary for the Feynman-Kac representation; that's why we postpone this issue. Next, we formulate the Feynman-Kac lemma, which is valid for arbitrary random walks $\left(\xi^{i,N}_r\right)_{r\geq 0}$ and $\left(\bar{\xi}^N_r\right)_{r\geq 0}$ satisfying (7.73), (7.74) and (7.75).

Lemma 7 $w^N_{\beta_N T}(i)$ $\left(i \in \mathbb{T}^d_N\right)$ *and* $\frac{w^N_{\beta_N t}\left(\mathbb{T}^d_N\right)}{\beta_N}$ *can be expressed by the following Feynman-Kac representations:*

$$w^N_{\beta_N T}(i) = \sum_{k=1}^{n} \frac{a_k}{\beta_N} \mathbb{E}\left[e^{-\int_0^{\beta_N t_k} w^N_{\beta_N T-r}\left(\xi^{i,N}_r\right) Y^N_r\left(\xi^{i,N}_r\right) dr} \right] \quad , \tag{7.76}$$

$$\frac{w^N_{\beta_N T}\left(\mathbb{T}^d_N\right)}{\beta_N} = \sum_{k=1}^{n} \frac{a_k}{\beta_N} \mathbb{E}\left[e^{-\int_0^{\beta_N t_k} w^N_{\beta_N T-r}\left(\bar{\xi}^N_r\right) Y^N_r\left(\bar{\xi}^N_r\right) dr} \right] \quad . \tag{7.77}$$

Before we prove this lemma, we discuss the coupling we are going to impose on the random walks – we have already mentioned that we some freedom to introduce a coupling. In the sequel let $\left(\xi^{i,N}_r\right)_{r\geq 0}$ and $\left(\bar{\xi}^N_r\right)_{r\geq 0}$ be defined such that the following holds:

1. *Processes stay together as soon as they meet:* The process $\xi^{i,N}$ coincides with $\bar{\xi}^N$ after the time they first meet. This stopping time is denoted by $\tau^{i,N}$.

2. *Processes meet fast*: The coupling of the processes is chosen such that
$$\mathbb{E}\left[\tau^{i,N}\right] \leq O\left(N^2\right) \quad (N \to \infty) \quad . \tag{7.78}$$

The following lemma makes sure that this requirement make sense:

Lemma 8 *It is possible to construct continuous time random walks* $\xi^{i,N}$ *and* $\bar{\xi}^N$ *with (7.73), (7.74) and (7.75) s. t.*

$$\mathbb{E}\left[\tau^{i,N}\right] \leq O\left(N^2\right) \quad (N \to \infty) \quad . \tag{7.79}$$

Proof. We construct a coupling for which (7.79) is valid.

Clearly, it is sufficient to prove the claim in a discrete time setting: we prove that there exists a family of Markov chains[7]

$$\left(z^{i,N}_n\right)_{n\in\mathbb{N}_0} = \left(z^{i,N}_n(1), ..., z^{i,N}_n(d)\right)_{n\in\mathbb{N}_0} \quad , \tag{7.80}$$

$$\left(\bar{z}^N_n\right)_{n\in\mathbb{N}_0} = \left(\bar{z}^N_n(1), ..., \bar{z}^N_n(d)\right)_{n\in\mathbb{N}_0} \quad , \tag{7.81}$$

defined on a common probability space $(\mathbb{O}, \mathfrak{A}, \mathbb{P})$, having values in $\mathbb{T}^d_N$ which satisfy the following:

- $$\mathbb{L}\left\{\left(z^{i,N}_n\right)_{n\in\mathbb{N}_0}\right\} = \mathbb{L}\left\{\left(\xi^{i,N}_n\right)_{n\in\mathbb{N}_0}\right\}, \quad \mathbb{L}\left\{\left(\bar{z}^N_n\right)_{n\in\mathbb{N}_0}\right\} = \mathbb{L}\left\{\left(\bar{\xi}^N_n\right)_{n\in\mathbb{N}_0}\right\} \quad , \tag{7.82}$$

- $$\vartheta^{i,N} := \min\left\{n \in \mathbb{N} : z^{i,N}_n = \bar{z}^N_n\right\} = O\left(N^2\right) \quad (N \to \infty) \quad . \tag{7.83}$$

Define the following $d+1$ stopping times $\left(\vartheta^{i,N}_k\right)_{k\in\{0,1,...,d\}}$:

$$\vartheta^{i,N}_0 := -1 \quad , \tag{7.84}$$

$$\vartheta^{i,N}_k := \min\left\{n \in \mathbb{N} : \forall g \in \{1, ..., k\} : z^{i,N}_n(g) = \bar{z}^N_n(g)\right\} \quad . \tag{7.85}$$

[7] implicitely, we introduce the notation for the coordinate processes

Define α as minimum of the jump rates from 0 to one of its $2d$ neighbors at time 1.

$$\alpha := \min_{i:|i|=1} a_1(0,i) \quad . \tag{7.86}$$

Since d is finite and a is irreducible by assumption, we have $\alpha > 0$.

For $n \in \left[\vartheta_k^{i,N}, \vartheta_{k+1}^{i,N}\right)$, let $\left(z_n^{i,N}, \bar{z}_n^N\right)_{n\in\mathbb{N}_0}$ be characterized by the following jump probabilities:

Let $x = (x_1, ..., x_d) \in \mathbb{Z}^d$ denote the jump vector.

- For $|x| \neq 1$ or ($|x| = 1$ and $x_{k+1} = 0$)

$$\mathbb{P}\left[\left(z_n^{i,N}, \bar{z}_n^N\right) = \left(z_{n-1}^{i,N} + x, \bar{z}_{n-1}^N + x\right)\right] := a_1^N(0,x) \quad . \tag{7.87}$$

- For $|x| = 1$ and $x_{k+1} = 1$:

$$\mathbb{P}\left[\left(z_n^{i,N}, \bar{z}_n^N\right) = \left(z_{n-1}^{i,N} + x, \bar{z}_{n-1}^N + x\right)\right] := a_1^N(0,x) - \frac{\alpha}{2} \quad , \tag{7.88}$$

$$\mathbb{P}\left[\left(z_n^{i,N}, \bar{z}_n^N\right) = \left(z_{n-1}^{i,N} + x, \bar{z}_{n-1}^N - x\right)\right] := \frac{\alpha}{2} \quad . \tag{7.89}$$

We are now in the following situation: if $\left[\vartheta_k^{i,N}, \vartheta_{k+1}^{i,N}\right)$, only the $(k+1)$th coordinate of $z_n^{i,N} - \bar{z}_n^N$ fluctuates as an irreducible next-neighbor random walk on the one-dimensional torus, whereas the other coordinates of the difference remain fixed. Therefore, it is sufficient for (7.83) that the maximal expectation of the time such a random walk needs to hit the origin is of order N^2. But this is an elementary fact. ∎

Proof of Lemma 7. Consider

$$\tilde{w}_t^N(i)$$

$$:= \sum_{k=1}^{n} a_k \mathbb{I}_{\{t\geq\beta_N(T-t_k)\}} \mathbb{E}\left[e^{-\int_{\beta_N(T-t_k)}^{t} w_r^N\left(\zeta_r^{k,N}\right) Y_{\beta_N T-r}^N\left(\zeta_r^{k,N}\right) dr} \mathbb{I}_{\left\{\zeta_t^{k,N}=i\right\}}\left(\zeta_t^{k,N}\right)\right] \quad , \tag{7.90}$$

where the random walks $\left(\zeta_t^{k,N}\right)_{t\geq\beta_N(T-t_k)}$ $(k \in \{1, ..., n\})$ are given by their transition rates, namely $c_x \bar{a}^N$, and initial condition

$$\mathcal{L}\left\{\zeta_{\beta_N(T-t_k)}^{k,N}\right\} = \frac{\mathfrak{n}}{\beta_N} \quad . \tag{7.91}$$

This way, $\left(\zeta_t^{k,N}\right)_{t\geq\beta_N(T-t_k)}$ is stationary. The evolution equations of $\tilde{w}_t^N(i)$ can be calculated as follows:

- for $t = \beta_N(T - t_k)$, $k \in \{1, ..., n\}$ we get

$$\begin{aligned}\tilde{w}_t^N(i) &= \lim_{s\nearrow t} \tilde{w}_s^N(i) + a_k \mathbb{E}\left[\mathbb{I}_{\left\{\zeta_t^{k,N}=i\right\}}\left(\zeta_t^{k,N}\right)\right] \\ &= \lim_{s\nearrow t} \tilde{w}_s^N(i) + a_k \cdot \frac{1}{\beta_N} \quad .\end{aligned} \tag{7.92}$$

- otherwise, by differentiation:

$$
\begin{aligned}
&\frac{\partial}{\partial t}\tilde{w}_t^N(i) \\
&= \frac{\partial}{\partial t}\left(\sum_{k=1}^{n} a_k \mathbb{I}_{\{t\geq\beta_N(T-t_k)\}}\mathbb{E}\left[e^{-\int_{\beta_N(T-t_k)}^{t} w_r^N\left(\zeta_r^{k,N}\right) Y_{\beta_N T-r}^N\left(\zeta_r^{k,N}\right) dr}\mathbb{I}_{\left\{\zeta_t^{k,N}=i\right\}}\right]\right) \\
&\overset{\text{Product rule}}{=} \sum_{k=1}^{n} a_k \mathbb{I}_{\{t\geq\beta_N(T-t_k)\}} \\
&\quad\cdot\mathbb{E}\left[-w_t^N\left(\zeta_t^{k,N}\right) Y_{\beta_N T-t}^N\left(\zeta_t^{k,N}\right) e^{-\int_{\beta_N(T-t_k)}^{t} w_r^N\left(\zeta_r^{k,N}\right) Y_{\beta_N T-r}^N\left(\zeta_r^{k,N}\right) dr}\mathbb{I}_{\left\{\zeta_t^{k,N}=i\right\}}\right] \\
&\quad+\sum_{k=1}^{n} a_k \mathbb{I}_{\{t\geq\beta_N(T-t_k)\}} \\
&\quad\cdot\mathbb{E}\left[e^{-\int_{\beta_N(T-t_k)}^{t} w_r^N\left(\zeta_r^{k,N}\right) Y_{\beta_N T-r}^N\left(\zeta_r^{k,N}\right) dr} c_x\sum_{j\in\mathbb{T}_N^d} a(j,i)\left(\mathbb{I}_{\left\{\zeta_t^{k,N}=j\right\}}-\mathbb{I}_{\left\{\zeta_t^{k,N}=i\right\}}\right)\right] \\
&= -w_t^N(i)\,\tilde{w}_t^N(i)\,Y_{\beta_N T-t}^N(i) \\
&\quad+\sum_{k=1}^{n} a_k \mathbb{I}_{\{t\geq\beta_N(T-t_k)\}} \\
&\quad\cdot\mathbb{E}\left[e^{-\int_{\beta_N(T-t_k)}^{t} w_r^N\left(\zeta_r^{k,N}\right) Y_{\beta_N T-r}^N\left(\zeta_r^{k,N}\right) dr} c_x\sum_{j\in\mathbb{T}_N^d} a(j,i)\left(\mathbb{I}_{\left\{\zeta_t^{k,N}=j\right\}}-\mathbb{I}_{\left\{\zeta_t^{k,N}=i\right\}}\right)\right] \\
&= -w_t^N(i)\,\tilde{w}_t^N(i)\,Y_{\beta_N T-t}^N(i) + c_x\sum_{j\in\mathbb{T}_N^d} a(j,i)\left(\tilde{w}_t^N(j)-\tilde{w}_t^N(i)\right) \quad . \qquad (7.93)
\end{aligned}
$$

Comparison of (7.93) with (7.45) and (7.46) and the uniqueness of the solution of equation (7.45) yield that $\tilde{w}_t^N = w_t^N$. In order to get (7.76), we consider at first (7.90) for $t = \beta_N T$:

$$
\tilde{w}_{\beta_N T}^N(i) = \sum_{k=1}^{n} a_k \mathbb{E}\left[e^{-\int_{\beta_N(T-t_k)}^{\beta_N T} w_r^N\left(\zeta_r^{k,N}\right) Y_{\beta_N T-r}^N\left(\zeta_r^{k,N}\right) dr}\mathbb{I}_{\left\{\zeta_{\beta_N T}^{k,N}=i\right\}}\left(\zeta_{\beta_N T}^{k,N}\right)\right] \quad . \tag{7.94}
$$

Substituting $r \mapsto \beta_N T - r$, we get:

$$
\tilde{w}_{\beta_N T}^N(i) = \sum_{k=1}^{n} a_k \mathbb{E}\left[e^{-\int_{0}^{\beta_N t_k} w_{\beta_N T-r}^N\left(\zeta_{\beta_N T-r}^{k,N}\right) Y_r^N\left(\zeta_{\beta_N T-r}^{k,N}\right) dr}\mathbb{I}_{\left\{\zeta_{\beta_N T}^{k,N}=i\right\}}\left(\zeta_{\beta_N T}^{k,N}\right)\right] \tag{7.95}
$$

Consider the random walk $\left(\bar{\xi}_t^N\right)_{t\geq 0}$ defined by (7.73). Having transition rates $c_x a^N(\cdot,\cdot)$ and starting with initial law $\frac{1}{\beta_N}\mathfrak{n}$, it is stationary. Taking account that the transition rates of $\zeta^{k,N}$ are given by $c_x\bar{a}^N(\cdot,\cdot)$we get:

$$\mathcal{L}\left\{\left(\bar{\xi}_t^N\right)_{t\in[0,\beta_N t_k]}\right\}=\mathcal{L}\left\{\left(\zeta_{\beta_N T-t}^{k,N}\right)_{t\in[0,\beta_N t_k]}\right\} \quad . \tag{7.96}$$

Therefore we get in continuation of (7.95):

$$=\sum_{k=1}^{n} a_k\mathbb{E}\left[e^{-\int\limits_0^{\beta_N t_k} w_{\beta_N T-r}^N\left(\bar{\xi}_r^N\right)\, Y_r^N\left(\bar{\xi}_r^N\right)dr}\,\mathbb{I}_{\left\{\bar{\xi}_0^N=i\right\}}\left(\bar{\xi}_0^N\right)\right] \tag{7.97}$$

Since $\left(\xi_t^{i,N}\right)_{t\geq 0}$ can be described as $\left(\bar{\xi}_t^N\right)_{t\geq 0}$ conditioned on starting in i,we get:

$$=\sum_{k=1}^{n} a_k\mathbb{E}\left[e^{-\int\limits_0^{\beta_N t_k} w_{\beta_N T-r}^N\left(\xi_r^{i,N}\right) Y_r^N\left(\xi_r^{i,N}\right)dr}\cdot\frac{1}{\beta_N}\right] \quad . \tag{7.98}$$

Thus, we have obtained (7.76). Equation (7.77) follows from similar calculations.

■

Using the Feynman-Kac representations (7.76), (7.77) and with

$$\beta_N w_t^N(i)\leq\sum_{k=1}^{n} a_k=:A \qquad \left(t\geq 0, i\in\mathbb{T}_N^d\right) \tag{7.99}$$

we get: (remember that it is our objective to prove (7.72))

$$\begin{aligned}
&\limsup_{N\to\infty}\mathbf{E}\left[\int_0^T \sum_{i\in\mathbb{T}_N^d} \beta_N Y^N_{\beta_N(T-t)}(i)\left|\left(\frac{w^N_{\beta_N t}(\mathbb{T}^d_N)}{\beta_N}\right)^2-\left(w^N_{\beta_N t}(i)\right)^2\right| dt\right]\\
&=\limsup_{N\to\infty}\mathbf{E}\left[\int_0^T \sum_{i\in\mathbb{T}_N^d} \beta_N Y^N_{\beta_N(T-t)}(i)\left|\frac{w^N_{\beta_N t}(\mathbb{T}^d_N)}{\beta_N}-w^N_{\beta_N t}(i)\right|\right.\\
&\qquad\qquad\left.\cdot\left(\frac{w^N_{\beta_N t}(\mathbb{T}^d_N)}{\beta_N}+w^N_{\beta_N t}(i)\right) dt\right]\\
&\leq\limsup_{N\to\infty}\mathbf{E}\left[\int_0^T 2A\sum_{i\in\mathbb{T}_N^d}\left|Y^N_{\beta_N(T-t)}(i)\left(w^N_{\beta_N t}(i)-\frac{w^N_{\beta_N t}(\mathbb{T}^d_N)}{\beta_N}\right)\right| dt\right]\\
&=\limsup_{N\to\infty}\mathbf{E}\left[\int_0^T 2A\sum_{i\in\mathbb{T}_N^d}\left|Y^N_{\beta_N(T-t)}(i)\right.\right.\\
&\qquad\cdot\sum_{k=1}^n\frac{a_k}{\beta_N}\mathbb{E}\left[e^{-\int_0^{\beta_N t_k} w^N_{\beta_N T-r}(\bar{\xi}^N_r)\, Y^N_r(\bar{\xi}^N_r)\,dr}\right.\\
&\qquad\qquad\left.\left.\left.-e^{-\int_0^{\beta_N t_k} w^N_{\beta_N T-r}(\xi^{i,N}_r)\, Y^N_r(\xi^{i,N}_r)\,dr}\right]\right| dt\right]\\
&\leq\limsup_{N\to\infty}\mathbf{E}\left[\int_0^T 2A\sum_{i\in\mathbb{T}_N^d} Y^N_{\beta_N(T-t)}(i)\,\cdot\right.\\
&\qquad\left.\cdot\sum_{k=1}^n\frac{a_k}{\beta_N}\mathbb{E}\left[\int_0^{\beta_N t_k}\left|w^N_{\beta_N T-r}\left(\bar{\xi}^N_r\right)\, Y^N_r\left(\bar{\xi}^N_r\right)-w^N_{\beta_N T-r}\left(\xi^{i,N}_r\right)\, Y^N_r\left(\xi^{i,N}_r\right)\right| dr\right] dt\right]
\end{aligned}\tag{7.100}$$

The processes $\bar{\xi}^N$ and $\xi^{i,N}$ coincide after time $\tau^{i,N}$. Therefore we get (in continuation of the sequence of inequalities begun above:)

$$\leq \limsup_{N\to\infty} \mathbf{E}\left[\int_0^T 2A \sum_{i\in\mathbb{T}_N^d} Y^N_{\beta_N(T-t)}(i) \cdot \right.$$

$$\left. \cdot \sum_{k=1}^n \frac{a_k}{\beta_N} \mathbb{E}\left[\int_0^{\tau^{i,N}} \left| w^N_{\beta_N T-r}\left(\bar{\xi}^N_r\right) Y^N_r\left(\bar{\xi}^N_r\right) - w^N_{\beta_N T-r}\left(\xi^{i,N}_r\right) Y^N_r\left(\xi^{i,N}_r\right)\right| dr\right] dt\right]$$

$$\overset{(7.57)}{\leq} \limsup_{N\to\infty} \mathbf{E}\left[\int_0^T 2A \sum_{i\in\mathbb{T}_N^d} Y^N_{\beta_N(T-t)}(i) \cdot \right.$$

$$\left. \cdot \left(\sum_{k=1}^n \frac{a_k}{\beta_N}\right) \mathbb{E}\left[\int_0^{\tau^{i,N}} \left(\sum_{k=1}^n \frac{a_k}{\beta_N}\right)\left(Y^N_r\left(\bar{\xi}^N_r\right) + Y^N_r\left(\xi^{i,N}_r\right)\right) dr\right] dt\right]$$

$$\leq \limsup_{N\to\infty} A \left(\sum_{k=1}^n \frac{a_k}{\beta_N}\right)^2$$

$$\cdot \sum_{i\in\mathbb{T}_N^d} \int_0^T \mathbb{E}\left[\int_0^{\tau^{i,N}} \mathbf{E}\left[2\left(Y^N_{\beta_N(T-t)}(i)\right)^2 + \left(Y^N_r\left(\bar{\xi}^N_r\right)\right)^2 + \left(Y^N_r\left(\xi^{i,N}_r\right)\right)^2\right] dr\right] dt \quad (7.101)$$

Let

$$\varrho_N := \mathbb{E}\left[\frac{\tau^{i,N}}{\beta_N}\right] \quad . \qquad (7.102)$$

By (7.79) we have the convergence

$$\lim_{N\to\infty} \varrho_N = 0 \quad . \qquad (7.103)$$

In addition to that, we know from [CGS95], Lemma 2.2c that the second moments of the processes on the torus have an uniform upper bound in $\beta_N T$:

$$\exists K \in \mathbb{R} : \sup_{N\in N} \sup_{t\in[0,\beta_N T]} \mathbf{E}\left[\left(Y^N_t(0)\right)^2\right] < K < \infty \quad . \qquad (7.104)$$

Thus, we get in countinuation of (7.101):

$$\leq \limsup_{N\to\infty} A \left(\sum_{k=1}^n a_k\right)^2 \frac{1}{\beta_N} \sum_{i\in\mathbb{T}_N^d} T\mathbb{E}\left[\frac{\tau^{i,N}}{\beta_N}\right] 4K \qquad (7.105)$$

$$= \limsup_{N\to\infty} A \left(\sum_{k=1}^n a_k\right)^2 T\varrho_N 4K = 0 \quad . \qquad (7.106)$$

We can conclude that the right hand side of (7.101) converges to zero.

This completes the proof of the first part of Theorem 1.

7.2 Proof of the second part of Theorem 1 – Local behavior

We have in mind to compare

- the local behavior of large torus systems observed at times $\beta_N s$. Note that at this time, the empirical averages have already begun to fluctuate.
- the law of a mixture of laws of stationary processes on the lattice, where the mixing law corresponds to the limit law Ψ_s of time-rescaled empirical averages.

We claim:

$$\mathcal{L}\left\{\left(X^N_{\beta_N s+t}, Y^N_{\beta_N s+t}\right)_{t\geq 0}\right\} \quad \overset{N\to\infty}{\Longrightarrow} \quad \int \nu_{(\tilde{\theta}_x,\tilde{\theta}_y)}\, \Psi_s\left(d\left(\tilde{\theta}_x,\tilde{\theta}_y\right)\right) \quad . \tag{7.107}$$

The main step of the proof contains two approximations. Roughly speaking, these two steps of approximation are

- approximate the mixing law on the right hand side of (7.107) by the empirical densities of time-rescaled empirical averages it is the limit of.
- compare the local distribution for given intensity with the equilibrium distribution.

There is a diagram on page 125 that gives a visualization of the approximation steps. First, we give a survey containing all steps the proof contains.

7.2.0 Survey of the proof

At first, we give a short overview of the structure of the proof. The following diagram tries to visualize the various steps. Namely, the tree structure is visualized (for example, to get the main result, one has to verify "fdd" and "tightness". For "fdd", "I" and "II and III" is needed etc.). The arrows give the the order in which we discuss the different points. Note: Distinguish between "**I**" and "I": "**I**" is bold, "I" not.

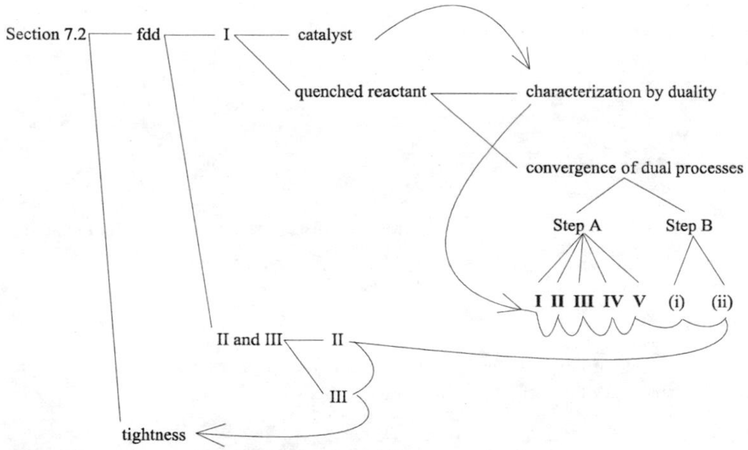

Structure of the proof of 7.2

The proof consists of the following steps:

7.2.1. Proof of the convergence result in terms of finite dimensional marginal distributions. Applying the triangular inequality, this falls into three conergence results

I. First step of the approximation (equation (7.115), p. 126). Making use of the catalytic setting, the proofs falls into two parts:

C Proof of convergence of the catalyst

QR Proof of convergence of the quenched reactant
This proof makes extensive use of the dual approach. It contains:

A. Characterization by dual processes

B. Convergence of dual processes
This part of the proof is rather technical: one has to follow a two-step program.

Step A, where the five summands **I-V** of a sum are discussed

Step B, which is subdivided into two parts (i) and (ii).

II., III. Second and third step of the approximation (equations (7.116) and (7.117), p. 126, proof on pages 7.2.1pp). In particular, a continuity lemma is needed.

7.2.2. A tightness argument completes the proof.

Next we describe what exactly is proven in the two main parts of the proof.

- Proof of convergence of finite dimensional marginal distributions
Let $n \in \mathbb{N}$. Consider an arbitrary sequence $(t_k)_{k\in\{1,\dots,n\}}$ of *observation times* satisfying

$$0 < t_1 < t_2 < \dots < t_n \tag{7.108}$$

and

$$\exists T < \infty : t_n < T \quad . \tag{7.109}$$

Let $\Upsilon := \{t_1, t_2, \dots, t_n\}$.
We claim

$$\mathcal{L}\left\{\left(X^N_{\beta_N s+t}, Y^N_{\beta_N s+t}\right)_{t\in\Upsilon}\right\} \overset{N\to\infty}{\Longrightarrow} \int \mu^{\Upsilon}_{(\tilde\theta_x,\tilde\theta_y)}\,\Psi_s\left(d\left(\tilde\theta_x,\tilde\theta_y\right)\right) \quad : \tag{7.110}$$

(Remember that $\mu^{\Upsilon}_{(\tilde\theta_x,\tilde\theta_y)}$ denotes the law of finite dimensional marginal distributions corresponding to Υ of the stationaly process with on the lattice with intensity $\left(\tilde\theta_x,\tilde\theta_y\right)$).

- Proof of tightness
The convergence of finite dimensional marginals (7.110) provides us with a characterization of the weak limit point of the sequence

$$\left(\mathcal{L}\left\{\left(X^N_{\beta_N s+t}, Y^N_{\beta_N s+t}\right)_{t\in\mathbb{R}_+}\right\}\right)_{N\in\mathbb{N}} \tag{7.111}$$

– if such a limit point exists.

We need to prove that the sequence (7.111) is tight in order to be able to guarantee that it has at least one accumulation point. The tightness property is proven in the second part of the proof (Subsection 7.2.2, p. 151pp).

7.2.1 Proof of convergence of finite dimensional marginal distributions

Notation 5 *Let ϱ^{Υ} denote the Prohorov metric on the space of laws of finite dimensional distributions*

$$\mathcal{M}^1\left(\left(\mathcal{M}\left(\mathbb{Z}^d\right)\times\mathcal{M}\left(\mathbb{Z}^d\right)\right)^{\Upsilon}\right) .$$

We can reformulate (7.110) as follows

$$\lim_{N\to\infty}\varrho^{\Upsilon}\left(\mathcal{L}\left\{\left(X^N_{\beta_N s+t},Y^N_{\beta_N s+t}\right)_{t\geq 0}\right\},\int\mu^{\Upsilon}_{(\tilde{\theta}_x,\tilde{\theta}_y)}\,\Psi_s\left(d\left(\tilde{\theta}_x,\tilde{\theta}_y\right)\right)\right)=0 \quad . \tag{7.112}$$

This has the advantage that the triangular inequality is applicable. In fact, we introduce two more processes as intermediate steps between the left hand side and the right hand side of (7.110).

These intermediate steps are visualized in the following diagram. As notation, we need the abbreviation of the laws of processes of empirical averages:

$$\Psi^N_s:=\mathcal{L}\left\{\left(\hat{X}^N_s,\hat{Y}^N_s\right)\right\} \quad , \quad (s\geq 0) \tag{7.113}$$

$$\begin{array}{ccc}
\mathcal{L}\left\{\left(X^N_{\beta_N s+t},Y^N_{\beta_N s+t}\right)_{t\in\Upsilon}\right\} & \xrightarrow{N\to\infty} & \int\mu^{\Upsilon}_{(\tilde{\theta}_x,\tilde{\theta}_y)}\,\Psi_s\left(d\left(\tilde{\theta}_x,\tilde{\theta}_y\right)\right) \\
{\scriptstyle (I)}\big\downarrow & & \big\uparrow{\scriptstyle (III)} \\
\int\mu^{\Upsilon}_{(\tilde{\theta}^N_x,\tilde{\theta}^N_y)}\,\Psi^N_{s-\frac{\ell_N}{\beta_N}}\left(d\left(\tilde{\theta}^N_x,\tilde{\theta}^N_y\right)\right) & \xrightarrow[(II)]{} & \int\mu^{\Upsilon}_{(\tilde{\theta}^N_x,\tilde{\theta}^N_y)}\,\Psi_{s-\frac{\ell_N}{\beta_N}}\left(d\left(\tilde{\theta}^N_x,\tilde{\theta}^N_y\right)\right)
\end{array}$$

In particular, we claim:

I. First step of approximation

(a) Heuristic interpretation: *For large N, the process at and after time $\beta_N s$ is well approximated by a mixture of stationary laws of the infinite lattice processes, where the law of intensities is given by the values of empirical averages on the torus at time $\beta_N s-\ell_N$. To be more precise, this approximation is an approximation in terms of finite dimensional marginal distributions.*

(b) Precise formulation: There exists an increasing sequence $(\ell_N)_{N\in\mathbb{N}}$ with

$$\lim_{N\to\infty}\ell_N=\infty,\quad \limsup_{N\to\infty}\frac{\ell_N}{\ln N}<1 \tag{7.114}$$

such that the following holds holds

$$\lim_{N\to\infty} \varrho^{\Upsilon}\Bigg(\mathcal{L}\left\{\left(X^N_{\beta_N s+t}, Y^N_{\beta_N s+t}\right)_{t\in\Upsilon}\right\},$$
$$\int \mu^{\Upsilon}_{\left(\tilde{\theta}^N_x,\tilde{\theta}^N_y\right)} \Psi^N_{s-\frac{\ell_N}{\beta_N}}\left(d\left(\tilde{\theta}^N_x,\tilde{\theta}^N_y\right)\right)\Bigg) = 0 \quad . \tag{7.115}$$

II. *It is of no importance as $N\to\infty$ if the mixing law is $\Psi^N_{s-\frac{\ell_N}{\beta_N}}$ (empirical averages on the torus) or $\Psi_{s-\frac{\ell_N}{\beta_N}}$ (limit process of empirical averages).*
Macroscopic limit[8]

$$\lim_{N\to\infty} \varrho^{\Upsilon}\Bigg(\int \mu^{\Upsilon}_{\left(\tilde{\theta}^N_x,\tilde{\theta}^N_y\right)} \Psi^N_{s-\frac{\ell_N}{\beta_N}}\left(d\left(\tilde{\theta}^N_x,\tilde{\theta}^N_y\right)\right),$$
$$\int \mu^{\Upsilon}_{\left(\tilde{\theta}^N_x,\tilde{\theta}^N_y\right)} \Psi_{s-\frac{\ell_N}{\beta_N}}\left(d\left(\tilde{\theta}^N_x,\tilde{\theta}^N_y\right)\right)\Bigg) = 0 \quad . \tag{7.116}$$

III. The term "$-\frac{\ell_N}{\beta_N}$" does not change the limiting behavior

$$\lim_{N\to\infty} \varrho\Bigg(\int \mu^{\Upsilon}_{\left(\tilde{\theta}^N_x,\tilde{\theta}^N_y\right)} \Psi_{s-\frac{\ell_N}{\beta_N}}\left(d\left(\tilde{\theta}^N_x,\tilde{\theta}^N_y\right)\right),$$
$$\int \mu^{\Upsilon}_{\left(\tilde{\theta}^N_x,\tilde{\theta}^N_y\right)} \Psi_s\left(d\left(\tilde{\theta}_x,\tilde{\theta}_y\right)\right)\Bigg) = 0 \tag{7.117}$$

By the triangular inequality, (7.115), (7.116) and (7.117) are sufficient for (7.107). We shall now prove in two seperate subsections "Step I" (7.115) and then "Step II" (7.116) together with "Step III" (7.117) (see pages p. 145pp)

7.2.1.1 Proof of Step I (7.115)

We consider finite $\mathbb{T}^d_N$-torus systems at times $\beta_N s$. Our claim is that at these times the catalytic interacting Feller diffusion process on the torus can be well approximated by stationary lattice processes; this approximation is the better the larger N is. To recall what we claim, we want to prove

$$\lim_{N\to\infty} \varrho^{\Upsilon}\Bigg(\mathcal{L}\left\{\left(X^N_{\beta_N s+t}, Y^N_{\beta_N s+t}\right)_{t\in\Upsilon}\right\},$$
$$\int \mu^{\Upsilon}_{\left(\tilde{\theta}^N_x,\tilde{\theta}^N_y\right)} \Psi^N_{s-\frac{\ell_N}{\beta_N}}\left(d\left(\tilde{\theta}^N_x,\tilde{\theta}^N_y\right)\right)\Bigg) = 0 \tag{7.118}$$

Let

$$\left(\bar{\theta}^N_x,\bar{\theta}^N_y\right) := \left(\hat{X}^N_{s-\frac{\ell_N}{\beta_N}}, \hat{Y}^N_{s-\frac{\ell_N}{\beta_N}}\right) \quad . \tag{7.119}$$

[8] Note that on the l.h.s we have $\Psi^N_{s-\frac{\ell_N}{\beta_N}}$, on the r.h.s $\Psi_{s-\frac{\ell_N}{\beta_N}}$.

We define $\mathcal{M}(\mathbb{Z}^d) \times \mathcal{M}(\mathbb{Z}^d)$-valued processes $\left(X_t^{N,\infty}, Y_t^{N,\infty}\right)_{t\geq 0}$ on the same probability space on which the sequence of processes $\left(X^N, Y^N\right)_{N\in\mathbb{N}}$ is defined. Let the law of $\left(X_t^{N,\infty}, Y_t^{N,\infty}\right)_{t\geq 0}$ be given by the right side of (7.118), namely

$$\mathcal{L}\left\{\left(X_t^{N,\infty}, Y_t^{N,\infty}\right)_{t\geq 0}\right\} = \int \nu_{\left(\bar{\theta}_x^N, \bar{\theta}_y^N\right)} \Psi^N_{s-\frac{\ell_N}{\beta_N}}\left(d\left(\tilde{\theta}_x^N, \tilde{\theta}_y^N\right)\right) \quad . \tag{7.120}$$

In addition to that, we require

$$\mathcal{L}\left\{\left(X_t^{N,\infty}, Y_t^{N,\infty}\right)_{t\geq 0}\middle|\left(\bar{\theta}_x^N, \bar{\theta}_y^N\right)\right\} := \nu_{\left(\bar{\theta}_x^N, \bar{\theta}_y^N\right)} \quad . \tag{7.121}$$

This makes sense, since one could construct the process $\left(X_t^{N,\infty}, Y_t^{N,\infty}\right)_{t\geq 0}$ by a two-step sampling, first constructing $\left(\bar{\theta}_x^N, \bar{\theta}_y^N\right)$ as random variable, then $\nu_{\left(\bar{\theta}_x^N, \bar{\theta}_y^N\right)}$ for given $\left(\bar{\theta}_x^N, \bar{\theta}_y^N\right)$ as strong solution of a stochastic differential equation. Making use of the catalyst-reactant setting, especially Proposition 1 (page 49), we can prove the following two facts.

1. The claim is true for the catalyst (even pathwise); let $\nu_{\tilde{\theta}_y^N}$ denotes the equilibrium of the catalyst process with intensity $\tilde{\theta}_y^N$. We have

$$\lim_{N\to\infty} \varrho\Bigg(\mathcal{L}\left\{\left(Y^N_{\beta_N s+t}\right)_{t\geq 0}\right\},$$
$$\int \nu_{\tilde{\theta}_y^N} \mathcal{L}\left\{\left(\hat{Y}^N_{s-\frac{\ell_N}{\beta_N}}\right)\right\}\left(d\tilde{\theta}_y^N\right)\Bigg) = 0 \quad , \tag{7.122}$$

   At this point, we only prove convergence with respect to finite dimensional distributions, as the tightness proof is postponed to Subsection 7.2.2.

2. Consider the laws of quenched laws of reactants – that in particular means that we consider reactants conditioned on pathwise converging catalysts. It is possible to assume this convergence without loss of generality since we have Skorohod's representation; compare the remark given below, especially equation (7.125).

   We have to prove

$$\mathcal{L}_{cat}\left\{\varrho^\Upsilon\left(\mathcal{L}^Y_{reac}\left\{\left(X^N_{\beta_N s+t}\right)_{t\in\Upsilon}\right\}, \mathcal{L}^Y_{reac}\left\{\left(X^{N,\infty}_{t+\ell_N}\right)_{t\in\Upsilon}\right\}\right)\right\} = 0 \quad . \tag{7.123}$$

   Again, weonly prove convergence with respect to finite dimensional distributions at this point; the tightness proof is postponed to Subsection 7.2.2.

In the proof of 2. (Convergence of the quenched reactant) we make use of the fact that we can define the catalyst-reactant processes on a joint probability space – preserving marginal distributions for fixed N – such that we have almost sure convergence of the catalyst. The following remark contains a discussion of this topic. We will refer to it later on.

Suppose that (7.122) is true. According to the Skorohod representation theorem, versions of the catalyst processes can be defined on a common probability space (with probability measure $\mathbb{P}$) such that

$$\lim_{N\to\infty} d_{Sk}\left(\left(Y^N_{\beta_N s-\ell_N+t}\right)_{t\geq 0},\left(Y^{N,\infty}_t\right)_{t\geq 0}\right)=0 \quad \mathbb{P}\text{-a.s.} \tag{7.124}$$

This is the reason why it is no loss of generality[9] if we assume

$$\lim_{N\to\infty} d_{Sk}\left(\left(Y^N_{\beta_N s-\ell_N+t}\right)_{t\geq 0},\left(Y^{N,\infty}_t\right)_{t\geq 0}\right)=0 \quad P_{cat}\text{-a.s.} \tag{7.125}$$

We want the sequence (ℓ_N) to grow sufficiently slowly to infinity, such that with (7.125) also

$$\lim_{N\to\infty} d_{Sk}\left(\left(Y^N_{\beta_N s+t}\right)_{t\geq 0},\left(Y^{N,\infty}_{t+\ell_N}\right)_{t\geq 0}\right)=0 \quad P_{cat}\text{-a.s.} \tag{7.126}$$

is true. In the following box, we give a justification that (7.126) is true?

Why is (7.126) true?
Recall that by (7.125)

$$\lim_{N\to\infty} d_{Sk}\left(\left(Y^N_{\beta_N s-\ell_N+t}\right)_{t\in[0,\Theta]},\left(Y^{N,\infty}_t\right)_{t\in[0,\Theta]}\right)=0 \tag{7.127}$$

holds for all $\Theta\in R_+$.
Let

$$\Theta_1=1 \quad , \tag{7.128}$$

$$\varepsilon_1=\sup_{N\in\mathbb{N}} d_{Sk}\left(\left(Y^N_{\beta_N s-\ell_N+t}\right)_{t\in[0,1]},\left(Y^{N,\infty}_t\right)_{t\in[0,1]}\right) \quad . \tag{7.129}$$

Define $(\Theta_N)_{N\in\mathbb{N}}$, $(\varepsilon_N)_{N\in\mathbb{N}}$ by

$$\Theta_{N+1}:=\Theta_N,\varepsilon_{N+1}:=\varepsilon_N$$
$$\text{if}$$
$$\sup_{M\geq N+1} d_{Sk}\left(\left(Y^M_{\beta_M s-\ell_M+t}\right)_{t\in[0,\Theta_N+1]},\left(Y^{M,\infty}_t\right)_{t\in[0,\Theta_N+1]}\right)>\tfrac{\varepsilon_N}{2}$$

and

$$\Theta_{N+1}:=\Theta_N+1,\varepsilon_{N+1}:=\tfrac{\varepsilon_N}{2}$$
$$\text{if}$$
$$\sup_{M\geq N+1} d_{Sk}\left(\left(Y^M_{\beta_M s-\ell_M+t}\right)_{t\in[0,\Theta_N+1]},\left(Y^{M,\infty}_t\right)_{t\in[0,\Theta_N+1]}\right)\leq\tfrac{\varepsilon_N}{2}$$

By (7.127), $(\Theta_N)_{N\in\mathbb{N}}$ jumps infinitely often, hence $\Theta_N \overset{N\to\infty}{\longrightarrow} \infty$.

[9] This is similar to the considerations carried out on page 88

Proof of 1. (Convergence of the catalyst)

In [CGS95] Cox, Greven and Shiga prove the finite system scheme for the catalyst process. A closer inspection of their prove yields that (7.122) holds. For those feeling incomfortable with this kind of argumentation, there is a second way to obtain (7.122): One simply has to carry out the same considerations and calculations as done below, but with a constant catalyst process instead of the fluctuating catalyst considered here. *Roughly speaking: the catalyst process is a reactant catalysed by a constant process.*

Proof of 2. (Convergence of the quenched reactant)

Characterization by duality

To obtain (7.123) we compare the Laplace transforms of the finite dimensional distributions characterizing the quenched laws

$$\mathcal{L}^Y_{reac}\left\{\left(X^N_{\beta_N s+t}\right)_{t\in\Upsilon}\right\}\ ,\ \mathcal{L}^Y_{reac}\left\{\left(X^{N,\infty}_{t+\ell_N}\right)_{t\in\Upsilon}\right\}\quad . \tag{7.130}$$

At our observation times $t_1, t_2, ..., t_n$ (the elements of Υ) we make observations applying $\varphi_k \in \mathbb{R}_+^{\mathbb{Z}^d}$ $(k \in \{1, ..., n\})$. We can assume

$$card\left\{i\in\mathbb{Z}^d : \sum_{k=1}^{n}\varphi_k(i)\right\} < \infty \quad . \tag{7.131}$$

We have to show the following that for any choice of test functions (observations) we have

$$\begin{aligned}\lim_{N\to\infty}\mathbf{E}\Bigg|\mathbf{E}^Y_{reac}&\left[\exp\left(-\sum_{k=1}^{n}\left\langle\varphi_k, X^N_{\beta_N s+t_k}\right\rangle\right)\right]\\ -\mathbf{E}^Y_{reac}&\left[\exp\left(-\sum_{k=1}^{n}\left\langle\varphi_k, X^{N,\infty}_{\ell_N+t_k}\right\rangle\right)\right]\Bigg|\\ =0\quad .\end{aligned} \tag{7.132}$$

By the so-called Cramér-Wold device[10] this is sufficient for convergence with respect to finite dimensional distributions. We can provide a dual representation of the terms occuring in (7.132):

$$\begin{aligned}\mathbf{E}^Y_{reac}&\left[\exp\left(-\sum_{k=1}^{n}\left\langle\varphi_k, X^N_{\beta_N s+t_k}\right\rangle\right)\right]\\ &=\mathbf{E}^Y_{reac}\left[\exp\left(-\left\langle u^N_{T+\ell_N}, X^N_{\beta_N s-\ell_N}\right\rangle\right)\right]\end{aligned} \tag{7.134}$$

[10] Remember that the main goal is to prove

$$\begin{aligned}\lim_{N\to\infty}\varrho\Bigg(\mathcal{L}&\left\{\left(X^N_{\beta_N s+t}, Y^N_{\beta_N s+t}\right)_{t\in\Upsilon}\right\},\\ &\int\mu^{\Upsilon}_{\left(\tilde\theta^N_x,\tilde\theta^N_y\right)}\,\Psi^N_{s-\frac{\ell_N}{\beta_N}}\left(d\left(\tilde\theta^N_x,\tilde\theta^N_y\right)\right)\Bigg)=0\quad .\end{aligned} \tag{7.133}$$

Remember that $\mu^{\Upsilon}_{\left(\tilde\theta^N_x,\tilde\theta^N_y\right)}$ denotes the joint law of the process at times $\Upsilon = \{t_1, t_2, ..., t_n\}$

and

$$\mathbf{E}^{Y}_{reac}\left[\exp\left(-\sum_{k=1}^{n}\left\langle\varphi_k, X^{N,\infty}_{\ell_N+t_k}\right\rangle\right)\right] \tag{7.135}$$

$$= \mathbf{E}^{Y}_{reac}\left[\exp\left(-\left\langle u^{N,\infty}_{T+\ell_N}, X^{N,\infty}_{0}\right\rangle\right)\right] \quad . \tag{7.136}$$

The dual processes $\left(u^N_t\right)_{t\geq 0}$ and $\left(u^{N,\infty}_t\right)_{t\geq 0}$ are given by

- initial condition

$$u^N_0 \equiv 0 \quad . \tag{7.137}$$

$$u^{N,\infty}_0 \equiv 0 \quad . \tag{7.138}$$

- evolution at non-jump times ($T - t \notin \{t_1, ..., t_n\}$)

$$\frac{\partial}{\partial t}u^N_t(i) = c_x \sum_{j\in\mathbb{T}^d_N} a^N(j,i)\left[u^N_t(j) - u^N_t(i)\right] - \gamma_x Y^N_{\beta_N s+T-t}(i)\left(u^N_t(i)\right)^2 \quad . \tag{7.139}$$

$$\frac{\partial}{\partial t}u^{N,\infty}_t(i) = c_x \sum_{j\in\mathbb{Z}^d} a(j,i)\left(u^{N,\infty}_t(j) - u^{N,\infty}_t(i)\right) - \gamma_x Y^{N,\infty}_{\ell_N+T-t}(i)\left(u^{N,\infty}_t(i)\right)^2 \quad . \tag{7.140}$$

- jumps at times t with $T - t \in \{t_1, ..., t_n\}$

$$u^N_{T-t_k}(i) = \lim_{\varsigma\to T-t_k-} u^N_\varsigma(i) + \varphi_k(i) \quad . \tag{7.141}$$

$$u^{N,\infty}_{T-t_k}(i) = \lim_{\varsigma\to T-t_k-} u^{N,\infty}_\varsigma(i) + \varphi_k(i) \quad . \tag{7.142}$$

We need the following supersolution:
Define $(v_t)_{t\geq 0}$ as unique solution of

- initial condition

$$v_0 \equiv 0 \quad . \tag{7.143}$$

- evolution at non-jump times ($T - t \notin \{t_1, ..., t_n\}$)

$$\frac{\partial}{\partial t}v_t(i) = c_x \sum_{j\in\mathbb{T}^d_N} a(j,i)\left(v_t(j) - v_t(i)\right) \tag{7.144}$$

- jumps at times t with $T - t \in \{t_1, ..., t_n\}$

$$v_{T-t_k}(i) = \lim_{\varsigma\to T-t_k-} v_\varsigma(i) + \varphi_k(i) \quad . \tag{7.145}$$

We can regard $(v_t)_{t\geq 0}$ as $\left(u_t^{N,\infty}\right)_{t\geq 0}$ with vanishing catalyst. Therefore we have

$$v_t(i) \geq u_t^{N,\infty}(i) \quad \left(i \in \mathbb{Z}^d\right). \tag{7.146}$$

The latter could be proven using a Feynman-Kac approach, for example.

Remark 17 *As a consequence of (7.131), there exists $K < \infty$ with*

$$K > \sup_{t\geq 0} \langle v_t, \mathfrak{n} \rangle \tag{7.147}$$

Comparison of dual processes

To verify (7.132) we have to check[11] if

$$\lim_{N\to\infty} \mathbf{E}\left[\left|\left\langle u_{T+\ell_N}^N, X_{\beta_N s-\ell_N}^N\right\rangle - \left\langle u_{T+\ell_N}^{N,\infty}, X_0^{N,\infty}\right\rangle\right| \wedge 1\right] = 0 \tag{7.148}$$

holds. Consider

$$\delta : \begin{cases} \mathbb{R}_+ \times \mathbb{R}_+ & \to & \mathbb{R}_+ \\ (x,y) & \mapsto & |x-y| \wedge 1 \end{cases} . \tag{7.149}$$

It is elementary that δ is a metric. Reformulating (7.148) using δ, we have to prove

$$\lim_{N\to\infty} \mathbf{E}\left[\delta\left(\left\langle u_{T+\ell_N}^N, X_{\beta_N s-\ell_N}^N\right\rangle, \left\langle u_{T+\ell_N}^{N,\infty}, X_0^{N,\infty}\right\rangle\right)\right] = 0 \quad . \tag{7.150}$$

We apply the triangular inequality:

$$\begin{aligned} &\delta\left(\left\langle u_{T+\ell_N}^N, X_{\beta_N s-\ell_N}^N\right\rangle, \left\langle u_{T+\ell_N}^{N,\infty}, X_0^{N,\infty}\right\rangle\right) \\ &\leq \delta\left(\left\langle u_{T+\ell_N}^N, X_{\beta_N s-\ell_N}^N\right\rangle, \left\langle u_{T+\ell_N}^N, \bar{\theta}_x^N \mathfrak{n}\right\rangle\right) \\ &\quad + \bar{\theta}_x^N \cdot \delta\left(\left\langle u_{T+\ell_N}^N, \mathfrak{n}\right\rangle, \left\langle u_{T+\ell_N}^{N,\infty}, \mathfrak{n}\right\rangle\right) \\ &\quad + \delta\left(\left\langle u_{T+\ell_N}^{N,\infty}, \bar{\theta}_x^N \mathfrak{n}\right\rangle, \left\langle u_{T+\ell_N}^{N,\infty}, X_0^{N,\infty}\right\rangle\right) \quad . \end{aligned} \tag{7.151}$$

The following steps have to be carried out:

Step A: We have to show

$$\lim_{N\to\infty} \mathbf{E}\left[\delta\left(\left\langle u_{T+\ell_N}^N, \mathfrak{n}\right\rangle, \left\langle u_{T+\ell_N}^{N,\infty}, \mathfrak{n}\right\rangle\right)\right] = 0 \quad . \tag{7.152}$$

Note first of all that $\left\langle u_{T+\ell_N}^N, \mathfrak{n}\right\rangle$ is a sum over the torus $\mathbb{T}_N^d$, whereas $\left\langle u_{T+\ell_N}^{N,\infty}, \mathfrak{n}\right\rangle$ is a sum over the whole lattice $\mathbb{Z}^d$.

There is the following interpretation of (7.152): The "losses of mass" (arising from the "$-u^2$"-terms) the log-Laplace transforms of the torus system suffer from converge to those of the infinite system as $N \to \infty$.

[11]Note that for $x, y \geq 0$ we have $\left|e^{-x} - e^{-y}\right| \leq |x-y| \wedge 1$.

Step B: We have to verify the following two equations:

1.
$$\lim_{N\to\infty} \mathbf{E}\left[\delta\left(\left\langle u^N_{T+\ell_N}, X^N_{\beta_N s-\ell_N}\right\rangle, \bar{\theta}^N_x \left\langle u^N_{T+\ell_N}, \mathfrak{n}\right\rangle\right)\right] = 0 \tag{7.153}$$

2.
$$\lim_{N\to\infty} \mathbf{E}\left[\delta\left(\left\langle u^{N,\infty}_{T+\ell_N}, X^{N,\infty}_0\right\rangle, \bar{\theta}^N_x \left\langle u^{N,\infty}_{T+\ell_N}, \mathfrak{n}\right\rangle\right)\right] = 0 \tag{7.154}$$

The key tool of the proof of 1. and 2. is that between times T and $T+\ell_N$ an averaging effect takes place.

Proof of Step A: First of all, note that

$$\lim_{N\to\infty} \mathbf{E}\left[\sum_{i\in\mathbb{Z}^d\setminus\mathbb{T}^d_N} u^{N,\infty}_{T+\ell_N}(i)\right] = 0 \quad . \tag{7.155}$$

This becomes obvious if we recall (7.146). Then (7.155) is an immediate consequence of

$$\lim_{N\to\infty} \mathbf{E}\left[\sum_{i\in\mathbb{Z}^d\setminus\mathbb{T}^d_N} v_{T+\ell_N}(i)\right] = 0 \quad . \tag{7.156}$$

It suffices to verify

$$\lim_{N\to\infty} \mathbf{E}\left[\sum_{i\in\mathbb{T}^d_N} \left|u^N_{T+\ell_N}(i) - u^{N,\infty}_{T+\ell_N}(i)\right| \wedge 1\right] = 0 \quad . \tag{7.157}$$

This is obviously sufficient for (7.153).

First of all, we want to describe the dynamics of the difference

$$u^N_t(i) - u^{N,\infty}_t(i) \quad , \left(i\in\mathbb{T}^d_N\right) \tag{7.158}$$

Since the jumps of u^N and $u^{N,\infty}$ coincide, the differences are – for arbitrary $i\in\mathbb{T}^d_N$ – continuous in time and piecewise differentiable. We will describe the difference processes in terms of differential equations, where by convention $\frac{\partial}{\partial t}$ denotes the right-sided derivatives at the points of non-differentiability. The following differential equation can easily be obtained from (7.139) and (7.140):

$$\begin{aligned}\frac{\partial}{\partial t} &= c_x \sum_{j\in\mathbb{T}^d_N} a^N(j,i)\left(u^N_t(j) - u^N_t(i)\right) \\ &- c_x \sum_{j\in\mathbb{T}^d_N} a(j,i)\left(u^{N,\infty}_t(j) - u^{N,\infty}_t(i)\right) \\ &- c_x \sum_{j\in\mathbb{Z}^d\setminus\mathbb{T}^d_N} a(j,i)\left(u^{N,\infty}_t(j) - u^{N,\infty}_t(i)\right) \\ &- \gamma_x Y^N_{\ell_N+T-t}(i)\left(u^N_t(i)\right)^2 \\ &+ \gamma_x Y^{N,\infty}_{\ell_N+T-t}(i)\left(u^{N,\infty}_t(i)\right)^2 \quad .\end{aligned} \tag{7.159}$$

Expanding this equation yields[12]:

$$
\begin{aligned}
&\frac{\partial}{\partial t}\left(u_t^N(i) - u_t^{N,\infty}(i)\right) \\
&= c_x \sum_{j\in\mathbb{T}_N^d} a^N(j,i)\left(u_t^N(j) - u_t^{N,\infty}(j) - \left(u_t^N(i) - u_t^{N,\infty}(i)\right)\right) \\
&+ c_x \sum_{j\in\mathbb{T}_N^d} \left(a(j,i) - a^N(j,i)\right)\left(u_t^{N,\infty}(j) - u_t^{N,\infty}(i)\right) \\
&- c_x \sum_{j\in\mathbb{Z}^d\setminus\mathbb{T}_N^d} a(j,i)\left(u_t^{N,\infty}(j) - u_t^{N,\infty}(i)\right) \\
&+ \frac{1}{2}\left(Y^{\infty}_{\ell_N+T-t}(i) - Y^N_{\ell_N+T-t}(i)\right)\left(\left(u_t^N(i)\right)^2 + \left(u_t^{N,\infty}(i)\right)^2\right) \\
&- \frac{1}{2}\left(Y^{\infty}_{\ell_N+T-t}(i) + Y^N_{\ell_N+T-t}(i)\right)\left(\left(u_t^N(i)\right)^2 - \left(u_t^{N,\infty}(i)\right)^2\right) \quad . \qquad (7.160)
\end{aligned}
$$

We focus on $\sum_{i\in\mathbb{T}_N^d}\left|u^N_{T+\ell_N}(i) - u^{N,\infty}_{T+\ell_N}(i)\right|$, which contains $|\bullet|$. Since there is non-differentiability at zero, it is not possible to get a differential equation for the evolution of $\sum_{i\in\mathbb{T}_N^d}\left|u^N_{T+\ell_N}(i) - u^{N,\infty}_{T+\ell_N}(i)\right|$ by simply applying the chain rule. Instead, we have to appproximate $|\bullet|$ by the sequence $(g_n(\bullet))_{n\in\mathbb{N}}$ defined by

$$
g_n : \left\{ \begin{array}{ccc} \mathbb{R} & \longrightarrow & \mathbb{R}_+ \\ x & \longmapsto & \frac{1}{n}\ln\cosh(nx) \end{array} \right. \quad . \qquad (7.161)
$$

Remark 18 *It is elementary (see page 93) to prove the following properties:*

1. $g_n'(x) = \tanh(nx)\,; \quad (x\in\mathbb{R})\,.$
2. $g_n(x) \le x g_n(x) \le |x| \le g_n(x) + \frac{\ln 2}{n}; \quad (x\in\mathbb{R})\,.$

We get from (7.160)[13]

$$
\begin{aligned}
&\frac{\partial}{\partial t}\sum_{i\in\mathbb{T}_N^d} g_{n\exp(|i|)}\left(u_t^N(i) - u_t^{N,\infty}(i)\right) \\
&= \mathbf{I} + \mathbf{II} + \mathbf{III} + \mathbf{IV} + \mathbf{V} \quad , \qquad (7.162)
\end{aligned}
$$

[12] We use $ab - a'b' = \frac{(a-a')(b+b')+(a+a')(b-b')}{2}$.

[13] We use $g_{n\exp(|i|)}$ instead of g_n. This gives us all uniform convergence properties we need to interchange limits as we like to.

with

$$\mathbf{I} := c_x \sum_{i\in\mathbb{T}_N^d} \sum_{j\in\mathbb{T}_N^d} a^N(j,i)\, g'_{n\exp(|i|)} \left(u_t^N(i) - u_t^{N,\infty}(i)\right) \left(u_t^N(j) - u_t^{N,\infty}(j)\right)$$
$$- c_x \sum_{i\in\mathbb{T}_N^d} \sum_{j\in\mathbb{T}_N^d} a^N(j,i)\, g'_{n\exp(|i|)} \left(u_t^N(i) - u_t^{N,\infty}(i)\right) \left(u_t^N(i) - u_t^{N,\infty}(i)\right) \quad , \tag{7.163}$$

$$\mathbf{II} := c_x \sum_{i\in\mathbb{T}_N^d} g'_{n\exp(|i|)} \left(u_t^N(i) - u_t^{N,\infty}(i)\right) \cdot \sum_{j\in\mathbb{T}_N^d} \left(a(j,i) - a^N(j,i)\right) \left(u_t^{N,\infty}(j) - u_t^{N,\infty}(i)\right) \quad , \tag{7.164}$$

$$\mathbf{III} := -c_x \sum_{i\in\mathbb{T}_N^d} g'_{n\exp(|i|)} \left(u_t^N(i) - u_t^{N,\infty}(i)\right) \cdot \sum_{j\in\mathbb{Z}^d\setminus\mathbb{T}_N^d} a(j,i) \left(u_t^{N,\infty}(j) - u_t^{N,\infty}(i)\right) \quad , \tag{7.165}$$

$$\mathbf{IV} := \frac{1}{2} \sum_{i\in\mathbb{T}_N^d} \left(Y^{\infty}_{\ell_N+T-t}(i) - Y^{N}_{\ell_N+T-t}(i)\right) \cdot g'_{n\exp(|i|)} \left(u_t^N(i) - u_t^{N,\infty}(i)\right) \left(\left(u_t^N(i)\right)^2 + \left(u_t^{N,\infty}(i)\right)^2\right) \quad , \tag{7.166}$$

$$\mathbf{V} := -\frac{1}{2} \sum_{i\in\mathbb{T}_N^d} \left(Y^{\infty}_{\ell_N+T-t}(i) + Y^{N}_{\ell_N+T-t}(i)\right) \cdot g'_{n\exp(|i|)} \left(u_t^N(i) - u_t^{N,\infty}(i)\right) \left(\left(u_t^N(i)\right)^2 - \left(u_t^{N,\infty}(i)\right)^2\right) \quad . \tag{7.167}$$

In the sequel, we show:

$$\forall t \in [0,T] \limsup_{N\to\infty} \mathbf{I} \le 0 \quad , \tag{7.168}$$

$$\lim_{N\to\infty} \int_0^{T+\ell_N} |\mathbf{II}|\, dt = 0 \quad , \tag{7.169}$$

$$\lim_{N\to\infty} \int_0^{T+\ell_N} |\mathbf{III}|\, dt = 0 \quad , \tag{7.170}$$

$$\int_0^{T+\ell_N} |\mathbf{IV}|\, dt \to 0 \quad P_{cat}\text{-sto.} \tag{7.171}$$

$$\forall N \in \mathbb{N}, \forall t \in [0,T] : \mathbf{V} \le 0 \quad . \tag{7.172}$$

I: We consider

$$\mathbf{I} = c_x \sum_{i\in\mathbb{T}_N^d} \sum_{j\in\mathbb{T}_N^d} a^N(j,i)\, g'_{n\exp(|i|)} \left(u_t^N(i) - u_t^{N,\infty}(i)\right)\left(u_t^N(j) - u_t^{N,\infty}(j)\right)$$
$$- c_x \sum_{i\in\mathbb{T}_N^d} \sum_{j\in\mathbb{T}_N^d} a^N(j,i)\, g'_{n\exp(|i|)} \left(u_t^N(i) - u_t^{N,\infty}(i)\right)\left(u_t^N(i) - u_t^{N,\infty}(i)\right) \quad . \tag{7.173}$$

With $-1 \le g'_{n\exp(|i|)} \le 1$ obtain:

$$\mathbf{I} \le c_x \sum_{i\in\mathbb{T}_N^d} \sum_{j\in\mathbb{T}_N^d} a^N(j,i) \left|u_t^N(j) - u_t^{N,\infty}(j)\right|$$
$$- c_x \sum_{i\in\mathbb{T}_N^d} \sum_{j\in\mathbb{T}_N^d} a^N(j,i)\, g'_{n\exp(|i|)} \left(u_t^N(i) - u_t^{N,\infty}(i)\right)$$
$$\cdot \left(u_t^N(i) - u_t^{N,\infty}(i)\right) \quad . \tag{7.174}$$

The limit (as $n \to \infty$) of the second line equals

$$-c_x \sum_{i\in\mathbb{T}_N^d} \sum_{j\in\mathbb{T}_N^d} a^N(j,i) \left|u_t^N(i) - u_t^{N,\infty}(i)\right| \quad . \tag{7.175}$$

This follows in particular from

$$0 \le g'_n(x) \cdot x \le g_n(x) \le \frac{\ln 2}{n} \quad . \tag{7.176}$$

Finally we get

$$\limsup_{N\to\infty} \mathbf{I} \le c_x \sum_{i\in\mathbb{T}_N^d} \sum_{j\in\mathbb{T}_N^d} a^N(j,i) \left(\left|u_t^N(j) - u_t^{N,\infty}(j)\right| - \left|u_t^N(i) - u_t^{N,\infty}(i)\right|\right) = 0 \quad . \tag{7.177}$$

II: We consider

$$\mathbf{II} = c_x \sum_{i\in\mathbb{T}_N^d} \sum_{j\in\mathbb{T}_N^d} g'_{n\exp(|i|)} \left(u_t^N(i) - u_t^{N,\infty}(i)\right)$$
$$\cdot \left(a(j,i) - a^N(j,i)\right)\left(u_t^{N,\infty}(j) - u_t^{N,\infty}(i)\right) \quad . \tag{7.178}$$

We get

$$|\mathbf{II}| \le c_x \sum_{i\in\mathbb{T}_N^d} \sum_{j\in\mathbb{T}_N^d} \left|a(j,i) - a^N(j,i)\right| \left(u_t^{N,\infty}(j) + u_t^{N,\infty}(i)\right)$$
$$\le c_x \sum_{i\in\mathbb{T}_N^d} \sum_{j\in\mathbb{T}_N^d} \left|\hat{a}(j,i) - \hat{a}^N(j,i)\right| \left(u_t^{N,\infty}(i)\right)$$
$$\le c_x \sum_{i\in\mathbb{T}_N^d} \sum_{j\in\mathbb{T}_N^d} \left|\hat{a}(j,i) - \hat{a}^N(j,i)\right| \cdot v_t(i) \quad . \tag{7.179}$$

Let $\varepsilon > 0$. By standard arguments (large deviations etc.) there exists $K_\varepsilon < \infty$ s. t.

$$\int_0^\infty \sum_{i\in\mathbb{Z}^d} v_t(i)\, \mathbb{I}_{\{|i|\geq K_\varepsilon(1+t)\}}(i)\, dt < \varepsilon \quad . \tag{7.180}$$

We get

$$\begin{aligned}
&\int_0^{T+\ell_N} |\mathbf{II}|\, dt \\
&\leq \int_0^{T+\ell_N} c_x \sum_{i\in\mathbb{T}_N^d} \sum_{j\in\mathbb{T}_N^d} \left|\hat{a}(j,i) - \hat{a}^N(j,i)\right| \cdot v_t(i)\, dt \\
&\leq \int_0^{T+\ell_N} c_x \sum_{i\in\mathbb{T}_N^d} \sum_{j\in\mathbb{T}_N^d} \left|\hat{a}(j,i) - \hat{a}^N(j,i)\right| \cdot v_t(i)\, \mathbb{I}_{\{|i|\geq K_\varepsilon(1+t)\}}(i)\, dt \\
&\quad + \int_0^{T+\ell_N} c_x \sum_{i\in\mathbb{T}_N^d} \sum_{j\in\mathbb{T}_N^d} \left|\hat{a}(j,i) - \hat{a}^N(j,i)\right| \cdot v_t(i)\, \mathbb{I}_{\{|i|< K_\varepsilon(1+t)\}}(i)\, dt \\
&\leq c_x\varepsilon \\
&\quad + \int_0^{T+\ell_N} c_x \sum_{i\in\mathbb{T}_N^d} \sum_{j\in\mathbb{T}_N^d} \left|\hat{a}(j,i) - \hat{a}^N(j,i)\right| \cdot v_t(i)\, \mathbb{I}_{\{|i|< K_\varepsilon(1+t)\}}(i)\, dt \quad .
\end{aligned} \tag{7.181}$$

If N is sufficiently large, we have $N \geq K_\varepsilon(1+T+\ell_N)$. Applying (7.147) we get

$$\begin{aligned}
&\leq c_x\varepsilon + \int_0^{T+\ell_N} c_x \sum_{i:|i|<K_\varepsilon(1+T+\ell_N)} \sum_{j\in\mathbb{T}_N^d} \left(\hat{a}^N(j,i) - \hat{a}(j,i)\right) \cdot K\, dt \\
&\leq c_x\varepsilon + c_x(T+\ell_N) \sum_{i:|i|<K_\varepsilon(1+T+\ell_N)} \sum_{j\in\mathbb{T}_N^d} \sum_{k\in(2\mathbb{Z})^d\setminus\{0,\ldots,0\}} \hat{a}(j+kN,i)\cdot K \quad .
\end{aligned} \tag{7.182}$$

In the equation above, the difference between i and $j+kN$ always exceeds N. In particular, we get

$$\begin{aligned}
&\limsup_{N\to\infty} \int_0^{T+\ell_N} |\mathbf{II}|\, dt \\
&\leq c_x\varepsilon \\
&\quad + \limsup_{N\to\infty} c_x(T+\ell_N)\left(2K_\varepsilon(1+T+\ell_N)\right)^d K \sum_{\ell\in\mathbb{Z}^d:|\ell|\geq N} \hat{a}(0,\ell) \quad .
\end{aligned} \tag{7.183}$$

As assumed in 1.2.4, we have

$$\tilde{K} := \sum_{i\in\mathbb{Z}^d} i^2 \hat{a}(0,i) < \infty \quad . \tag{7.184}$$

Together with (7.114) this yields

$$\limsup_{N\to\infty} \int_0^{T+\ell_N} |\mathbf{II}| \, dt$$
$$\leq c_x\varepsilon + \limsup_{N\to\infty} c_x \left(T+\ln N\right)\left(2K_\varepsilon\left(1+T+\ln N\right)\right)^d K\frac{\tilde{K}}{N^2}$$
$$\leq c_x\varepsilon \quad . \tag{7.185}$$

Finally, let $\varepsilon \to 0$.

III: We consider

$$\mathbf{III} := -c_x \sum_{i\in\mathbb{T}_N^d} \sum_{j\in\mathbb{Z}^d\setminus\mathbb{T}_N^d} g'_{n\exp(|i|)}\left(u_t^N(i) - u_t^{N,\infty}(i)\right) \cdot a(j,i)\left(u_t^{N,\infty}(j) - u_t^{N,\infty}(i)\right) \quad . \tag{7.186}$$

We have to verify

$$\lim_{N\to\infty} \int_0^{T+\ell_N} |\mathbf{III}| \, dt = 0 \quad . \tag{7.187}$$

We have

$$\limsup_{N\to\infty} \int_0^{T+\ell_N} |\mathbf{III}| \, dt$$
$$\leq \limsup_{N\to\infty} \int_0^{T+\ell_N} c_x \sum_{i\in\mathbb{T}_N^d} \sum_{j\in\mathbb{Z}^d\setminus\mathbb{T}_N^d} a(j,i)\left(u_t^{N,\infty}(j) + u_t^{N,\infty}(i)\right) dt$$

We have replaced a " $-$ " by a "$+$" – this of course generates an upper bound

$$\leq \limsup_{N\to\infty} \int_0^{T+\ell_N} c_x \sum_{i\in\mathbb{T}_N^d} \sum_{j\in\mathbb{Z}^d\setminus\mathbb{T}_N^d} a(j,i)\, v_t(j)\, dt$$
$$+ \limsup_{N\to\infty} \int_0^{T+\ell_N} c_x \sum_{i\in\mathbb{T}_N^d} \sum_{j\in\mathbb{Z}^d\setminus\mathbb{T}_N^d} a(j,i)\, v_t(i)\, dt$$

$$\begin{aligned}
&\leq \limsup_{N\to\infty} \int_0^{T+\ell_N} c_x \sum_{j\in\mathbb{Z}^d\setminus\mathbb{T}_N^d} v_t(j)\, \mathbb{I}_{\{|j|<K_\varepsilon(1+t)\}}(j)\, dt \\
&+ \limsup_{N\to\infty} \int_0^{T+\ell_N} c_x \sum_{j\in\mathbb{Z}^d\setminus\mathbb{T}_N^d} v_t(j)\, \mathbb{I}_{\{|j|\geq K_\varepsilon(1+t)\}}(j)\, dt \\
&+ \limsup_{N\to\infty} \int_0^{T+\ell_N} c_x \sum_{i\in\mathbb{T}_N^d} \sum_{j\in\mathbb{Z}^d\setminus\mathbb{T}_N^d} a(j,i)\, v_t(i)\, \mathbb{I}_{\{|i|\geq K_\varepsilon(1+t)\}}(i)\, dt \\
&+ \limsup_{N\to\infty} \int_0^{T+\ell_N} c_x \sum_{i\in\mathbb{T}_N^d} \sum_{j\in\mathbb{Z}^d\setminus\mathbb{T}_N^d} a(j,i)\, v_t(i)\, \mathbb{I}_{\{|i|<K_\varepsilon(1+t)\}}(i)\, dt
\end{aligned}$$

As the sets $\mathbb{Z}^d\setminus\mathbb{T}_N^d$ and $\{|j| < K_\varepsilon(1+T)\}$ are disjoint if N is large, the first lie vanishes.

The second and third line are both bounded above by $c_x\varepsilon$. This follows from the definition of K_ε – compare (7.180).

For the last line, the distance between i and j exceeds N if N is large enough. The summation (over i) contains $(2K_\varepsilon(1+T+\ln N))^d$ terms. Applying (7.184), we get:

$$\begin{aligned}
&\limsup_{N\to\infty} \int_0^{T+\ell_N} |\mathbf{III}|\, dt \\
&\leq 2c_x\varepsilon \\
&\quad + \limsup_{N\to\infty} (T+\ell_N)\cdot c_x \frac{\tilde{K}}{(N-K_\varepsilon(1+T+\ell_N))^2} K\,(2K_\varepsilon(1+T+\ln N))^d \\
&\leq 2c_x\varepsilon \quad . \qquad (7.188)
\end{aligned}$$

As finite step, let $\varepsilon\to 0$. Obviously, II. and III. have been proven using similar methods.

IV: We have to prove that

$$\int_0^{T+\ell_N} |\mathbf{IV}|\, dt \to 0 \quad P_{cat} - sto. \qquad (7.189)$$

holds, where **IV** is given by

$$\begin{aligned}
\mathbf{IV} &= \frac{1}{2}\sum_{i\in\mathbb{T}_N^d} \left(Y^{\infty}_{\ell_N+T-t}(i) - Y^{N}_{\ell_N+T-t}(i)\right) \\
&\quad \cdot g'_{n\exp(|i|)}\left(u_t^N(i) - u_t^{N,\infty}(i)\right)\left(\left(u_t^N(i)\right)^2 + \left(u_t^{N,\infty}(i)\right)^2\right) \quad . \qquad (7.190)
\end{aligned}$$

It is sufficient to prove

$$\lim_{N\to\infty} \mathbf{E}\left[\sum_{i\in\mathbb{Z}^d} \int_0^{T+\ell_N} \frac{1}{2}\left|Y^{N,\infty}_{\ell_N+T-t}(i) - Y^{N}_{\ell_N+T-t}(i)\right| \cdot \left(\left(u^N_t(i)\right)^2 + \left(u^{N,\infty}_t(i)\right)^2\right) dt \wedge 1\right] = 0 \quad . \tag{7.191}$$

is true.

We need the following:

$$\limsup_{N\to\infty} \int_0^{T+\ell_N} \sum_{i\in\mathbb{T}^d_N} \left(\left(u^N_t(i)\right)^2 + \left(u^{N,\infty}_t(i)\right)^2\right) dt < \infty \quad . \tag{7.192}$$

Define $\left(v^N_t\right)_{t\geq 0}$ as unique solution of

- initial condition

$$v^N_0 \equiv 0 \quad . \tag{7.193}$$

- evolution at non-jump times ($T-t \notin \{t_1, ..., t_n\}$)

$$\frac{\partial}{\partial t} v^N_t(i) = c_x \sum_{j\in\mathbb{T}^d_N} a^N(j,i)\left(v^N_t(j) - v^N_t(i)\right) \tag{7.194}$$

- jumps at times t with $T-t \in \{t_1, ..., t_n\}$

$$v^N_{T-t_k}(i) = \lim_{\varsigma\to T-t_k-} v^N_\varsigma(i) + \varphi_k(i) \quad . \tag{7.195}$$

Clearly, $\left(v^N_t\right)_{t\geq 0}$ exceeds $\left(u^N_t\right)_{t\geq 0}$ for arbitrary $N\in\mathbb{N}$. Hence it is sufficient to show

$$\limsup_{N\to\infty} \int_0^{T+\ell_N} \sum_{i\in\mathbb{T}^d_N} \left(\left(v^N_t(i)\right)^2 + \left(v_t(i)\right)^2\right) dt < \infty \quad . \tag{7.196}$$

We know from transience that

$$\int_0^{\infty} \sum_{i\in\mathbb{Z}^d} v_t(i)^2 \; dt < \infty \quad . \tag{7.197}$$

As a result of Cox, Greven and Shiga ([CGS95], Lemma 2.1c) we have

$$\lim_{N\to\infty} \int_0^{\beta_N s} \hat{a}^N_t(0,0)\, dt = \lim_{N\to\infty} \int_0^{\beta_N s} \hat{a}_t(0,0)\, dt + s \quad (s\geq 0) \quad . \tag{7.198}$$

This yields

$$\limsup_{N\to\infty} \int_0^{T+\ell_N} \sum_{i\in\mathbb{T}^d_N} \left(v^N_t(i)\right)^2 dt < \infty \quad .$$

Hence, for any $\varepsilon > 0$ there exists a bounded set $B_\varepsilon \subset \mathbb{R}_+ \times \mathbb{Z}^d$ with

$$\limsup_{N\to\infty} \int_0^\infty \sum_{i\in\mathbb{T}_N^d} \left(\left(u_t^N(i)\right)^2 + \left(u_t^{N,\infty}(i)\right)^2 \right) \mathbb{1}_{\{(t,i)\notin B_\varepsilon\}}(t,i)\ dt < \varepsilon \quad . \tag{7.199}$$

Decomposing the left hand side of (7.191) with the help of B_ε, there remains to prove:

(i)

$$\begin{aligned} &\lim_{N\to\infty} \mathbf{E}\left[\frac{1}{2} \left\langle \mathbf{n}, \int_0^{T+\ell_N} \left| Y^{N,\infty}_{\ell_N+T-t}(\cdot) - Y^N_{T-t}(\cdot) \right| \right.\right. \\ &\qquad \left.\left. \left(\left(u_t^N(\cdot)\right)^2 + \left(u_t^\infty(\cdot)\right)^2 \right) \mathbb{1}_{\{(t,\cdot)\notin B_\varepsilon\}}(t,\cdot)\ dt \wedge 1 \right\rangle \right] \\ &< \theta_y \varepsilon \quad . \end{aligned} \tag{7.200}$$

(ii)

$$\begin{aligned} &\lim_{N\to\infty} \mathbf{E}\left[\frac{1}{2} \left\langle \mathbf{n}, \int_0^{T+\ell_N} \left| Y^{N,\infty}_{\ell_N+T-t}(\cdot) - Y^N_{T-t}(\cdot) \right| \right.\right. \\ &\qquad \left.\left. \left(\left(u_t^N(\cdot)\right)^2 + \left(u_t^\infty(\cdot)\right)^2 \right) \mathbb{1}_{\{(t,\cdot)\in B_\varepsilon\}}(t,\cdot)\ dt \right\rangle \right] \wedge 1 \\ &= 0 \quad . \end{aligned} \tag{7.201}$$

Equation (7.200) follows by the definition of B_ε :

$$\begin{aligned} &\lim_{N\to\infty} \mathbf{E}\left[\sum_{i\in\mathbb{Z}^d} \int_0^{T+\ell_N} \frac{1}{2} \left| Y^{N,\infty}_{\ell_N+T-t}(i) - Y^N_{T-t}(i) \right| \right. \\ &\qquad \left. \cdot \left(\left(u_t^N(i)\right)^2 + \left(u_t^\infty(i)\right)^2 \right) \mathbb{1}_{\{(t,i)\notin B_\varepsilon\}}\ dt \wedge 1 \right] \\ &\le \lim_{N\to\infty} \sum_{i\in\mathbb{Z}^d} \int_0^{T+\ell_N} \theta_y \left(\left(u_t^N(i)\right)^2 + \left(u_t^\infty(i)\right)^2 \right) \mathbb{1}_{\{(t,i)\notin B_\varepsilon\}}(t,i)\ dt \\ &\le \theta_y \varepsilon \end{aligned} \tag{7.202}$$

To check (7.201) we use that

$$\begin{aligned} &\sum_{i\in\mathbb{Z}^d} \int_0^{T+\ell_N} \frac{1}{2} \left| Y^{N,\infty}_{\ell_N+T-t}(i) - Y^N_{T-t}(i) \right| \\ &\qquad \cdot \left(\left(u_t^N(i)\right)^2 + \left(u_t^\infty(i)\right)^2 \right) \mathbb{1}_{\{(t,i)\in B_\varepsilon\}}(t,i)\ dt \end{aligned} \tag{7.203}$$

converges P_{cat}-a.s. to zero.

This fact is an immediate consequence of (7.126). Note that this is the point where the Skorohod representation comes in. Due to the fact that the right hand side of (7.201) is bounded above by 1 and that B_ε, which is the domain of integration and summation, is a bounded set we can apply the dominated convergence theorem: therefore, (7.203) is sufficient for (7.201).

V: There remains to prove

$$\forall N \in \mathbb{N}, \forall t \in [0,T] : \mathbf{V} \le 0 \quad . \tag{7.204}$$

where $\mathbf{V}$ was given by

$$\mathbf{V} = -\frac{1}{2} \sum_{i \in \mathbb{T}_N^d} \left(Y^{\infty}_{\ell_N+T-t}(i) + Y^{N}_{\ell_N+T-t}(i) \right) \cdot g'_{n \exp(|i|)} \left(u_t^N(i) - u_t^{N,\infty}(i) \right) \left(\left(u_t^N(i)\right)^2 - \left(u_t^{N,\infty}(i)\right)^2 \right) \quad . \tag{7.205}$$

This follows immediately from

$$\begin{aligned} & g'_{n \exp(|i|)} \left(u_t^N(i) - u_t^{N,\infty}(i) \right) \left(\left(u_t^N(i)\right)^2 - \left(u_t^{N,\infty}(i)\right)^2 \right) \\ &= g'_{n \exp(|i|)} \left(u_t^N(i) - u_t^{N,\infty}(i) \right) \left(u_t^N(i) - u_t^{N,\infty}(i) \right) \cdot \left(u_t^N(i) + u_t^{N,\infty}(i) \right) \end{aligned} \tag{7.206}$$

This is nonnegative, since we have

$$\tanh\left(n \exp\left(|i|\right) x\right) \cdot x \ge 0 \tag{7.207}$$

for all $x \in \mathbb{R}$.

This completes the proof of (7.157).

Proof of Step B: We have to prove

$$\lim_{N\to\infty} \mathbf{E}\left[\left| \left\langle u^N_{T+\ell_N}, X^N_{\beta_N s-\ell_N} \right\rangle - \bar{\theta}^N_x \left\langle u^N_{T+\ell_N}, \mathfrak{n} \right\rangle \right| \right] = 0 \quad . \tag{7.208}$$

and

$$\lim_{N\to\infty} \mathbf{E}\left[\left| \left\langle u^{N,\infty}_{T+\ell_N}, X^{N,\infty}_0 \right\rangle - \bar{\theta}^N_x \left\langle u^{N,\infty}_{T+\ell_N}, \mathfrak{n} \right\rangle \right| \right] = 0 \quad . \tag{7.209}$$

We only give a proof of (7.208) – the proof of (7.209) is similar.

Note that for arbitrary $T > 0$

$$\mathbf{E}\left[\left\langle u^N_{T+\ell_N}, X^N_{\beta_N s-\ell_N} \right\rangle - \bar{\theta}^N_x \left\langle u^N_{T+\ell_N}, \mathfrak{n} \right\rangle \right] = 0$$

holds by construction. Therefore, we only have to verify

$$\lim_{N\to\infty} var\left(\left\langle u^N_{T+\ell_N}, X^N_{\beta_N s-\ell_N} \right\rangle \right) = 0 \quad P_{cat}\text{-a.s..} \tag{7.210}$$

Let

$$\varrho^N := u^N_{T+\ell_N} \quad . \tag{7.211}$$

By definition, we have

$$\varrho^N(i) \le \sum_{k=1}^{n} \sum_{j\in\mathbb{T}^d_N} a^N_{T-t_k+\ell_N}(j,i)\,\varphi_k(j) =: \hat{\varrho}^N(i) \quad . \tag{7.212}$$

We know that the coordinate processes of the reactant are positively correlated (compare Lemma 1, p. 12). This yields:

$$\lim_{N\to\infty} var\left(\left\langle u^N_{T+\ell_N}, X^N_{\beta_N s-\ell_N}\right\rangle\right) \le \lim_{N\to\infty} var\left(\sum_{i\in\mathbb{Z}^d} \hat{\varrho}^N(i)\, X^N_{\beta_N s-\ell_N}(i)\right) \tag{7.213}$$

Calculating the quenched variance, we hence obtain

$$\begin{aligned} & var\left(\sum_{i\in\mathbb{Z}^d} \hat{\varrho}^N(i)\, X^N_{\beta_N s-\ell_N}(i)\right) \\ &= \mathbf{E}\left[\left(\sum_{i\in\mathbb{T}^d_N} \hat{\varrho}^N(i)\, X^N_{\beta_N s-\ell_N}(i)\right)^2 - \left(\bar{\theta}^N_x \langle \hat{\varrho}^N, \mathfrak{n}\rangle\right)^2\right] \quad . \end{aligned} \tag{7.214}$$

At this point, we use that

$$\begin{aligned} X^N_{\beta_N s-\ell_N}(i) &= \sum_{j\in\mathbb{T}^d_N} a^N_{c_x(\beta_N s-\ell_N)}(j,i)\, X^N_0(j) \\ &\quad + \sum_{j\in\mathbb{T}^d_N} a^N_{c_x(\beta_N s-\ell_N-t)}(j,i)\,\sqrt{2\gamma_x X^N_t(j)\, Y^N_t(j)}\, dB^x_t(j) \quad . \end{aligned} \tag{7.215}$$

(compare Lemma 4 on page 69). Inserting (7.215) in (7.214) we get:

$$\begin{aligned} & var\left(\sum_{i\in\mathbb{Z}^d} \hat{\varrho}^N(i)\, X^N_{\beta_N s-\ell_N}(i)\right) \\ &= \mathbf{E}\left[\left(\sum_{i\in\mathbb{T}^d_N}\sum_{j\in\mathbb{T}^d_N} \hat{\varrho}^N(i)\, a^N_{c_x(\beta_N s-\ell_N)}(j,i)\, X^N_0(j)\right)^2 - \left(\bar{\theta}^N_x \langle \hat{\varrho}^N, \mathfrak{n}\rangle\right)^2\right] \\ &+ 2\gamma_x \mathbf{E}\left[\sum_{i\in\mathbb{Z}^d} \left(\hat{\varrho}^N(i)\right)^2 \cdot \sum_{j\in\mathbb{T}^d_N} \int_0^{\beta_N s-\ell_N} \left(a^N_{c_x(\beta_N s-\ell_N-t)}(j,i)\right)^2 X^N_t(j)\, Y^N_t(j)\, dt\right] \quad . \end{aligned} \tag{7.216}$$

We have to prove:

(i)

$$\lim_{N\to\infty} \mathbf{E}\left[\left(\sum_{i\in\mathbb{T}_N^d}\sum_{j\in\mathbb{T}_N^d} \hat{\varrho}^N(i)\, a^N_{c_x(\beta_N s-\ell_N)}(j,i)\, X_0^N(j)\right)^2 - \left(\bar{\theta}_x^N \langle \hat{\varrho}^N, \mathfrak{n}\rangle\right)^2\right] = 0 \quad .$$

(ii)

$$\lim_{N\to\infty} \mathbf{E}\left[\sum_{i\in\mathbb{Z}^d} \left(\hat{\varrho}^N(i)\right)^2 \cdot \sum_{j\in\mathbb{T}_N^d} \int\limits_0^{\beta_N s-\ell_N} \left(a^N_{c_x(\beta_N s-\ell_N-t)}(j,i)\right)^2 X_t^N(j)\, Y_t^N(j)\, dt\right] = 0 \quad .$$

Proof of (i) We have assumed $d \geq 3$. This makes Lemma 2.1 from [CGS95] applicable, namely

$$\lim_{N\to\infty} a^N_{c_x(\beta_N s-\ell_N)}(j,i) = \frac{1}{\beta_N} \quad . \tag{7.217}$$

For this reason and with the uniform summability of $\left(\hat{\varrho}^N\right)_{N\in\mathbb{N}}$, we can write

$$\begin{aligned}
&\lim_{N\to\infty} \mathbf{E}\left[\left(\sum_{i\in\mathbb{T}_N^d}\sum_{j\in\mathbb{T}_N^d} \hat{\varrho}^N(i)\, a^N_{c_x(\beta_N s-\ell_N)}(j,i)\, X_0^N(j)\right)^2\right] \\
&= \lim_{N\to\infty} \mathbf{E}\left[\left(\sum_{j\in\mathbb{T}_N^d}\sum_{i\in\mathbb{T}_N^d} \frac{1}{\beta_N} X_0^N(j)\, \hat{\varrho}^N(i)\right)^2\right] \\
&= \lim_{N\to\infty} \mathbf{E}\left[\left(\bar{\theta}_x^N \langle \hat{\varrho}^N, \mathfrak{n}\rangle\right)^2\right]
\end{aligned} \tag{7.218}$$

Hence the first summand of (7.216), namely

$$\left|\mathbf{E}^Y_{reac}\left[\left(\sum_{j\in\mathbb{T}_N^d}\sum_{i\in\mathbb{T}_N^d} a^N_{c_x\beta_N s}(j,i)\, X_0^N(j)\, \hat{\varrho}^N(i)\right)^2 - \left(\bar{\theta}_x^N \langle \varrho^N, \mathfrak{n}\rangle\right)^2\right]\right| \quad , \tag{7.219}$$

converges to zero as N tends to infinity.

Proof of (ii) We have

$$\mathbf{E}\left[\sum_{i\in\mathbb{T}_N^d}\left(\hat{\varrho}^N(i)\right)^2\sum_{j\in\mathbb{T}_N^d}\int\limits_0^{\beta_N s-\ell_N}\left(a^N_{c_x(\beta_N s-\ell_N-t)}(j,i)\right)^2 X_t^N(j)\,Y_t^N(j)\;dt\right]$$
$$\leq\sum_{i\in\mathbb{T}_N^d}\left(\sum_{k=1}^n\sum_{j\in\mathbb{T}_N^d}a^N_{T-t_k+\ell_N}(j,i)\,\varphi_k(j)\right)^2$$
$$\cdot\sum_{j\in\mathbb{T}_N^d}\int\limits_0^{\beta_N s-\ell_N}\left(a^N_{c_x(\beta_N s-\ell_N-t)}(j,i)\right)^2\mathbf{E}\left[X_t^N(j)\,Y_t^N(j)\right]\;dt$$
$$=\sum_{i\in\mathbb{T}_N^d}\left(\sum_{k=1}^n\sum_{j\in\mathbb{T}_N^d}a^N_{T-t_k+\ell_N}(j,i)\,\varphi_k(j)\right)^2\int\limits_0^{\beta_N s-\ell_N}\hat{a}^N_{2c_xt}(i,i)\;dt\cdot\mathbf{E}\left[X_t^N(0)\,Y_t^N(0)\right]$$
$$=\left(\sum_{i\in\mathbb{T}_N^d}\left(\sum_{k=1}^n\sum_{j\in\mathbb{T}_N^d}a^N_{T-t_k+\ell_N}(j,i)\,\varphi_k(j)\right)^2\right)$$
$$\cdot\int\limits_0^{\beta_N s-\ell_N}\hat{a}^N_{2c_xt}(0,0)\;dt$$
$$\cdot\sup_{t\in[0,\beta_N s-\ell_N]}\mathbf{E}\left[X_t^N(0)\,Y_t^N(0)\right] \tag{7.220}$$

- The first factor converges to zero for $N\to\infty$ as the following calculation shows:

$$\sum_{i\in\mathbb{T}_N^d}\left(\sum_{k=1}^n\sum_{j\in\mathbb{T}_N^d}a^N_{T-t_k+\ell_N}(j,i)\,\varphi_k(j)\right)^2$$
$$=\sum_{i\in\mathbb{T}_N^d}\sum_{k_1=1}^n\sum_{j_1\in\mathbb{T}_N^d}\sum_{k_2=1}^n\sum_{j_2\in\mathbb{T}_N^d}a^N_{T-t_{k_1}+\ell_N}(j_1,i)\,a^N_{T-t_{k_2}+\ell_N}(j_2,i)\,\varphi_{k_1}(j_1)\,\varphi_{k_2}(j_2)$$
$$=\sum_{k_1=1}^n\sum_{j_1\in\mathbb{T}_N^d}\sum_{k_2=1}^n\sum_{j_2\in\mathbb{T}_N^d}\hat{a}^N_{T-t_{k_1}+\ell_N+T-t_{k_2}+\ell_N}(j_1,j_2)\,\varphi_{k_1}(j_1)\,\varphi_{k_2}(j_2)\quad. \tag{7.221}$$

Recall that we have assumed (7.131), namely

$$card\left\{i\in\mathbb{Z}^d:\sum_{k=1}^n\varphi_k(i)\right\}<\infty\quad. \tag{7.222}$$

With

$$F:=\left\{i\in\mathbb{Z}^d:\sum_{k=1}^n\varphi_k(i)\right\}\quad, \tag{7.223}$$

we can write (7.221) as a finite sum:

$$\text{“(7.221)”} = \sum_{k_1=1}^{n} \sum_{j_1 \in F} \sum_{k_2=1}^{n} \sum_{j_2 \in F} \hat{a}^N_{T-t_{k_1}+\ell_N+T-t_{k_2}+\ell_N}(j_1,j_2)\,\varphi_{k_1}(j_1)\,\varphi_{k_2}(j_2) \quad . \tag{7.224}$$

Hence it suffices to verify

$$\lim_{N\to\infty} \sup\left\{\hat{a}^N_{2T+2\ell_N-t_{k_1}-t_{k_2}}(j_1,j_2), \quad k_1=1,...,n;\ k_2=1,...,n;\ j_1,j_2\in F\right\} = 0 \quad . \tag{7.225}$$

But this follows immediately from

$$\lim_{N\to\infty} \ell_N = \infty \quad .$$

- The second factor is bounded, since we have (see [CGS95], lemma 2.1c):

$$\begin{aligned}
&\lim_{N\to\infty} \int_0^{\beta_N s} a^N_{2c_x t}(0,0)\ dt \\
&= \left(\frac{1}{c_x}\int_0^{\beta_N c_x s} a_{2t}(0,0)\ dt + s\right) \quad .
\end{aligned} \tag{7.226}$$

- The third factor is bounded since catalyst and reactant are uncorrelated, both having uniformly bounded first moment.

We can conclude that

$$\mathbf{E}\left[\sum_{i\in\mathbb{T}^d_N} \left(\hat{\varrho}^N(i)\right)^2 \sum_{j\in\mathbb{T}^d_N} \int_0^{\beta_N s-\ell_N} \left(a^N_{c_x(\beta_N s-\ell_N-t)}(j,i)\right)^2 X^N_t(j)\,Y^N_t(j)\ dt\right] \tag{7.227}$$

converges to zero. This completes the proof of (7.153).

The proof of equation (7.154) is similar and omitted here. This completes the proof of Step I.

7.2.1.2 Proof of step II and step III ((7.116) and (7.117))

The main work for the argument is the proof of the following property:

Lemma 9 *Given $\Upsilon = \{t_1,...,t_n\}$, the mapping*

$$\left\{\begin{array}{ccc} \mathbb{R}^2_+ & \longrightarrow & \mathcal{M}^1\left(\mathcal{C}\left(\mathbb{R}_+, \mathcal{M}\left(\mathbb{Z}^d\right)\times\mathcal{M}\left(\mathbb{Z}^d\right)\right)\right) \\ (\theta_x,\theta_y) & \longmapsto & \mu^{\Upsilon}_{(\theta_x,\theta_y)} \end{array}\right\} \tag{7.228}$$

is uniformly continuous with respect to the metric ϱ^{Υ}.

Proof. We shall first consider the catalyst by itself and then discuss the combined catalyst-reactant system. The family $\left\{\mathcal{L}\left\{\left(\tilde{Y}_t^{(\theta_y)}\right)_{t\in\mathbb{R}}\right\}\right\}_{\theta_y\in\mathbb{R}_+}$ of laws of stationary catalyst processes. Applying the branching property of Feller's branching diffusion process, we can define all these processes on a common probability space, such that

- the difference processes are again Feller interacting branching diffusions:

$$\mathcal{L}\left\{\left|\tilde{Y}^{\left(\theta_y^{(2)}\right)} - \tilde{Y}^{\left(\theta_y^{(1)}\right)}\right|\right\} = \mathcal{L}\left\{\tilde{Y}^{\left(\left|\theta_y^{(2)}-\theta_y^{(1)}\right|\right)}\right\} \tag{7.229}$$

- there is monotonicity:

$$\theta_y^{(1)} \le \theta_y^{(2)} \quad \Longrightarrow \quad \tilde{Y}^{\left(\theta_y^{(1)}\right)} \le \tilde{Y}^{\left(\theta_y^{(2)}\right)} \quad . \tag{7.230}$$

In particular, we have

$$\mathbf{E}\left|\tilde{Y}^{\left(\theta_y^{(2)}\right)} - \tilde{Y}^{\left(\theta_y^{(1)}\right)}\right| = \left|\theta_y^{(2)} - \theta_y^{(1)}\right| \quad . \tag{7.231}$$

Now we return to the catalyst-reactant system. We recall one of the main results of Section 6.2. Namely, the stationary process $\nu_{(\theta_x,\theta_y)}$ can be characterized as limit of processes with starting times $T<0$ and initial law in $S^{(\theta_x,\theta_y)}$, letting $T\to-\infty$:
For $(\theta_x,\theta_y)\in\mathbb{R}_+^2$, let $\left(\tilde{Y}_t^{(\theta_x,\theta_y),T}\right)_{t\in[T,\infty)}$ be defined by

$$\tilde{Y}_t^{(\theta_x,\theta_y),T} := \tilde{Y}_t^{(\theta_y)} \quad (t\in[T,\infty)) \tag{7.232}$$

and $\left(X_t^{(\theta_x,\theta_y),T}\right)_{t\in[T,\infty)}$ as solution of

$$X_T^{(\theta_x,\theta_y),T} := \theta_x \mathfrak{n} \quad , \tag{7.233}$$

$$\begin{aligned} dX_t^{(\theta_x,\theta_y),T}(i) &= c_x \sum_{j\in\mathbb{Z}^d} a(i,j)\left[X_t^{(\theta_x,\theta_y),T}(j) - X_t^{(\theta_x,\theta_y),T}(i)\right] \\ &\quad + \sqrt{2\gamma_x X_t^{(\theta_x,\theta_y),T}(i)\,\tilde{Y}_t^{(\theta_x,\theta_y),T}(i)}dB_t \quad . \end{aligned} \tag{7.234}$$

Consider a sequence $\left(\theta_x^m,\theta_y^m\right)_{m\in\mathbb{N}}\in(\mathbb{R}_+\times\mathbb{R}_+)^{\mathbb{N}}$ with $\lim_{m\to\infty}\left(\theta_x^m,\theta_y^m\right)=(\theta_x,\theta_y)$. We prove (7.228) by means of convergence of Laplace transforms. We have to show that for all increasing sequences $0<t_1<...<t_n<\varpi<\infty$ and test functions

$$\varphi_1^x,...,\varphi_n^x,\varphi_1^y,...\varphi_n^y \in \left\{\varphi\in\mathbb{R}_+^{\mathbb{Z}^d} : card\left\{\varphi(i)>0\right\}<\infty\right\} \tag{7.235}$$

the following holds:

$$\begin{aligned}&\lim_{m\to\infty}\left(\lim_{T\to-\infty}\mathbf{E}\left[\exp\left(-\sum_{k=1}^{n}\left\langle\varphi_k^y,\tilde{Y}_{t_k}^{(\theta_x^m,\theta_y^m),T}\right\rangle\right)\right.\right.\\&\qquad\left.\left.\cdot\mathbf{E}_{reac}^{Y}\left[\exp\left(-\sum_{k=1}^{n}\left\langle\varphi_k^x,X_{t_k}^{(\theta_x^m,\theta_y^m),T}\right\rangle\right)\right]\right]\right)\\&=\lim_{T\to-\infty}\mathbf{E}\left[\exp\left(-\sum_{k=1}^{n}\left\langle\varphi_k^y,\tilde{Y}_{t_k}^{(\theta_x,\theta_y),T}\right\rangle\right)\right.\\&\qquad\left.\cdot\mathbf{E}_{reac}^{Y}\left[\exp\left(-\sum_{k=1}^{n}\left\langle\varphi_k^x,X_{t_k}^{(\theta_x,\theta_y),T}\right\rangle\right)\right]\right]\quad.\qquad(7.236)\end{aligned}$$

Taking (7.231) into account, it is enough to prove

$$\begin{aligned}&\lim_{m\to\infty}\lim_{T\to-\infty}\mathbf{E}\left[\mathbf{E}_{reac}^{Y}\left[\exp\left(-\sum_{k=1}^{n}\left\langle\varphi_k^x,X_{t_k}^{(\theta_x^m,\theta_y^m),T}\right\rangle\right)\right]\right.\\&\qquad\left.-\mathbf{E}_{reac}^{Y}\left[\exp\left(-\sum_{k=1}^{n}\left\langle\varphi_k^x,X_{t_k}^{(\theta_x,\theta_y),T}\right\rangle\right)\right]\right]=0\quad.\qquad(7.237)\end{aligned}$$

We again characterize the Laplace transforms by a dual representation. Let $\left(u_t^{T,m}\right)_{t\geq0}\in\mathbb{R}_+^{\mathbb{Z}^d}$, $\left(u_t^{T,\infty}\right)_{t\geq0}\in\mathbb{R}_+^{\mathbb{Z}^d}$ be defined by

$$u_0^{T,m}\equiv0\qquad(7.238)\qquad\qquad u_0^{T,\infty}\equiv0\qquad(7.239)$$

if $\varpi-t\notin\{t_1,...,t_n\}$:

$$\begin{aligned}&\frac{\partial}{\partial t}u_t^{T,m}(i)\\&=c_x\sum_{j\in\mathbb{Z}^d}a(j,i)\left[u_t^{T,m}(j)-u_t^{T,m}(i)\right]\\&\quad-\gamma_x\tilde{Y}_{\varpi-t}^{(\theta_x^m,\theta_y^m),T}\left(u_t^{T,m}(i)\right)^2\qquad(7.240)\end{aligned}$$

$$\begin{aligned}&\frac{\partial}{\partial t}u_t^{T,\infty}(i)\\&=c_x\sum_{j\in\mathbb{Z}^d}a(j,i)\left[u_t^{T,\infty}(j)-u_t^{T,\infty}(i)\right]\\&\quad-\gamma_x\tilde{Y}_{\varpi-t}^{(\theta_x,\theta_y),T}\left(u_t^{T,\infty}(i)\right)^2\qquad(7.241)\end{aligned}$$

if $\varpi-t=t_k$:

$$u_t^{T,m}(i)=\lim_{\varsigma\to t-}u_\varsigma^{T,m}(i)+\varphi_k^x\quad,\qquad(7.242)$$

$$u_t^{T,\infty}(i)=\lim_{\varsigma\to t-}u_\varsigma^{T,\infty}(i)+\varphi_k^x\quad.\qquad(7.243)$$

Having introduced $\left(u_t^{T,m}\right)_{t\geq 0}$ and $\left(u_t^{T,\infty}\right)_{t\geq 0}$, we can write

$$\mathbf{E}^Y_{reac}\left[\exp\left(-\sum_{k=1}^{n}\left\langle \varphi_k^x, X_{t_k}^{(\theta_x^m,\theta_y^m),T}\right\rangle\right)\right] = \exp\left(-\left\langle u^{T,m}_{\varpi+|T|}, X_T^{(\theta_x^m,\theta_y^m),T}\right\rangle\right)$$
$$= \exp\left(-\left\langle u^{T,m}_{\varpi+|T|}, \theta_x^m \mathfrak{n}\right\rangle\right) \tag{7.244}$$

and

$$\mathbf{E}^Y_{reac}\left[\exp\left(-\sum_{k=1}^{n}\left\langle \varphi_k^x, X_{t_k}^{(\theta_x,\theta_y),T}\right\rangle\right)\right] = \exp\left(-\left\langle u^{T,\infty}_{\varpi+|T|}, X_T^{(\theta_x,\theta_y),T}\right\rangle\right)$$
$$= \exp\left(-\left\langle u^{T,\infty}_{\varpi+|T|}, \theta_x \mathfrak{n}\right\rangle\right) \quad . \tag{7.245}$$

Taking (7.232) into account, we realize that $\left(u_t^{T,m}\right)_{t\geq 0}$ and $\left(u_t^{T,\infty}\right)_{t\geq 0}$ do not depend on T :

$$(u_t^m)_{t\geq 0} = \left(u_t^{T,m}\right)_{t\geq 0}, \quad (u_t^\infty)_{t\geq 0} = \left(u_t^{T,\infty}\right)_{t\geq 0} \quad . \tag{7.246}$$

It remains to prove:

$$\lim_{m\to\infty}\lim_{\mathcal{T}\to\infty}\mathbf{E}\left[\left|\exp\left(-\left\langle u_\mathcal{T}^m, \theta_x^m\mathfrak{n}\right\rangle\right) - \exp\left(-\left\langle u_\mathcal{T}^\infty, \theta_x^\infty\mathfrak{n}\right\rangle\right)\right|\right] = 0 \quad . \tag{7.247}$$

We prove this in two steps:

1. $\lim_{m\to\infty}\frac{1}{|\theta_x^m-\theta_x^\infty|}\lim_{\mathcal{T}\to\infty}\mathbf{E}\left[\left|\langle u_\mathcal{T}^\infty, \theta_x^m\mathfrak{n}\rangle - \langle u_\mathcal{T}^\infty, \theta_x^\infty\mathfrak{n}\rangle\right|\right] < \infty.$
2. $\lim_{m\to\infty}\frac{1}{|\theta_y^m-\theta_y^\infty|}\lim_{\mathcal{T}\to\infty}\mathbf{E}\left[\left|\langle u_\mathcal{T}^m, \mathfrak{n}\rangle - \langle u_\mathcal{T}^\infty, \mathfrak{n}\rangle\right|\right] < \infty.$

ad 1: This follows from

$$\lim_{m\to\infty}\frac{1}{|\theta_x^m-\theta_x^m|}\lim_{\mathcal{T}\to\infty}\mathbf{E}\left[\left|\langle u_\mathcal{T}^\infty, \theta_x^m\mathfrak{n}\rangle - \langle u_\mathcal{T}^\infty, \theta_x^\infty\mathfrak{n}\rangle\right|\right]$$
$$= \lim_{m\to\infty}\frac{1}{|\theta_x^m-\theta_x^m|}\lim_{\mathcal{T}\to\infty}\mathbf{E}\left[|\theta_x^m-\theta_x^\infty|\cdot\langle u_\mathcal{T}^\infty, \mathfrak{n}\rangle\right]$$
$$\leq \sum_{k=1}^{n}\langle\varphi_k^x, \mathfrak{n}\rangle < \infty \quad . \tag{7.248}$$

ad 2: We can even prove

$$\lim_{m\to\infty}\frac{1}{|\theta_y^m-\theta_y^\infty|}\lim_{\mathcal{T}\to\infty}\mathbf{E}\left[\langle|u_\mathcal{T}^m - u_\mathcal{T}^\infty|, \mathfrak{n}\rangle\right] < \infty \quad . \tag{7.249}$$

We have carried out similar calculations before. By argumemts similar to the derivation of (6.161) (page 95), we get:

$$\lim_{\mathcal{T}\to\infty}\mathbf{E}\left[\langle|u_\mathcal{T}^m - u_\mathcal{T}^\infty|, \mathfrak{n}\rangle\right]$$
$$\leq \mathbf{E}\left[\int_0^\infty\frac{\gamma_x}{2}\sum_{i\in\mathbb{Z}^d}\left|\tilde{Y}_{\varpi-t}^{(\theta_x,\theta_y)} - \tilde{Y}_{\varpi-t}^{(\theta_x^m,\theta_y^m)}\right|\right.$$
$$\left.\cdot\left((u_t^m(i))^2 + (u_t^\infty(i))^2\right)dt\right] \quad . \tag{7.250}$$

Let $(v_t)_{t\geq 0}$ be defined by

$$v_0 = 0 \tag{7.251}$$

if $\varpi - t \notin \{t_1, ..., t_n\}$:

$$\frac{\partial}{\partial t} v_t(i) = c_x \sum_{j\in\mathbb{Z}^d} a(j,i)\left[v_t(j) - v_t(i)\right] \tag{7.252}$$

if $\varpi - t = t_k$:

$$v_\varsigma(i) = \lim_{s\to t-} v_\varsigma(i) + \varphi_k^x \quad . \tag{7.253}$$

Then we have:

- $v_t \geq u_t^m$ $(\forall m \in \mathbb{N}\cup\{\infty\})$ ([14])
- $$\int_0^\infty \sum_{i\in\mathbb{Z}^d} (v_t(i))^2\, dt =: C < \infty \tag{7.254}$$

  by transience.

This yields

$$\lim_{\tau\to\infty} \mathbf{E}\left[\langle |u_\tau^m - u_\tau^\infty|, \mathfrak{n}\rangle\right] \leq C\gamma_x \left|\theta_y^m - \theta_y^m\right|, \tag{7.255}$$

completing the proof of (7.249).

This completes the proof of the continuity lemma. ■

We need this lemma to prove II and III; we recall what we claim:

II. Macroscopic limit

$$\lim_{N\to\infty} \varrho^\Upsilon \left(\int \mu^\Upsilon_{\left(\tilde\theta_x^N, \tilde\theta_y^N\right)} \Psi^N_{s-\frac{\ell_N}{\beta_N}} \left(d\left(\tilde\theta_x^N, \tilde\theta_y^N\right)\right), \right.$$
$$\left. \int \mu^\Upsilon_{\left(\tilde\theta_x^N, \tilde\theta_y^N\right)} \Psi_{s-\frac{\ell_N}{\beta_N}} \left(d\left(\tilde\theta_x^N, \tilde\theta_y^N\right)\right)\right) = 0 \quad . \tag{7.256}$$

III. The term "$-\frac{\ell_N}{\beta_N}$" does not change the limiting behavior

$$\lim_{N\to\infty} \varrho^\Upsilon \int \mu^\Upsilon_{\left(\tilde\theta_x^N, \tilde\theta_y^N\right)} \Psi_{s-\frac{\ell_N}{\beta_N}} \left(d\left(\tilde\theta_x^N, \tilde\theta_y^N\right)\right),$$
$$\left. \int \mu^\Upsilon_{\left(\tilde\theta_x, \tilde\theta_y\right)} \Psi_s \left(d\left(\tilde\theta_x, \tilde\theta_y\right)\right)\right) = 0 \tag{7.257}$$

[14] This could be proven by means of a Feynman-Kac representation, for example.

Proof of II We know from our investigations concerning fluctuations of rescaled empirical averages that we have the weak convergence property

$$\mathcal{L}\left\{\hat{X}^N, \hat{Y}^N\right\} \overset{N\to\infty}{\Longrightarrow} \mathcal{L}\left\{\hat{X}^\infty, \hat{Y}^\infty\right\} \quad . \tag{7.258}$$

Applying Skorohod's representation once again, it turns out that we can define versions of the considered processes on a common probability space $\left(\mathring{\Omega}, \mathring{A}, \mathring{\mathbf{P}}\right)$ (we denote the law with respect to this probability space by $\mathring{\mathcal{L}}$) such that we have

$$\mathcal{L}\left\{\hat{X}^N, \hat{Y}^N\right\} = \mathring{\mathcal{L}}\left\{\mathring{X}^N, \mathring{Y}^N\right\} \quad \forall N \in \mathbb{N} \cup \{\infty\} \tag{7.259}$$

and

$$\lim_{N\to\infty} \sup_{t\in[0,1]} \left(\left|\mathring{X}^N_{s-t} - \mathring{X}^\infty_{s-t}\right| + \left|\mathring{Y}^N_{s-t} - \mathring{Y}^\infty_{s-t}\right|\right) = 0 \quad \mathring{\mathbf{P}}\text{-a.s.} \tag{7.260}$$

Let $(\Theta_N)_{N\in\mathbb{N}}$ be a sequence of random variables defined as follows

$$\Theta_N := \left(\mathring{X}^N_{s-\frac{\ell_N}{\beta_N}}, \mathring{Y}^N_{s-\frac{\ell_N}{\beta_N}}\right) \tag{7.261}$$

and let $(\Theta_N^\infty)_{N\in\mathbb{N}}$ be defined by

$$\Theta_N^\infty := \left(\mathring{X}^\infty_{s-\frac{\ell_N}{\beta_N}}, \mathring{Y}^\infty_{s-\frac{\ell_N}{\beta_N}}\right) \quad . \tag{7.262}$$

By (7.260), we obtain

$$\lim_{N\to\infty} |\Theta_N - \Theta_N^\infty| = 0 \quad \mathring{\mathbf{P}}\text{-a.s..} \tag{7.263}$$

We insert this in (7.256). We have:

$$\int \mu^{\Upsilon}_{\left(\tilde{\theta}^N_x, \tilde{\theta}^N_y\right)} \mathcal{L}\left\{\left(\hat{X}^N_{s-\frac{\ell_N}{\beta_N}}, \hat{Y}^N_{s-\frac{\ell_N}{\beta_N}}\right)\right\}\left(d\left(\tilde{\theta}^N_x, \tilde{\theta}^N_y\right)\right) = \mathring{\mathbf{E}}\left[\mu^{\Upsilon}_{\Theta_N}\right] \tag{7.264}$$

and

$$\int \mu^{\Upsilon}_{\left(\tilde{\theta}^N_x, \tilde{\theta}^N_y\right)} \Psi_{s-\frac{\ell_N}{\beta_N}}\left(d\left(\tilde{\theta}^N_x, \tilde{\theta}^N_y\right)\right) = \mathring{\mathbf{E}}\left[\mu^{\Upsilon}_{\Theta_N^\infty}\right] \quad . \tag{7.265}$$

An application of Lemma 2 (p. 44) yields

$$\varrho^{\Upsilon}\left(\mathring{\mathbf{E}}\left[\mu^{\Upsilon}_{\Theta_N}\right], \mathring{\mathbf{E}}\left[\mu^{\Upsilon}_{\Theta_N^\infty}\right]\right) \le \mathring{\mathbf{E}}\left[\varrho^{\Upsilon}\left(\mu^{\Upsilon}_{\Theta_N}, \mu^{\Upsilon}_{\Theta_N^\infty}\right)\right] \tag{7.266}$$

and applying that

- it can be assumed that ϱ^{Υ} is bounded above by 1
- there is uniform continuity as proven in Lemma 9 (p.145) we get

$$\lim_{N\to\infty} \mathring{\mathbf{E}}\left[\varrho^{\Upsilon}\left(\mu^{\Upsilon}_{\Theta_N}, \mu^{\Upsilon}_{\Theta_N^\infty}\right)\right] = 0 \quad . \tag{7.267}$$

This completes the proof of II.

Proof of III The proof of III is quite similar to the proof of II. We therefore only provide it in a short version. The main steps are:

1.	$s - \frac{\ell_N}{\beta_N} \overset{N\to\infty}{\longrightarrow} s$	by definition of (ℓ_N)
2.	$\Psi_{s-\frac{\ell_N}{\beta_N}} \overset{N\to\infty}{\Longrightarrow} \Psi_s$	continuity of sample paths
3.	$\hat{X}^\infty_{s-\frac{\ell_N}{\beta_N}} \overset{N\to\infty}{\longrightarrow} \hat{X}^\infty_s \quad a.s.$	this are the sample paths
4.	$\mu^\Upsilon_{\hat{X}^\infty_{s-\frac{\ell_N}{\beta_N}}} \overset{N\to\infty}{\Longrightarrow} \mu^\Upsilon_{\hat{X}^\infty_s} \quad a.s.$	convergence of stationary processes due to continuity lemma
5.	$\mathbf{E}\left[\mu^\Upsilon_{\hat{X}^\infty_{s-\frac{\ell_N}{\beta_N}}}\right] \overset{N\to\infty}{\Longrightarrow} \mathbf{E}\left[\mu^\Upsilon_{\hat{X}^\infty_s}\right]$	weak convergence of mixtures of measures
6.	(7.257)	5. is the same as (7.257)

7.2.2 Tightness

As we only have to verify tightness locally[15], it suffices to prove tightness of the sequence

$$\left(\mathcal{L}\left\{X^N_{|\mathbb{T}^d_N|s+\cdot}(0), Y^N_{|\mathbb{T}^d_N|s+\cdot}(0)\right\}\right)_{N\in\mathbb{N}}. \tag{7.268}$$

The proof of tightness consists of two parts: firstly, we have to show that the initial laws are tight, secondly, we have to establish tightness from regularity properties of the paths. The first point is easy: since expectations are preserved, we have

$$\mathbf{E}\left[X^N_{|\mathbb{T}^d_N|s}(0)\right] = \theta_x \quad , \quad \mathbf{E}\left[Y^N_{|\mathbb{T}^d_N|s}(0)\right] = \theta_y \quad . \tag{7.269}$$

Since the considered processes are $\mathbb{R}_+$-valued, this is sufficient.

We apply the criterion of Kolmogorov again (see [EK86], Proposition 3.6.3). We have to verify that there exist $C < \infty$, $\tilde{p} > 2$ such that for $u + 1 > t > u > 0$ we have

$$\limsup_{N\to\infty} \mathbf{E}\left[\left|X^N_{|\mathbb{T}^d_N|s+t}(0) - X^N_{|\mathbb{T}^d_N|s+u}(0)\right|^{\tilde{p}}\right] \le C\,|t-u|^{\frac{\tilde{p}}{2}} \tag{7.270}$$

and

$$\limsup_{N\to\infty} \mathbf{E}\left[\left|Y^N_{|\mathbb{T}^d_N|s+t}(0) - Y^N_{|\mathbb{T}^d_N|s+u}(0)\right|^{\tilde{p}}\right] \le C\,|t-u|^{\frac{\tilde{p}}{2}} \quad . \tag{7.271}$$

We only provide a proof of (7.270) – (7.271) can be verified by similar - but more simple - calculations.

Using the decomposition verified in Lemma 4 (p. 69) and in the sequel the inequality

$$(\alpha + \beta)^{\tilde{p}} \le (\tilde{p} - 1)\,\alpha^{\tilde{p}} + C_{\tilde{p}}\beta^{\tilde{p}} \quad \forall \alpha, \beta \ge 0 \quad . \tag{7.272}$$

[15] This is a consequence of Jakubowski's tightness criterion. (e.g., compare [Pf03], Section 5.2)

(this is Lemma B.11, which is proven in the appendix on page 249), we get:

$$\mathbf{E}\left[\left|X^N_{|\mathbb{T}^d_N|s+t}(0) - X^N_{|\mathbb{T}^d_N|s+u}(0)\right|^{\tilde{p}}\right]$$

$$= \mathbf{E}\left[\left(\int_{|\mathbb{T}^d_N|s+u}^{|\mathbb{T}^d_N|s+t} \sum_{j\in\mathbb{T}^d_N} c_x a^N(0,j)\left[X^N_\varrho(j) - X^N_\varrho(0)\right] d\varrho\right.\right.$$

$$\left.\left. + \int_{|\mathbb{T}^d_N|s+u}^{|\mathbb{T}^d_N|s+t} \sqrt{2\gamma_x X^N_\varrho(0)\, Y^N_\varrho(0)} dB^x_\varrho(0)\right)^{\tilde{p}}\right]$$

$$\leq C_{\tilde{p}}\mathbf{E}\left[(t-u)^{\tilde{p}}\left|\frac{1}{t-u}\int_{|\mathbb{T}^d_N|s+u}^{|\mathbb{T}^d_N|s+t} \sum_{j\in\mathbb{T}^d_N} c_x a^N(0,j)\left[X^N_\varrho(j) - X^N_\varrho(0)\right] d\varrho\right|^{\tilde{p}}\right]$$

$$+ (\tilde{p}-1)\,\mathbf{E}\left[\left|\int_{|\mathbb{T}^d_N|s+u}^{|\mathbb{T}^d_N|s+t} \sqrt{2\gamma_x X^N_\varrho(0)\, Y^N_\varrho(0)} dB^x_\varrho(0)\right|^{\tilde{p}}\right] \tag{7.273}$$

We apply Jensen's inequality[16] and a martingale inequality by Burkholder[17] [Bu]:

$$\leq c_x^{\tilde{p}}\mathbf{E}\left[C_{\tilde{p}}(t-u)^{\tilde{p}}\frac{1}{t-u}\int_{|\mathbb{T}^d_N|s+u}^{|\mathbb{T}^d_N|s+t} \sum_{j\in\mathbb{T}^d_N} a^N(0,j)\left|X^N_\varrho(j) - X^N_\varrho(0)\right|^{\tilde{p}} d\varrho\right]$$

$$+ (\tilde{p}-1)(\tilde{p}-1)^{\tilde{p}}\,\mathbf{E}\left[\left|\int_{|\mathbb{T}^d_N|s+u}^{|\mathbb{T}^d_N|s+t} 2\gamma_x X^N_\varrho(0)\, Y^N_\varrho(0)\, d\varrho\right|^{\frac{\tilde{p}}{2}}\right] \tag{7.275}$$

---

[16] Jensen's inequality is applied as follows:

$$\left(\frac{1}{t-u}\int_u^t \sum_{j\in\mathbb{T}^d_N} a^N(0,j)(\cdot(j))\, d\varrho\right)^p$$

$$\leq \frac{1}{t-u}\int_u^t \sum_{j\in\mathbb{T}^d_N} a^N(0,j)(\cdot(j))^p\, d\varrho \tag{7.274}$$

Note that $(\cdot)^p$ is convex.

[17] Let M denote a L^p-martingale, $p > 2$. Then there is the following inequality

$$\sup_{0\leq r\leq R}\mathbf{E}\left[|M_r|^p\right] \leq (p-1)^p\,\mathbf{E}\left[\left(\langle M\rangle_R\right)^{\frac{p}{2}}\right]$$

In the first part of the sum, we apply Jensen's inequality (with respect to $\sum_{j\in\mathbb{T}_N^d} a^N(0,j)$). The second part of the sum is prepared in order to make Jensen's inequality applicable.

$$\leq c_x^{\tilde{p}} C_{\tilde{p}} (t-u)^{\tilde{p}} \frac{1}{t-u} \int_{|\mathbb{T}_N^d|s+u}^{|\mathbb{T}_N^d|s+t} 2\mathbf{E}\left[\left|X_\varrho^N(0)\right|^{\tilde{p}}\right] d\varrho$$

$$+ (2\gamma_x)^{\frac{\tilde{p}}{2}} (\tilde{p}-1)^{\tilde{p}+1} (t-u)^{\frac{\tilde{p}}{2}} \mathbf{E}\left[\left|\frac{1}{t-u} \int_{|\mathbb{T}_N^d|s+u}^{|\mathbb{T}_N^d|s+t} X_\varrho^N(0) Y_\varrho^N(0)\, d\varrho\right|^{\frac{\tilde{p}}{2}}\right] \tag{7.276}$$

The integral can be bounded above by taking the supremum over a set containing the domain of integration. We apply Jensen's inequality to the second part of the sum. With

$$X_\varrho^N(0) Y_\varrho^N(0) \leq \left(X_\varrho^N(0)\right)^{\tilde{p}} + \left(Y_\varrho^N(0)\right)^{\tilde{p}} \tag{7.277}$$

we get:

$$\leq c_x^{\tilde{p}} C_{\tilde{p}} (t-u)^{\tilde{p}} \sup_{\varrho\in[0,|\mathbb{T}_N^d|s+t]} \left(2\mathbf{E}\left[\left|X_\varrho^N(0)\right|^{\tilde{p}}\right]\right)$$

$$+ (2\gamma_x)^{\frac{\tilde{p}}{2}} (\tilde{p}-1)^{\tilde{p}+1} (t-u)^{\frac{\tilde{p}}{2}} \frac{1}{t-u} \int_{|\mathbb{T}_N^d|s+u}^{|\mathbb{T}_N^d|s+t} \mathbf{E}\left[\left(\left(X_\varrho^N(0)\right)^{\tilde{p}} + \left(Y_\varrho^N(0)\right)^{\tilde{p}}\right) d\varrho\right]$$

$$\leq (t-u)^{\frac{\tilde{p}}{2}}$$

$$\cdot \left(c_x^{\tilde{p}} C_{\tilde{p}} (t-u)^{\frac{\tilde{p}}{2}} \sup_{\varrho\in[0,|\mathbb{T}_N^d|s+t]} \left(2\mathbf{E}\left[\left|X_\varrho^N(0)\right|^{\tilde{p}}\right]\right)\right.$$

$$\left. + (2\gamma_x)^{\frac{\tilde{p}}{2}} (\tilde{p}-1)^{\tilde{p}+1} \sup_{\varrho\in[0,|\mathbb{T}_N^d|s+t]} \mathbf{E}\left[\left(X_\varrho^N(0)\right)^{\tilde{p}} + \left(Y_\varrho^N(0)\right)^{\tilde{p}}\right]\right) \tag{7.278}$$

We prove in the appendix (Lemma 14, p. 247) that there exists $\tilde{p} > 2$ such that for given s

$$\sup_{\varrho\in[0,|\mathbb{T}_N^d|s]} \left(\mathbf{E}\left[\left(X_\varrho^N(0)\right)^{\tilde{p}} + \left(Y_\varrho^N(0)\right)^{\tilde{p}}\right]\right)$$

is finite. It is not hard to see that for any finite t also

$$\sup_{\varrho\in[0,|\mathbb{T}_N^d|s+t]} \left(\mathbf{E}\left[\left(X_\varrho^N(0)\right)^{\tilde{p}} + \left(Y_\varrho^N(0)\right)^{\tilde{p}}\right]\right) < \infty \tag{7.279}$$

holds. Together with (7.278), this completes the proof of (7.270).

Chapter 8

Proof of Theorem 2

8.0 Introduction and survey of the proof

8.0.1 Introduction

In this chapter we give a proof of Theorem 2, namely the Mean Field Limit of the $\mathcal{M}(\{0,...,N-1\}) \times \mathcal{M}(\{0,...,N-1\})$-valued processes given by the system of stochastic differential equations

$$dX_t^N(i) = c_x \left(\frac{1}{N} \sum_{j=0}^{N-1} X_t^N(j) - X_t^N(i) \right) dt + \sqrt{2\gamma_x X_t^N(i)\, Y_t^N(i)}\, dB_t^x(i), \tag{8.1}$$

$$dY_t^N(i) = c_y \left(\frac{1}{N} \sum_{j=0}^{N-1} Y_t^N(j) - Y_t^N(i) \right) dt + \sqrt{2\gamma_y Y_t^N(i)}\, dB_t^y(i) \quad . \tag{8.2}$$

with initial conditions as described in 4.1.2. In particular, we have

- Independence properties of the initial state:
  The initial configuration is constructed as a projection of a product measure on $(\mathbb{R}_+ \times \mathbb{R}_+)^{\mathbb{N}_0}$. In particular this implies
  - catalyst and reactant are independent
  - configurations at different sites are independent
- The initial intensity is finite:

$$\mathbf{E}\left[\left(X_0^N(0), Y_0^N(0)\right)\right] = (\theta_x, \theta_y) \in \mathbb{R}_+ \times \mathbb{R}_+ \quad . \tag{8.3}$$

- At time zero, there exists $p > 2$ for which pth moments are uniformly bounded:[1]

$$\sup_{N \in \mathbb{N}} \mathbf{E}\left[\left(X_0^N(0)\right)^p + \left(Y_0^N(0)\right)^p\right] < \infty \quad . \tag{8.4}$$

We center our investigations on the limit $N \to \infty$.

[1] This is needed for the proof of tightness.

8.0.2 Preliminary remark

The proof we provide is based on Laplace transforms and the dual point of view.

Certainly, this is not the intuitive way to prove Theorem 2. Namely, it seems to be more natural to compare coefficients of stochastic differential equations. We provide more information on this approach in subsection 8.6.

The reason why we use the duality method is the following: Theorem 2 may be regarded as as special case of Theorem 3 (namely $L = 1$). Therefore, it makes sense to introduce the methods used to prove Theorem 3 in the more simple context of Theorem 2.

8.0.3 Survey of the proof

The proof is divided in six parts (subsections 8.1-8.6). We want to give a short description what is happening in each of these parts.

The catalytic process we discuss can be constructed two-step sampling. In subsection 8.1 we give a discussion of the consequences of this property. In particular, we lay out that this two-step sampling enables us to prove the theorem by a proof consisting of two steps: At first, we prove convergence of the catalyst; then we prove convergence of the reactant conditioned on the catalyst processes (the so-called quenched reactant), which are assumed to converge pathwise. This assumption is made possible by constructing all the processes on a suitable joint probability space. By Skorohod's representation this probability space can be chosen such that there is pathwise convergence of the catalyst.

In 8.2 we explain how the limit of the finite dimensional distributions of the quenched reactant processes can be described in a dual picture. What we get is a system of interacting differential equations, of which the solutions characterize finite dimensional distributions.

In 8.3 we describe the finite dimensional distributions of the limiting processes using duality.

In 8.4 we prove that the finite dimensional distributions of the considered processes converge weakly to those of the limit process as we let $N \to \infty$. This convergence property is verified by comparison of the corresponding dual equations.

In 8.5 it is shown that the family of laws of the considered processes is a tight family. This property has to be shown in order to obtain pathwise convergence from convergence of finite dimensional distributions.

As a digression we discuss in 8.6 another way of proving Theorem 2. Namely, we discuss the fluctuations of empirical averages. It turns out that the processes given by the empirical averages for different $N \in \mathbb{N}$ approximate the constant process given by the initial intensities (θ_x, θ_y) as $N \to \infty$. We explain how this result could be applied to prove the theorem.

8.1 Catalytic setting – Quenched approach

Due to the catalytic nature of the process the process corresponds to a two-step experiment. It is possible to provide a proof consisting of two steps which correspond to the two steps of the experiment.

- Convergence of the catalyst laws
- A possible interpretation of the catalyst-reactant law of the process conditioned on a given catalyst is the following: The conditioned law can be regarded as functional of the catalyst path. This is a consequence of the two-step construction.

  We assume that the catalyst paths converge a. s.. Skorohod's representation enables us to do so. We prove that the laws of the reactant processes converge after conditioning on these converging catalysts.

It is explained in Chapter 5 (in particular page 43) that it suffices to verify the two points we have just mentioned..

8.1.1 Convergence of the catalyst laws

It is not a new result that the laws of the catalyst converge: a proof can be found in ([DG96]).

We will use this fact to make the following assumption:

$$\forall i \in \mathbb{N}_0 : \lim_{N\to\infty} \sup_{0\le t\le T} \left\{ \left| Y_t^N(i) - Y_t^\infty(i) \right| \right\} = 0 \tag{8.5}$$

We know by Skorohod's representation (compare 5.1.1) that this assumption implies no loss of generality.[2]

8.1.2 Convergence of the quenched reactant (in law)

We will prove that the laws of conditioned laws of the reactant given the catalyst converge weakly: We claim that there exists a limit point of the sequence

$$\left(\mathcal{L}_{cat} \left\{ \mathcal{L}_{reac}^Y \left\{ X^N \right\} \right\} \right)_{N\in\mathbb{N}} . \tag{8.6}$$

Note at this point that $\mathcal{L}_{reac}^Y$ denotes the law of catalyst-reactant processes conditioned on the sequence of catalyst realizations $\left(Y^N\right)$. By construction, we have (8.5).

We prove the convergence of (8.6) using a dual approach.

[2]Namely, all processes can be defined on a common probability space such that (8.5) holds.

8.2 Dual representation of the finite dimensional distributions of the limit process

Our objective is to characterize the finite dimensional distributions of the limit process $(X_t^\infty)_{t\geq 0}$ conditioned on the catalyst. We do this by means of Laplace transforms using a dual representation.

We recall the stochastic differential equations governing the evolution of $\left\{(X_t^\infty(i))_{t\geq 0}\right\}_{i\in\mathbb{N}_0}$:

$$
\begin{aligned}
& dX_t^\infty(i) \\
& = c_x\theta_x dt && \text{(immigration)} \\
& \quad - c_x X_t^\infty(i)\, dt && \text{(emigration/killing)} \\
& \qquad + \sqrt{2\gamma_x X_t^\infty(i)\, Y_t^\infty(i)}\, dB_t^x(i) \quad . && \text{(branching)}
\end{aligned}
$$

Note that the family $\left\{(X_t^\infty(i))_{t\geq 0}\right\}_{i\in\mathbb{N}_0}$ is an independent family for given catalytic medium. We have already learned to describe the branching term in the dual picture - compare e. g. the duality arguments carried out when considering the Finite System Scheme in Chapter 7. We expect that the branching mechanism will be reflected by the term "$-u_t^2$" (in a catalytic setting "$-Y_{T-t}u_t^2$").

We have not yet considered the effects of immigration and killing in the dual picture. Before describing $(X_t^\infty)_{t\geq 0}$ by means of duality, we analyze which terms have to occur in the cumulant equation due to immigration and killing. This is the topic of the following two digressions.

Immigration and Duality

Consider a $\mathbb{R}_+^\Lambda$-valued process $(I_t)_{t\geq 0}$ defined by its initial condition I_0 and the evolution equation

$$\frac{\partial}{\partial t} I_t(i) = c > 0 \quad (i \in \Lambda) \quad . \tag{8.7}$$

This deterministic process is the diffusion limit of a particle process with immigration, where the immigration events happen with rate c at each site. It is clear by (8.7) that I is explicitely given by $I_t(i) = I_0(i) + ct$.

We want to give a dual representation of I. To this end, we introduce an additional site Δ (that's where all the immigrating particles come from) and define

$$I_t(\Delta) = 1 \quad \forall t \geq 0 \quad .$$

Consider $\varphi \in \mathbb{R}_+^{\Lambda \cup \{\Delta\}}$ with $card\{i \in \Lambda \cup \{\Delta\} : \varphi(i) > 0\} < \infty$.

We define the process $\left(\left\{u_t^{\checkmark}(i)\right\}_{i \in \Lambda \cup \{\Delta\}}\right)_{t \geq 0}$ by

$$u_0^{\checkmark} = \varphi \quad , \tag{8.8}$$

$$u_t^{\checkmark}(i) = u_0^{\checkmark}(i) \quad (t \geq 0) \text{ for } i \in \Lambda \quad , \tag{8.9}$$

$$\frac{\partial}{\partial t} u_t^{\checkmark}(\Delta) = \sum_{i \in \Lambda} c u_t^{\checkmark}(i) \quad . \tag{8.10}$$

(hence $u_t^{\checkmark}(\Delta) = \varphi(\Delta) + \sum_{i \in \Lambda} ct\varphi(i)$). This process can be considered as dual process of I since we have

$$\exp\left(-\sum_{i \in \Lambda} \varphi(i) I_t(i) - \varphi(\Delta) \cdot I_t(\Delta)\right) = \exp\left(-\sum_{i \in \Lambda} u_t^{\checkmark}(i) I_0(i) - u_t^{\checkmark}(\Delta) I_0(\Delta)\right) \tag{$*$}$$

as the following calculation shows:

$$\begin{aligned}
-\log(l.h.s(*)) &= \sum_{i \in \Lambda} \varphi(i) I_t(i) + \varphi(\Delta) I_t(\Delta) \\
&= \sum_{i \in \Lambda} \varphi(i) (I_0(i) + ct) + \varphi(\Delta) I_0(\Delta) \\
&= \sum_{i \in \Lambda} u_t^{\checkmark}(i) I_0(i) + \sum_{i \in \Lambda} \varphi(i) ct + \varphi(\Delta) I_0(\Delta) \\
&= \sum_{i \in \Lambda} u_t^{\checkmark}(i) I_0(i) + \sum_{i \in \Lambda} \varphi(i) ct I_0(\Delta) + \varphi(\Delta) I_0(\Delta) \\
&= \sum_{i \in \Lambda} u_t^{\checkmark}(i) I_0(i) + \sum_{i \in \Lambda} (\varphi(i) ct + \varphi(\Delta)) I_0(\Delta) \\
&= -\log(r.h.s(*)) \quad .
\end{aligned} \tag{8.11}$$

Emigration/Killing and Duality

Consider a $\mathbb{R}_+^\Lambda$-valued process $(E_t)_{t\geq 0}$ defined by its initial condition E_0 and the evolution equation

$$\frac{\partial}{\partial t} E_t(i) = -cE_t(i) \quad (i \in \Lambda) \quad , \tag{8.12}$$

where $c \geq 0$ is the killing rate.
Clearly,

$$E_t(i) = \exp(-ct) E_0(i) \quad (i \in \Lambda) \tag{8.13}$$

follows. The process $(E_t)_{t\geq 0}$ arises as diffusion limit of a particle process, where particles are killed with rate c. Again, we consider test functions $\varphi \in \mathbb{R}_+^{\not{Z}}$ with $card\{\varphi > 0\} < \infty$, claiming

$$\exp\left(-\sum_{i\in\Lambda} \varphi(i) E_t(i)\right) = \exp\left(-\sum_{i\in\Lambda} u_t^{\nearrow}(i) E_0(i)\right). \tag{$**$}$$

We define the process $\left(u_t^{\nearrow}\right)_{t\geq 0}$ by

$$u_0^{\nearrow} = \varphi \tag{8.14}$$

and

$$\frac{\partial}{\partial t} u_t^{\nearrow}(i) = -cu_t^{\nearrow}(i) \quad (i \in \Lambda). \tag{8.15}$$

This yields:

$$u_t^{\nearrow} = e^{-ct}\varphi \quad . \tag{8.16}$$

Equation $(**)$ follows from

$$
\begin{aligned}
-\log(l.h.s\,(**)) &= \sum_{i\in\Lambda} \varphi(i) E_t(i) && (8.17)\\
&= \sum_{i\in\Lambda} \varphi(i) e^{-ct} E_0(i) && (8.18)\\
&= \sum_{i\in\Lambda} u_t^{\nearrow}(i) E_0(i) = -\log(r.h.s\,(**)) \quad . && (8.19)
\end{aligned}
$$

We want to describe finite dimensional distributions of the process $(X_t^\infty)_{t\geq 0}$ [conditioned on the catalyst] from a dual point of view. We will see that the dual equation reflects all mechanisms of evolutions contributing to the evolution of the considered process, namely immigration, killing and branching.

In order to characterize finite dimensional distributions, it is natural to consider

$$X_t^\infty(\Sigma) := \sum_{k=1}^{n} \sum_{i\in\mathbb{N}_0} \varphi_k(i) X_{t_k}^\infty(i) \mathbb{1}_{\{t\geq t_k\}}(t) \quad (t \geq 0), \tag{8.20}$$

with

$$
\begin{aligned}
\varphi_k &\in \left\{\varphi \in \mathbb{R}_+^{\mathbb{N}_0} : card\{\varphi > 0\} < \infty\right\} \quad , && (8.21)\\
k &\in \{1, ..., n\}, 0 < t_1 < ... < t_n < T \quad . && (8.22)
\end{aligned}
$$

Since we have immigration, it is convenient for us to introduce an additional site "Δ" (this is somehow the site where the particles wait to appear). This *delivery room* used for birth processes is the counterpart of the *cemetery* frequently used for processes with killing. (compare [Sz], p. 51)

We introduce a process at site Δ by

$$X_{\cdot}^{\infty}(\Delta) \equiv \theta_x. \tag{8.23}$$

Having introduced two additional sites, Δ and Σ, the process X^{∞} we consider has become a $\mathcal{M}(\mathbb{N}_0 \cup \{\Delta, \Sigma\})$-valued process.

Notation 6 *We use brackets to simplify the notation of sums:*

$$\langle a, b\rangle = \sum_{i\in\mathbb{N}_0} a(i)\, b(i) \quad \text{when considering the infinite case,} \tag{8.24}$$

$$\langle a, b\rangle = \sum_{i\in\{0,\ldots,N-1\}} a(i)\, b(i) \quad \text{when considering the finite case,} \tag{8.25}$$

$$\langle a, b\rangle_{\Sigma} = \sum_{i\in\{0,\ldots,N-1\}\cup\{\Sigma\}} a(i)\, b(i) \quad , \tag{8.26}$$

$$\langle a, b\rangle_{\Delta,\Sigma} = \sum_{i\in\mathbb{N}_0\cup\{\Delta,\Sigma\}} a(i)\, b(i) \quad . \tag{8.27}$$

We want to construct a dual process $(u_t)_{t\geq 0}$ for which

$$\mathbf{E}_{reac}^{Y}\left[\exp\left(-X_T^{\infty}(\Sigma)\right)\right] = \mathbf{E}_{reac}^{Y}\left[\exp\left(-\langle u_T, X_0^{\infty}\rangle_{\Delta,\Sigma}\right)\right] \tag{8.28}$$

holds for $T > 0$. Clearly, the initial condition has to be chosen as follows:

<u>*initial condition*</u>

$$u_0(i) := 0 \quad (i \in \mathbb{N}_0) \quad , \tag{8.29}$$

$$u_0(\Delta) := 0 \quad , \tag{8.30}$$

$$u_0(\Sigma) := 1 \quad . \tag{8.31}$$

We hope that (8.28) is true for $(u_t)_{t\geq 0}$ if we define this process as follows, simply adding up the terms corresponding to catalytic branching, immigration, killing and observations at times (t_k) :

<u>*evolution:*</u> for $T - t \notin \{t_1, \ldots, t_n\}$

$$\frac{\partial}{\partial t} u_t(i) = -\gamma_x Y_{T-t}^{\infty}(i)\,(u_t(i))^2 \qquad \text{(catalytic branching)}$$

$$- c_x u_t(i) \quad , \qquad \text{(killing)}$$

$$\frac{\partial}{\partial t} u_t(\Delta) = \sum_{i\in\mathbb{N}_0} c_x u_t(i) \quad , \qquad \text{(immigration)}$$

$$\frac{\partial}{\partial t} u_t(\Sigma) = 0 \quad . \tag{8.32}$$

for $T-t=t_k$ $(k\in\{1,...,n\})$:

$$u_{T-t_k}(i) = \lim_{s\nearrow T-t_k} u_s(i) + \varphi_k(i) \quad \text{for } i\in\mathbb{N}_0 \qquad \text{(observations)}$$

$$u_{T-t_k}(\Delta) = \lim_{s\nearrow T-t_k} u_s(\Delta) \tag{8.33}$$

$$u_{T-t_k}(\Sigma) = \lim_{s\nearrow T-t_k} u_s(\Sigma) \quad . \tag{8.34}$$

For given catalyst, the process $(u_t)_{t\geq 0}$ is deterministic. It is a dual process of the process $(X_t^\infty)_{t\geq 0}$ with given catalyst $(Y_t^\infty)_{t\geq 0}$ in the sense of the following lemma:

Lemma 10 *For almost all realizations of the catalyst the following duality holds:*

$$\mathbf{E}^Y_{reac}\left[\exp\left(-X_T^\infty(\Sigma)\right)\right] = \exp\left(-\langle u_T, X_0^\infty\rangle_{\Delta,\Sigma}\right) \quad . \tag{8.35}$$

Proof. Another formulation of (8.35) is:

$$\mathbf{E}^Y_{reac}\left[\exp\left(-\langle u_0, X_T^\infty\rangle_{\Delta,\Sigma}\right)\right] = \exp\left(-\langle u_T, X_0^\infty\rangle_{\Delta,\Sigma}\right) \quad P_{cat}\text{.a.s..} \tag{8.36}$$

Therefore, it suffices to verify that for P_{cat}-almost all (converging) catalysts we have

1. where no jumps happen – for $t\notin\{t_1,...,t_n\}$:

$$\frac{\partial}{\partial t}\mathbf{E}^Y_{reac}\left[\exp\left(-\langle u_{T-t}, X_t^\infty\rangle_{\Delta,\Sigma}\right)\right] = 0;$$

2. at the jumps:

$$\lim_{\varsigma\searrow 0}\langle u_{T-t_k-\varsigma}, X^\infty_{t_k+\varsigma}\rangle_{\Delta,\Sigma} - \langle u_{T-t_k+\varsigma}, X^\infty_{t_k-\varsigma}\rangle_{\Delta,\Sigma} = 0 \quad (k=1,...,n)\,.$$

Proof of 1.
Using Itô's formula (product rule[3]), we get (for Y fixed)

$$\begin{aligned}
&d\exp\left(-\langle u_{T-t}, X_t^\infty\rangle_{\Delta,\Sigma}\right)\\
&= d\exp\left(-\sum_{i\in\mathbb{N}_0} u_{T-t}(i)\,X_t^\infty(i)\right)\\
&\qquad\cdot\exp\left(-u_{T-t}(\Delta)\,X_t^\infty(\Delta) - u_{T-t}(\Sigma)\,X_t^\infty(\Sigma)\right)\\
&+ d\exp\left(-u_{T-t}(\Delta)\,X_t^\infty(\Delta)\right)\\
&\qquad\cdot\exp\left(-\sum_{i\in\mathbb{N}_0} u_{T-t}(i)\,X_t^\infty(i) - u_{T-t}(\Sigma)\,X_t^\infty(\Sigma)\right)\\
&+ d\exp\left(-u_{T-t}(\Sigma)\,X_t^\infty(\Sigma)\right)\\
&\qquad\cdot\exp\left(-\sum_{i\in\mathbb{N}_0} u_{T-t}(i)\,X_t^\infty(i) - u_{T-t}(\Delta)\,X_t^\infty(\Delta)\right)
\end{aligned} \tag{8.37}$$

[3]Note that by (8.21) the summation $\langle u_{T-t}, X_t^\infty\rangle_{\Delta,\Sigma}$ has in fact only finitely many summands.

There do not occur more terms, because only the processes $(X_\cdot^\infty(i))$ $(i \in \mathbb{N}_0 \cup \{\Sigma\} \cup \{\Delta\})$ are processes of infinite variation, not the dual processes. For $t \notin \{t_1, ..., t_n\}$, we get:[4]

$$
\begin{aligned}
&\frac{\partial}{\partial t}\mathbf{E}^Y_{reac}\left[\exp\left(-\langle u_{T-t}, X_t^\infty\rangle_{\Delta,\Sigma}\right)\right] \\
= \;& \mathbf{E}^Y_{reac}\Bigg[\exp\left(-\langle u_{T-t}, X_t^\infty\rangle_{\Delta,\Sigma}\right)\cdot\Bigg(-\sum_{i\in\mathbb{N}_0} X_t^\infty(i)\frac{\partial}{\partial t}u_{T-t}(i) \\
& -\sum_{i\in\mathbb{N}_0}\Bigg[u_{T-t}(i)\,c_x\left(X_t^\infty(\Delta)-X_t^\infty(i)\right)+\underbrace{\gamma_x X_t^\infty(i)\,Y_t^\infty(i)\left(u_{T-t}(i)\right)^2}_{\text{quadr. variation term of Ito's formula}}\Bigg] \\
& -X_t^\infty(\Delta)\frac{\partial}{\partial t}u_{T-t}(\Delta)-u_{T-t}(\Sigma)\frac{\partial}{\partial t}X_t^\infty(\Sigma)\Bigg)\Bigg] \\
= \;& \mathbf{E}^Y_{reac}\Bigg[\exp\left(-\langle u_{T-t}, X_t^\infty\rangle_{\Delta,\Sigma}\right)\cdot\Bigg(\sum_{i\in\mathbb{Z}}\Bigg[X_t^\infty(i)\left(-\gamma_x Y_t^\infty(i)\left(u_{T-t}(i)\right)^2-c_x u_{T-t}(i)\right) \\
& -u_{T-t}(i)\left(c_x\left(X_t^\infty(\Delta)-X_t^\infty(i)\right)\right)+\left(u_{T-t}(i)\right)^2\gamma_x X_t^\infty(i)\,Y_t^\infty(i) \\
& +X_t^\infty(\Delta)\cdot c_x u_{T-t}(i)\Bigg]\Bigg)\Bigg] \\
= \;& \mathbf{E}^Y_{reac}\Bigg[\exp\left(-\langle u_{T-t}, X_t^\infty\rangle_{\Delta,\Sigma}\right) \\
& \cdot\Bigg(\sum_{i\in\mathbb{Z}}\Bigg[\underline{-\gamma_x X_t^\infty(i)\,Y_t^\infty(i)\left(u_{T-t}(i)\right)^2}-\underline{\underline{c_x X_t^\infty(i)\,u_{T-t}(i)}} \\
& -\underline{\underline{\underline{c_x X_t^\infty(\Delta)\,u_{T-t}(i)}}}+\underline{\underline{c_x X_t^\infty(i)\,u_{T-t}(i)}}+\underline{\gamma_x X_t^\infty(i)\,Y_t^\infty(i)\left(u_{T-t}(i)\right)^2} \\
& +\underline{\underline{\underline{c_x X_t^\infty(\Delta)\,u_{T-t}(i)}}}\Bigg]\Bigg)\Bigg] = 0 \quad .
\end{aligned}
\tag{8.38}
$$

<u>Proof of 2.</u>
Considering the case $t = t_k$ $(k \in \{0, ..., n\})$ we obtain:

$$
\begin{aligned}
&\lim_{\varsigma\searrow 0}\left\langle u_{T-t_k-\varsigma}, X_{t_k+\varsigma}^\infty\right\rangle_{\Delta,\Sigma}-\left\langle u_{T-t_k+\varsigma}, X_{t_k-\varsigma}^\infty\right\rangle_{\Delta,\Sigma} \\
&=\lim_{\varsigma\searrow 0}\left(\left\langle u_{T-t_k-\varsigma}, X_{t_k+\varsigma}^\infty\right\rangle-\left\langle u_{T-t_k+\varsigma}, X_{t_k-\varsigma}^\infty\right\rangle\right) \\
&\quad+\lim_{\varsigma\searrow 0}\left(u_{T-t_k-\varsigma}(\Delta)\,X_{t_k+\varsigma}^\infty(\Delta)-u_{T-t_k+\varsigma}(\Delta)\,X_{t_k-\varsigma}^\infty(\Delta)\right) \\
&\quad+\lim_{\varsigma\searrow 0}\left(u_{T-t_k-\varsigma}(\Sigma)\,X_{t_k+\varsigma}^\infty(\Sigma)-u_{T-t_k+\varsigma}(\Sigma)\,X_{t_k-\varsigma}^\infty(\Sigma)\right) \\
&=-\left\langle\varphi_k, X_{t_k}^\infty\right\rangle+0+u_{T-t_k-\varsigma}(\Sigma)\left\langle\varphi_k, X_{t_k}^\infty\right\rangle \\
&=-\left\langle\varphi_k, X_{t_k}^\infty\right\rangle+0+1\cdot\left\langle\varphi_k, X_{t_k}^\infty\right\rangle \\
&=0 \quad .
\end{aligned}
\tag{8.39}
$$

■

[4] We underline equal terms in order to improve readability.

8.3 Dual representation of the finite dimensional distributions of the limiting processes

So far, we have found means to describe the limit process from the point of view of duality. In the following, we will provide similar considerations for the limiting objects.

Given the catalyst processes (which we assume to live on a common probability space, compare (8.5)), we want to calculate the log-Laplace transform of the processes $\left(X^N\right)$. We define

$$X_t^N(\Sigma) := \sum_{k=1}^{n} \sum_{i \in \mathbb{N}_0} \varphi_k(i)\, X_{t_k}^N(i)\, \mathbb{1}_{\{t \geq t_k\}}(t) \quad (t \geq 0) \quad , \tag{8.40}$$

with $\varphi_k \in \left\{\varphi \in \mathbb{R}_+^{\mathbb{N}_0} : card\{\varphi > 0\} < \infty\right\}$, $k \in \{1, ..., n\}$, $0 < t_1 < ... < t_n < T$. In particular, we have

$$A := \int_0^T \sum_{k=1}^{n} \sum_{i \in \mathbb{N}_0} \varphi_k(i) < \infty \quad . \tag{8.41}$$

We introduce processes $\left(u^N\right)_{N \in \mathbb{N}}$, of which we will prove that they characterize the finite dimensional distributions of X^N (applying the Cramér-Wold device) by means of a duality equation. Let $\left(u^N\right)_{N \in \mathbb{N}}$ be defined by

- *initial condition*

$$u_0^N(i) := 0 \quad (i \in \{0, ..., N-1\}) \tag{8.42}$$

$$u_0^N(\Sigma) := 1 \tag{8.43}$$

- *evolution*

  – for $T - t \notin \{t_1, ..., t_n\}$

$$\frac{\partial}{\partial t} u_t^N(i) = c_x \frac{1}{N} \sum_{j=0}^{N-1} \left(u_t^N(j) - u_t^N(i)\right) - \gamma_x Y_{T-t}^N(i) \left(u_t(i)\right)^2 \tag{8.44}$$

$$\frac{\partial}{\partial t} u_t^N(\Sigma) = 0 \tag{8.45}$$

  – for $T - t = t_k$ $(k \in \{1, ..., n\})$

$$u_{T-t_k}^N(i) = \lim_{s \nearrow T-t_k} u_s^N(i) + \varphi_k(i) \quad \text{for } i \in \mathbb{N}_0 \tag{8.46}$$

$$u_{T-t_k}^N(\Sigma) = \lim_{s \nearrow T-t_k} u_s^N(\Sigma) \quad . \tag{8.47}$$

We need the following lemma:

Lemma 11 *The dual processes $\left(u_t^N\right)_{t \geq 0}$ $(N \in \mathbb{N})$ have the following property:*

$$\mathbf{E}_{reac}^Y\left[\exp\left(-X_T^N(\Sigma)\right)\right] = \exp\left(-\left\langle u_T^N, X_0^N\right\rangle_\Sigma\right) \quad P_{cat} - a.s. \tag{8.48}$$

Proof. The proof consists of two steps:

1. For $t \notin \{t_1, ..., t_n\}$ we have to prove

$$\frac{\partial}{\partial t}\mathbf{E}^{Y}_{reac}\left[\exp\left(-\left\langle u^N_{T-t}, X^N_t\right\rangle\right)_\Sigma\right] = 0 \tag{8.49}$$

2. If t is a time of observation ($t = t_k, k \in \{1, ..., n\}$) we have to prove

$$\lim_{\varsigma \searrow 0}\left(\left\langle u^N_{T-t_k+\varsigma}, X^N_{t_k-\varsigma}\right\rangle_\Sigma - \left\langle u^N_{T-t_k-\varsigma}, X^N_{t_k+\varsigma}\right\rangle_\Sigma\right) = 0 \tag{8.50}$$

We prove (8.49) using stochastic calculus. We get (for Y^N fixed)

$$\begin{aligned}
&d\exp\left(-\left\langle u^N_{T-t}, X^N_t\right\rangle\right)\\
&= d\exp\left(-\sum_{i\in\mathbb{Z}} u^N_{T-t}(i)\, X^N_t(i)\right)\\
&\quad\cdot\exp\left(-u^N_{T-t}(\Sigma)\, X^N_t(\Sigma)\right)\\
&+ d\exp\left(-u^N_{T-t}(\Sigma)\, X^N_t(\Sigma)\right)\\
&\quad\cdot\exp\left(-\sum_{i\in\mathbb{Z}} u^N_{T-t}(i)\, X^N_t(i)\right)
\end{aligned} \tag{8.51}$$

inserting the evolution equations of u^N and X^N:

$$\begin{aligned}
&\frac{\partial}{\partial t}\mathbf{E}^{Y}_{reac}\left[\exp\left(-\left\langle u^N_{T-t}, X^N_t\right\rangle\right)\right]\\
&= \mathbf{E}^{Y}_{reac}\Bigg[\exp\left(-\left\langle u^N_{T-t}, X^N_t\right\rangle\right)\cdot\Bigg(-\sum_{i\in\mathbb{Z}} X^N_t(i)\,\frac{\partial}{\partial t}u^N_{T-t}(i)\\
&\quad -\sum_{i\in\mathbb{Z}} u^N_{T-t}(i)\,\frac{c_x}{N}\sum_{j=0}^{N-1}\left(X^N_t(j) - X^N_t(i)\right) + \sum_{i\in\mathbb{Z}}\left(u^N_{T-t}(i)\right)^2\gamma_x X^\infty_t(i)\, Y^\infty_t(i)\\
&\quad - u^N_{T-t}(\Sigma)\,\frac{\partial}{\partial t}X^N_t(\Sigma)\Bigg)\Bigg]\\
&= \mathbf{E}^{Y}_{reac}\Bigg[\exp\left(-\left\langle u^N_{T-t}, X^\infty_t\right\rangle\right)\\
&\quad\cdot\Bigg(\sum_{i\in\mathbb{Z}} X^\infty_t(i)\left(-\gamma_x Y^N_t(i)\left(u^N_{T-t}(i)\right)^2 + \frac{c_x}{N}\sum_{j=0}^{N-1}\left(u^N_{T-t}(j) - u^N_{T-t}(i)\right) + 0\right)\\
&\quad -\sum_{i\in\mathbb{Z}} u^N_{T-t}(i)\,\frac{c_x}{N}\sum_{j=0}^{N-1}\left(X^N_t(j) - X^N_t(i)\right) + \sum_{i\in\mathbb{Z}}\left(u^N_{T-t}(i)\right)^2\gamma_x X^N_t(i)\, Y^N_t(i)\\
&\quad -0\Bigg)\Bigg]\\
&= 0\quad .
\end{aligned} \tag{8.52}$$

This is similar to (8.38).

Next, we have to prove (8.50). We get:

$$\begin{aligned}
&\lim_{\varsigma\searrow 0}\left(\langle u^N_{T-t_k+\varsigma}, X^N_{t_k-\varsigma}\rangle_\Sigma - \langle u^N_{T-t_k-\varsigma}, X^N_{t_k+\varsigma}\rangle_\Sigma\right)\\
&= \lim_{\varsigma\searrow 0}\sum_{i\in\mathbb{N}_0}\left(u^N_{T-t_k+\varsigma}(i)\, X^N_{t_k-\varsigma}(i) - u^N_{T-t_k-\varsigma}(i)\, X^N_{t_k+\varsigma}(i)\right)\\
&\quad + \lim_{\varsigma\searrow 0}\left(u^N_{T-t_k+\varsigma}(\Sigma)\, X^N_{t_k-\varsigma}(\Sigma) - u^N_{T-t_k-\varsigma}(\Sigma)\, X^N_{t_k+\varsigma}(\Sigma)\right)\\
&= \sum_{i\in\mathbb{N}_0} X^N_{t_k}(i)\left(-\varphi_k(i)\right)\\
&+ u^N_{T-t_k+\varsigma}(\Sigma)\left(\sum_{i\in\mathbb{T}^d_N}\varphi_k(i)\, X^N_{t_k-\varsigma}(i)\right)\\
&= \sum_{i\in\mathbb{N}_0} X^N_{t_k}(i)\left(-\varphi_k(i)\right) + 1\cdot\sum_{i\in\mathbb{N}_0}\varphi_k(i)\, X^N_{t_k-\varsigma}(i) = 0 \quad . \qquad (8.53)
\end{aligned}$$

This completes the proof. ■

8.4 Convergence of dual processes

We give a survey of our results on the log-Laplace transform of the finite and infinite model:

finite model	**infinite model**

initial condition

Finite model:

$$u_0^N(i) := 0$$
$$u_0^N(\Sigma) := 1$$

Infinite model:

$$u_0(i) := 0$$
$$u_0(\Delta) := 0$$
$$u_0(\Sigma) := 1$$

Differential equation at non-jump times

Finite model: for $T - t \notin \{t_1, ..., t_n\}$

$$\frac{\partial}{\partial t} u_t^N(i) = c_x \frac{1}{N} \sum_{j=0}^{N-1} \left(u_t^N(j) - u_t^N(i)\right) - \gamma_x Y_{T-t}^N(i) \left(u_t^N(i)\right)^2 \quad ,$$
$$\frac{\partial}{\partial t} u_t^N(\Sigma) = 0 \quad ,$$

Infinite model: for $T - t \notin \{t_1, ..., t_n\}$

$$\frac{\partial}{\partial t} u_t(i) = -\gamma_x Y_{T-t}^\infty(i) \left(u_t(i)\right)^2 - c_x u_t(i) \quad ,$$
$$\frac{\partial}{\partial t} u_t(\Delta) = \sum_{i \in \mathbb{Z}} c_x u_t(i) \quad ,$$
$$\frac{\partial}{\partial t} u_t(\Sigma) = 0 \quad .$$

Points of discontinuity

Finite model: for $T - t = t_k$ $(k \in \{1, ..., n\})$

$$u_{T-t_k}^N(i) = \lim_{s \nearrow T-t_k} u_s^N(i) + \varphi_k(i)$$
$$u_{T-t_k}^N(\Sigma) = \lim_{s \nearrow T-t_k} u_s^N(\Sigma) \quad .$$

Infinite model: for $T - t = t_k$ $(k \in \{1, ..., n\})$:

$$u_{T-t_k}(i) = \lim_{s \nearrow T-t_k} u_s(i) + \varphi_k(i)$$
$$u_{T-t_k}(\Delta) = \lim_{s \nearrow T-t_k} u_s(\Delta)$$
$$u_{T-t_k}(\Sigma) = \lim_{s \nearrow T-t_k} u_s(\Sigma) \quad .$$

Convergence of finite dimensional distributions in equivalent to convergence of sums of observations as considered here (Cramér-Wold device).

With (8.48) and (8.35), the claim of Theorem 2 may be reformulated as follows:

$$\mathcal{L}\left\{\exp\left(-\langle u_T^N, X_0^N\rangle_\Sigma\right)\right\} \overset{N\to\infty}{\Longrightarrow} \mathcal{L}\left\{\exp\left(-\langle u_T, X_0^\infty\rangle_{\Delta,\Sigma}\right)\right\} \quad . \tag{8.54}$$

With

$$X_0^N(i) = X_0^\infty(i) \quad (i \in \{0, ..., N-1\}) \tag{8.55}$$

by construction, (8.54) is equivalent to

$$\mathcal{L}\left\{\exp\left(-\langle u_T^N, X_0^\infty\rangle_\Sigma\right)\right\} \overset{N\to\infty}{\Longrightarrow} \mathcal{L}\left\{\exp\left(-\langle u_T, X_0^\infty\rangle_{\Delta,\Sigma}\right)\right\} \quad . \tag{8.56}$$

The most remarkable difference between the finite and the infinite model is the state Δ, which is missing in the finite model. In the infinite model Δ is a model for "the world outside". We try to make a parallel construction for the finite model. The idea is to divide $\mathbb{N}_0$ into

- the *world inside (window of observation)*

$$W := \left\{ i \in \mathbb{N}_0 : \quad \sum_{k=1}^{N} \varphi_k(i) > 0 \right\} \quad . \tag{8.57}$$

- the *world outside*

$$\hat{\Delta} := \{ i \in \mathbb{N}_0 : \quad \varphi_k(i) = 0 \quad \forall k \in \{1, ..., n\} \} \quad . \tag{8.58}$$

Obviously we have

$$W \cup \hat{\Delta} = \mathbb{N}_0 \quad . \tag{8.59}$$

It will turn out that in some sense $\hat{\Delta}$ and Δ correspond to each other. Namely, we prove (8.56) in three steps:

1. For $\varepsilon > 0$:

$$\lim_{N\to\infty} P\left[\left| \sum_{i \in W \cap \{0,...,N-1\}} \left(u_T^N(i)\, X_0^\infty(i) - u_T(i)\, X_0^\infty(i) \right) \right| > \varepsilon \right] = 0 \quad . \tag{8.60}$$

2. For $\varepsilon > 0$:

$$\lim_{N\to\infty} P\left[\left| \sum_{i \in \hat{\Delta} \cap \{0,...,N-1\}} \left(u_T^N(i)\, X_0^N(i) \right) - u_T(\Delta)\, X_0^\infty(\Delta) \right| > \varepsilon \right] = 0 \quad . \tag{8.61}$$

3. For $\varepsilon > 0$:

$$\lim_{N\to\infty} P\left[\left| \left(u_T^N(\Sigma)\, X_0^\infty(\Sigma) \right) - u_T(\Sigma)\, X_0^\infty(\Sigma) \right| > \varepsilon \right] = 0 \quad . \tag{8.62}$$

Proof of the first step (8.60)
By construction we have:

- W is a finite set
- $\mathbf{E}_{reac}^Y [X_0^\infty(i)] = \theta_x < \infty \quad (i \in \mathbb{N}_0)$

Therefore it is sufficient to verify

$$\forall i \in W : \lim_{N\to\infty} P\left[\left| u_T^N(i) - u_T(i) \right| > \varepsilon \right] = 0 \quad . \tag{8.63}$$

With $g_N(x) := \frac{1}{N} \ln\cosh(Nx) \geq |x|$ it suffices to prove[5]

$$\forall i \in W : \lim_{N\to\infty} P\left[g_N\left(u_T^N(i) - u_T(i) \right) > \varepsilon \right] = 0 \quad . \tag{8.64}$$

[5]The properties of (g_N) are discussed on page 93

Consider the differential equation describing $u_t^N(i) - u_t(i)$ for $i \in \{0, ..., N-1\} \cap W$, namely

$$\frac{\partial}{\partial t}\left(u_t^N(i) - u_t(i)\right)$$

$$= \frac{c_x}{N}\sum_{j=0}^{N-1} u_t^N(j) \tag{8.65}$$

$$- c_x\left(u_t^N(i) - u_t(i)\right) \tag{8.66}$$

$$- \frac{\gamma_x}{2}\left(\left(u_t^N(i)\right)^2 - \left(u_t(i)\right)^2\right)\left(Y_{T-t}^N(i) + Y_{T-t}^\infty(i)\right) \tag{8.67}$$

$$- \frac{\gamma_x}{2}\left(\left(u_t^N(i)\right)^2 + \left(u_t(i)\right)^2\right)\left(Y_{T-t}^N(i) - Y_{T-t}^\infty(i)\right) \quad . \tag{8.68}$$

Applying the chain rule, we get

$$\frac{\partial}{\partial t} g_N\left(u_t^N(i) - u_t(i)\right)$$

$$= \tanh\left(N \cdot \left(u_t^N(i) - u_t(i)\right)\right) \cdot \frac{c_x}{N}\sum_{j=0}^{N-1} u_t^N(j) \tag{8.69}$$

$$- \tanh\left(N \cdot \left(u_t^N(i) - u_t(i)\right)\right) \cdot c_x\left(u_t^N(i) - u_t(i)\right) \tag{8.70}$$

$$- \tanh\left(N \cdot \left(u_t^N(i) - u_t(i)\right)\right) \cdot \frac{\gamma_x}{2}\left(\left(u_t^N(i)\right)^2 - \left(u_t(i)\right)^2\right)\left(Y_{T-t}^N(i) + Y_{T-t}^\infty(i)\right) \tag{8.71}$$

$$- \tanh\left(N \cdot \left(u_t^N(i) - u_t(i)\right)\right) \cdot \frac{\gamma_x}{2}\left(\left(u_t^N(i)\right)^2 + \left(u_t(i)\right)^2\right)\left(Y_{T-t}^N(i) - Y_{T-t}^\infty(i)\right) \quad . \tag{8.72}$$

Since the terms (8.70) and (8.71) are negative and tanh is bounded above by 1, we get:

$$\frac{\partial}{\partial t} g_N\left(u_t^N(i) - u_t(i)\right)$$

$$\leq \frac{c_x}{N}\sum_{j=0}^{N-1} u_t^N(j) + \frac{\gamma_x}{2}\left(\left(u_t^N(i)\right)^2 + \left(u_t(i)\right)^2\right)\left|Y_{T-t}^N(i) - Y_{T-t}^\infty(i)\right| \quad . \tag{8.73}$$

and after integration (using $\frac{1}{N}\ln\cosh 0 = \frac{1}{N}$):

$$\left|u_T^N(i) - u_T(i)\right|$$

$$\leq \frac{1}{N} + \int_0^T \left[\frac{c_x}{N}\sum_{j=0}^{N-1} u_t^N(j) + \frac{\gamma_x}{2}\left(\left(u_t^N(i)\right)^2 + \left(u_t(i)\right)^2\right)\left|Y_{T-t}^N(i) - Y_{T-t}^\infty(i)\right|\right] dt \quad . \tag{8.74}$$

We can derive from the equations defining u^N and u and from (8.41) that we have

$$\frac{1}{N}\sum_{j=0}^{N-1} u_t^N(j) \leq \frac{A}{N} \quad , (t \in [0, T]) \tag{8.75}$$

and

$$u_t^N(i) \le A \quad , \quad u_t(i) \le A \quad (i \in \{0, ..., N-1\}) \quad . \tag{8.76}$$

This yields

$$\left|u_T^N(i) - u_T(i)\right| \le \frac{1}{N} + \int_0^T \left[\frac{A}{N} + \gamma_x A^2 \left|Y_{T-t}^N(i) - Y_{T-t}^\infty(i)\right|\right] dt \quad . \tag{8.77}$$

Since we have

$$\lim_{N\to\infty} \int_0^T \left|Y_{T-t}^N(i) - Y_{T-t}^\infty(i)\right| dt = 0 \quad P_{cat}\text{-a.s.} \tag{8.78}$$

by (8.5), we get (compare(8.74))

$$\forall i \in \mathbb{N}_0 : \lim_{N\to\infty} \left|u_T^N(i) - u_T(i)\right| = 0 \quad P_{cat}\text{-a.s.} \tag{8.79}$$

Proof of the second step (8.61)

Applying the triangular inequality to (8.61) (remember that we have $X^\infty(\Delta) \equiv \theta_x$ by (8.23)), it suffices to prove the following two points

a)

$$\lim_{N\to\infty} P\left[\left|\sum_{i\in\hat{\Delta}\cap\{0,...,N-1\}} u_T^N(i) - u_T(\Delta)\right| > \varepsilon\right] = 0 \quad . \tag{8.80}$$

b)

$$\lim_{N\to\infty} P\left[\left|\sum_{i\in\hat{\Delta}\cap\{0,...,N-1\}} u_T^N(i)\left(X_0^N(i) - X_0^\infty(\Delta)\right)\right| > \varepsilon\right] = 0 \quad . \tag{8.81}$$

Proof of (8.80):

We compare $u^N\left(\hat{\Delta}\right)$ with $u(\Delta)$. Since the "window of observation is finite"

$$M := card(W) < \infty, \tag{8.82}$$

we can assume without loss of generality

$$\{0, \ldots, N-1\} \supseteq W \quad . \tag{8.83}$$

We write down the differential equality describing $u_t^N\left(\hat{\Delta}\right)$:

$$\begin{aligned}\frac{\partial}{\partial t} u_t^N\left(\hat{\Delta}\right) &= c_x \frac{1}{N} \sum_{i\in\hat{\Delta}} \sum_{j\in W} \left(u_t^N(j) - u_t^N(i)\right) \\ &\quad - \sum_{i\in\hat{\Delta}} \gamma_x Y_{T-t}^N(i) \left(u_t(i)\right)^2 \quad .\end{aligned} \tag{8.84}$$

We get:

$$\begin{aligned}
&\frac{\partial}{\partial t}\left(u_t(\Delta) - u_t^N\left(\hat{\Delta}\right)\right) \\
&= \sum_{k\in\mathbb{N}_0} c_x u_t(k) - c_x \frac{1}{N} \sum_{k\in\{0,\dots,N-1\}\cap\hat{\Delta}} \sum_{j\in W} \left(u_t^N(j) - u_t^N(k)\right) \\
&\quad + \sum_{k\in\{0,\dots,N-1\}\cap\hat{\Delta}} \gamma_x Y_{T-t}^N(k)\left(u_t^N(k)\right)^2 \\
&= \sum_{k\in\mathbb{N}_0} c_x u_t(k) - \frac{c_x}{N}\left(card\left(\{0,\dots,N-1\}\cap\hat{\Delta}\right) \sum_{j\in W} u_t^N(j) \right. \\
&\qquad\qquad \left. - card(W) \sum_{k\in\{0,\dots,N-1\}\cap\hat{\Delta}} u_t^N(k)\right) \\
&\quad + \sum_{k\in\hat{\Delta}} \gamma_x Y_{T-t}^N(k)\left(u_t^N(k)\right)^2 \\
&= \sum_{k\in\mathbb{N}_0} c_x u_t(k) - \frac{c_x}{N}\left((N-M)\sum_{j\in W} u_t^N(j) - M \sum_{k\in\{0,\dots,N-1\}\cap\hat{\Delta}} u_t^N(k)\right) \\
&\quad + \sum_{k\in\{0,\dots,N-1\}\cap\hat{\Delta}} \gamma_x Y_{T-t}^N(k)\left(u_t^N(k)\right)^2
\end{aligned}$$

Since $u_t(k) = 0 \ \forall k \in \hat{\Delta}$, we can write

$$\begin{aligned}
&= \sum_{k\in\{0,\dots,N-1\}\setminus\hat{\Delta}} c_x u_t(k) - c_x\left(\sum_{k\in\{0,\dots,N-1\}\setminus\hat{\Delta}} u_t^N(k) \right. \\
&\quad \left. - \frac{M}{N} \sum_{k\in\{0,\dots,N-1\}} u_t^N(k)\right) \\
&\quad + \sum_{k\in\hat{\Delta}\cap\{0,\dots,N-1\}} \gamma_x Y_{T-t}^N(k)\left(u_t^N(k)\right)^2 \\
&= \sum_{k\in\{0,\dots,N-1\}\setminus\hat{\Delta}} c_x\left(u_t(k) - u_t^N(k)\right) \\
&\quad - c_x\frac{M}{N} \sum_{k\in\{0,\dots,N-1\}} u_t^N(k) + \sum_{k\in\hat{\Delta}\cap\{0,\dots,N-1\}} \gamma_x Y_{T-t}^N(k)\left(u_t^N(k)\right)^2 \quad . \qquad (8.85)
\end{aligned}$$

This yields:

$$\begin{aligned}
&\left|u_t(\Delta) - u_t^N\left(\hat{\Delta}\right)\right| \\
&\leq \int_0^T \sum_{k\in W} c_x \left|u_t(k) - u_t^N(k)\right| dt + c_x \frac{M}{N} \int_0^T \sum_{k\in\{0,\dots,N-1\}} u_t^N(k)\, dt \\
&+ \int_0^T \sum_{k\in\hat{\Delta}} \gamma_x Y_{T-t}^N(k) \left(u_t^N(k)\right)^2 dt
\end{aligned} \tag{8.86}$$

With

$$\int_0^T \sum_{k\in\{0,\dots,N-1\}} u_t^N(k)\, dt \leq A < \infty \quad , \tag{8.87}$$

we get:

$$\begin{aligned}
\left|u_t(\Delta) - u_t^N\left(\hat{\Delta}\right)\right| \leq &\int_0^T \sum_{k\in W} c_x \left|u_t(k) - u_t^N(k)\right| dt + c_x \frac{M}{N} A \\
&+ \int_0^T \sum_{k\in\hat{\Delta}} \gamma_x Y_{T-t}^N(k) \left(\frac{A}{N}\right)^2 dt \quad .
\end{aligned} \tag{8.88}$$

This converges to zero Y-stochastically as $N \to \infty$, in particular taking (8.79) into consideration.

Proof of (8.81)
We have to prove

$$\lim_{N\to\infty} P\left[\left|\sum_{i\in\hat{\Delta}\cap\{0,\dots,N-1\}} u_T^N(i)\left(X_0^N(i) - X_0^\infty(\Delta)\right)\right| > \varepsilon\right] = 0 \quad . \tag{8.89}$$

Since (by (8.23)) we have

$$X_0^\infty(\Delta) = \theta_x \tag{8.90}$$

it suffices to prove

$$\lim_{N\to\infty} var\left[\sum_{i\in\hat{\Delta}\cap\{0,\dots,N-1\}} u_T^N(i)\left(X_0^N(i) - \theta_x\right)\right] = 0 \quad . \tag{8.91}$$

Since the initial configuration of the reactant is a product measure by assumption, and applying

$$u_T^N(i) \leq \frac{A}{N} \quad (i \in \{0,\dots,N-1\}) \tag{8.92}$$

we get

$$\lim_{N\to\infty} var\left[\sum_{i\in\hat{\Delta}\cap\{0,\dots,N-1\}} u_T^N(i)\left(X_0^N(i)-\theta_x\right)\right]$$
$$\leq \lim_{N\to\infty} \sum_{i\in\hat{\Delta}\cap\{0,\dots,N-1\}} var\left[\frac{A}{N}\left(X_0^N(i)-\theta_x\right)\right]$$
$$\leq \lim_{N\to\infty} (N-M)\left(\frac{A}{N}\right)^2 var\left[\left(X_0^N(0)\right)\right]$$
$$= \lim_{N\to\infty} \frac{A^2}{N} var\left[\left(X_0^N(0)\right)\right] = 0 \quad . \tag{8.93}$$

In particular we have applied that the initial configuration is given by a product measure (compare 4.1.2).

Proof of the third step (8.62)

Since we have

$$X_0^\infty(\Sigma) = 0 \quad , \tag{8.94}$$

there is nothing to prove.

8.5 Tightness

So far, we have only verified convergence of finite dimensional distributions. To obtain pathwise convergence, one has to prove that the family consisting of the laws of the processes is a tight family. This can be proven by applying the Kolmogorov criterion of tightness. Therefore it is crucial to have uniform boundedness of pth moments for some $p > 2$.

Since Theorem 2 is a special case of Theorem 3 (namely $L = 1$), we do not carry out the complete proof of tightness here, but refer to the tightness argument carried out when proving Theorem 3.

8.6 Digression – Empirical averages

In the following digression we give a sketch of one possibility to prove Theorem 2. Actually, we provide afterwards a different proof using methods that also can be applied when proving Theorem 3.Therefore, the proof of Theorem 2 can be regarded as preparation of the proof of Theorem 3.

But first we line out how Theorem 2 could be proven making use of the convergence of empirical averages:

Inspecting equations (8.1) and (8.2) it is natural to think of a law of large numbers. The idea is to proceed a two-step program, namely:

1. For fixed $t > 0$, prove

$$\lim_{N\to\infty} \mathbf{E}\left[\left(\frac{1}{N}\sum_{j=0}^{N-1} X_t^N(j) - \theta_x\right)^2\right] = 0 \quad , \tag{8.95}$$

$$\lim_{N\to\infty} \mathbf{E}\left[\left(\frac{1}{N}\sum_{j=0}^{N-1} Y_t^N(j) - \theta_y\right)^2\right] = 0 \quad . \tag{8.96}$$

2. Show that this is sufficient for law of the limit process to coincide (with respect to finite dimensional distributions) with the process given by the following stochastic differential equations:

$$X_0^\infty(i) = \theta_x, \tag{8.97}$$

$$Y_0^\infty(i) = \theta_y, \tag{8.98}$$

$$dX_t^\infty(i) = c_x\left(\theta_x - X_t^\infty(i)\right) dt + \sqrt{2\gamma_x X_t^\infty(i)\, Y_t^\infty(i)}\, dB_t^x(i), \tag{8.99}$$

$$dY_t^\infty(i) = c_y\left(\theta_y - Y_t^\infty(i)\right) dt + \sqrt{2\gamma_y Y_t^\infty(i)}\, dB_t^y(i) \quad , i \in \mathbb{N}_0. \tag{8.100}$$

where $\left\{\left(B_t^x(i), B_t^y(i)\right)_{t\geq 0}\right\}_{i\in\mathbb{N}_0}$ is a family of independent bivariate Brownian motions.

In particular, this implies that in the limit we have independent evolutions at different sites.

ad 1. We prove that the empirical averages converge in L^2 to their expected values, namely θ_x and θ_y. By the strong law of large numbers the claim holds for the initial condition, which we have assumed to be a product law. We take a closer look at the evolution equation of the empirical averages. Adding up the equations for all N components, one gets:

$$\frac{\partial}{\partial t}\mathbf{E}\left[\left(\frac{1}{N}\sum_{j=0}^{N-1} X_t^N(j) - \theta_x\right)^2\right] = \frac{2\gamma_x}{N}\theta_x\theta_y \quad , \tag{8.101}$$

and

$$\frac{\partial}{\partial t}\mathbf{E}\left[\left(\frac{1}{N}\sum_{j=0}^{N-1} Y_t^N(j) - \theta_y\right)^2\right] = \frac{2\gamma_y}{N}\theta_y \quad . \tag{8.102}$$

Using the fact, that the empirical averages are martingales[6], we get immediately:

$$\lim_{N\to\infty} \sup_{s\in[0,t]} \mathbf{E}\left[\left(\frac{1}{N}\sum_{j=0}^{N-1} X_t^N(j) - \theta_x\right)^2\right] = 0 \tag{8.103}$$

$$\lim_{N\to\infty} \sup_{s\in[0,t]} \mathbf{E}\left[\left(\frac{1}{N}\sum_{j=0}^{N-1} Y_t^N(j) - \theta_y\right)^2\right] = 0 \quad . \tag{8.104}$$

ad 2. By 1., we have some kind of convergence of coefficients. Applying results of Joffe and Métivier - they provide in [JM], Proposition 3.2.3 a relationship between pathwise convergence and convergence of drift and fluctuation parameters - it is possible to obtain the considered convergence result.

Somehow, the paper of Joffe and Métivier comes in as a miraculous black box – this is another reason for providing a different proof.

[6] hence for all $t > 0$ we have

$$\mathbf{E}\left[\frac{1}{N}\sum_{j=0}^{N-1} X_t^N(j) - \theta_x\right] = 0 \quad ,$$

$$\mathbf{E}\left[\frac{1}{N}\sum_{j=0}^{N-1} Y_t^N(j) - \theta_y\right] = 0 \quad .$$

Chapter 9

Proof of Theorem 3

In this chapter, we prove Theorem 3. As a reminder, we provide the formulation of the Theorem at this place:

Theorem 3 We consider the processes $\left(X_t^N, Y_t^N\right)_{t\geq 0}$ started in a product measure with first moments given by $\theta_0 = (\theta_0^x, \theta_0^y)$. Let $i \in \Omega_{N,L}$. We claim:

1.
$$\mathcal{L}\left\{\left(\bar{X}^{N,k}_{N^k t}(i), \bar{Y}^{N,k}_{N^k t}(i)\right)_{t\geq 0}\right\} \overset{N\to\infty}{\Longrightarrow} \Psi^{\theta_0}_{c_k} \tag{9.1}$$

2. Let $k \leq j$. For $s : \mathbb{N} \to \mathbb{R}_+$ with $\lim_{N\to\infty} s(N) = \infty$, $\lim_{N\to\infty} \frac{s(N)}{N} = 0$ we have
$$\mathcal{L}\left\{\left(\bar{X}^{N,k}_{N^j s(N)+N^k t}(i), \bar{Y}^{N,k}_{N^j s(N)+N^k t}(i)\right)_{t\geq 0}\right\} \overset{N\to\infty}{\Longrightarrow} \int \hat{\Psi}^{\theta^*}_{c_k} \; \mu^{j,k+1}_{\theta_0}(d\theta^*) \tag{9.2}$$

In Section 9.1 we prove (9.1), in Section 9.2 we prove (9.2).

9.1 First part: Fluctuations of averages

An important part of work has already been done: Dawson and Greven prove in [DG96] that the claim is true for the catalyst. Namely, they prove weak convergence of the laws of catalyst averages. Defining all catalyst processes $\left(Y^N\right)$ on a common probability space, we enforce by Skorohod's representation that almost surely there is pathwise convergence of the catalyst. We denote the joint probability law of the catalyst processes by P_{cat}. Next, the reactant processes can be constructed for given catalyst processes.

By this construction, the reactant is a stochastic process in a random medium, which consists of the catalyst processes. As we already have described on page 108 when discussing the quenched approach in the context of fluctuating averages of processes defined on tori of growing size, we can decompose the probability measure P describing the catalyst-reactant system with pathwise converging catalyst averages:[1]

$$P[\cdot] = \int P^{Y}_{reac}[\cdot] \quad P_{cat}(dY). \tag{9.3}$$

It has turned out that all we have to do is to verify

1. that the laws catalyst processes converge weakly.
2. that the laws of the catalyst/reactant processes converge for a given sequence of catalyst processes. We are free to assume that this sequence converges pathwise almost surely by choosing a suitable probability space. This assumption is made possible by Skorohod's representation.

After this short survey of the proof, we want to go into details and describe the stucture of the proof. Namely, it contains the following steps; to each step there corresponds a subsection.

- *Weak convergence of the catalyst laws:* We know from [DG96] that the laws of the empirical k-level intensities of the catalyst converge after rescaling of time. We define all catalyst realizations on a common probability space, assuming that pathwise convergence holds almost surely. This is possible, since all processes live on Polish spaces and therefore Skorohod's representation can be applied (compare Section 5.1.1, p. 40).
- *Some words about the quenched approach:* To make notation more precise, we recall what is called the quenched approach.
- *Cramér-Wold device:* This gives us a criterion for convergence with respect to finite dimensional distributions. We characterize finite dimensional distributions by Laplace-transforms for a given catalytic medium.
- *Duality:* The Laplace transforms can be described in terms of systems of interacting differential equations. These differential equations are deterministic for given catalyst.

  We will make frequent use of this dual representation.
- *Convergence of dual processes*: Given the almost surely converging sequence $Y = (Y^N)_{N\in\mathbb{N}}$, we compare the dual processes of the reactants X^N, letting $N \to \infty$. It turns out that the Laplace transforms describing finite dimensional distributions of conditioned dual processes converge in a L^1 sense.
- *Tightness:* We need to prove tightness to fill the gap between convergence with respect to finite dimensional distributions and pathwise convergence.

[1]It is important that the mapping $Y \longmapsto \mathcal{L}^{Y}_{reac}$ is measurable. This can be proven similarly to Lemma 6, p. 109.

9.1.1 Weak convergence of the catalyst laws

Notation 7 *We introduce a notation for empirical averages in accelerated timescale*

$$\hat{X}_t^{N,k}(i) := \bar{X}_{N^k t}^{N,k}(i) \quad , \tag{9.4}$$
$$\hat{Y}_t^{N,k}(i) := \bar{Y}_{N^k t}^{N,k}(i) \quad . \tag{9.5}$$

Using this notation, we have to prove[2]

$$\mathcal{L}\left\{\left(\hat{X}_t^{N,k}(0), \hat{Y}_t^{N,k}(0)\right)_{t\geq 0}\right\} \quad \overset{N\to\infty}{\Longrightarrow} \quad \Psi_{c_k}^{\theta_0} \tag{9.6}$$

Let $\left(\hat{X}_t^{\infty,k}(0), \hat{Y}_t^{\infty,k}(0)\right)_{t\geq 0}$ be a process, which is a realization of the limit law $\Psi_{c_k}^{\theta_0}$, namely

$$\mathcal{L}\left\{\left(\hat{X}_t^{\infty,k}(0), \hat{Y}_t^{\infty,k}(0)\right)_{t\geq 0}\right\} = \Psi_{c_k}^{\theta_0} \quad . \tag{9.7}$$

The processes $\left(\hat{X}_t^{N,k}, \hat{Y}_t^{N,k}\right)_{t\geq 0}$ $(N \in \mathbb{N})$ attain values in a Polish space. Therefore, it is possible to apply Skorohod's representation (compare 5.1.1). This enables us to assume

Assumption: Let all the catalyst processes be defined on a common probability space, such that catalyst averages shall converge pathwise. Namely, we assume that

$$\lim_{N\to\infty} d_{Sk}\left(\left(\hat{Y}_t^{N,k}\right)_{t\geq 0}, \left(\hat{Y}_t^{\infty,k}\right)_{t\geq 0}\right) = 0 \tag{9.8}$$

holds for almost all realizations of the catalyst (using a notation to be introduced in the next paragraph: P_{cat}-almost surely.) In particular, we emphasize that we define the processes $\left(X_t^N, Y_t^N\right)_{t\geq 0}$ $(N \in \mathbb{N})$ on a common probability space.

Note that a similar consideration has been carried out on page 108 proving the convergence of time-rescaled fluctuating averages of processes defined on tori.

9.1.2 Some words about the quenched approach

We make use of the following decomposition:

$$P[\cdot] = \int P_{reac}^{Y}[\cdot] \quad P_{cat}(dY). \tag{9.9}$$

Such a decomposition is possible since the catalytic process can be realized by a two-step experiment: At first, we can sample the catalyst, then we can construct the catalyst-reactant process for given catalyst. We have assumed that the catalyst processes converge (compare (9.8)). Let the joint law of these coupled processes be denoted by P_{cat} :

$$P_{cat} = \mathcal{L}\left\{\left(Y_t^1\right)_{t\geq 0}, \left(Y_t^2\right)_{t\geq 0}, \dots\right\} \tag{9.10}$$

[2]For the definition of $\Psi_{c_k}^{\theta_0}$ recall (4.19).

In other words, P_{cat} describes all catalyst processes at once. Formally, Y in (9.9) describes nothing else but the vector

$$\left(\left(Y_t^1\right)_{t\geq 0}, \left(Y_t^2\right)_{t\geq 0}, \ldots\right) \tag{9.11}$$

Given the converging catalyst processes "Y", the reactant processes are sampled. The law of the catalyst-reactant processes is again obtained by a two-step construction: first sample the catalyst, then solve the evolution equations for given catalyst to get both, catalyst and reactant. In particular, the law P_{reac}^{Y} of the process constructed given the catalyst vector Y has the property

$$P_{reac}^{Y} = \mathcal{L}\left\{\left(X^N, Y^N\right)_{N\in\mathbb{N}} \middle| \left(Y^N\right)_{N\in\mathbb{N}}\right\} \quad .$$

It is our aim to prove

$$\mathcal{L}\left\{\left(\hat{X}^{N,k}, \hat{Y}^{N,k}\right)\right\} \quad \overset{N\to\infty}{\Longrightarrow} \quad \mathcal{L}\left\{\left(\hat{X}^{\infty,k}, \hat{Y}^{\infty,k}\right)\right\} \quad . \tag{9.12}$$

This can be related to the quenched approach: we have the pathwise convergence property (9.8); applying the strategy described in Chapter 5 – compare in particular Lemma 2, p. 44 – we have to prove

$$\forall \varepsilon > 0 : \lim_{N\to\infty} P_{cat}\left[\varrho\left(\mathcal{L}_{reac}^{Y}\left\{\left(\hat{X}^{N,k}\right)\right\}, \mathcal{L}_{reac}^{Y}\left\{\left(\hat{X}^{\infty,k}\right)\right\}\right) > \varepsilon\right] = 0 \quad . \tag{9.13}$$

which is sufficient for (9.12).

Note that this is similar to the discussion of the Finite System Scheme (compare page 108).

9.1.3 Cramér-Wold device

To prove (9.13), besides tightness we have to prove weak convergence of finite dimensional distributions. We can formulate this claim in terms of Laplace transforms as follows:

In the sequel, consider $0 < t_1 < t_2 < \ldots < t_n \leq T < \infty$ arbitrarily chosen.

Claim 7 *For any $(a_\ell)_{\ell\in\{1,\ldots,n\}} \in \mathbb{R}_+^n$, the following convergence holds:*

$$\lim_{N\to\infty} \mathbf{E}_{cat}\left|\mathbf{E}_{reac}^{Y}\left[\exp\left(-\sum_{\ell=1}^{n} a_\ell \hat{X}_{t_\ell}^{N,k}\right)\right] - \mathbf{E}_{reac}^{Y}\left[\exp\left(-\sum_{\ell=1}^{n} a_\ell \hat{X}_{t_\ell}^{\infty,k}\right)\right]\right| = 0 \quad . \tag{9.14}$$

9.1.4 Duality

Formulating the dual equations, we choose a slightly more general framework.

Consider test functions $\left(\varphi_\ell \in \mathbb{R}_+^{\Omega_{N,L}}\right)$, $\ell \in \{1, \ldots, n\}$) having bounded support

$$card\left\{i \in \Omega_{N,L} : \sum_{\ell=1}^{n} \varphi_\ell(i) > 0\right\} < \infty \quad . \tag{9.15}$$

We want the following duality equation to hold:

$$\mathbf{E}^{Y}_{reac}\left[\exp\left(-\sum_{\ell=1}^{n}\left\langle \varphi_\ell, X^N_{N^k t_\ell}\right\rangle\right)\right] = \mathbf{E}^{Y}_{reac}\left[\exp\left(-\left\langle u^N_{N^k T}, X^N_0\right\rangle\right)\right] \tag{9.16}$$

This can be realized (compare [Isc86] and the remarks on the topic of random occupation times in the "Methods and Tools" Section (Proposition 1, p. 49), defining the "càglàd" dual process $\left(u^N_t\right)_{t\geq 0}$ by

- initial condition:

$$u^N_0 \equiv 0 \quad , \tag{9.17}$$

- jumps:

$$\lim_{t'\to N^k(T-t_\ell)+} u^N_{N^k(T-t_\ell)} = \lim_{t'\to N^k(T-t_\ell)-} u^N_{t'} + \varphi_\ell \quad , \tag{9.18}$$

- evolution (for $N^kT - t \notin \{N^k t_1, ..., N^k t_n\}$)

$$\begin{aligned} \frac{\partial}{\partial t} u^N_t(i) &= \sum_{\varsigma=1}^{L} \frac{c_{\varsigma-1}}{N^{\varsigma-1}} \left(\frac{1}{N^\varsigma} \sum_{j\in\Omega_{N,L}: d(i,j)\leq\varsigma} u^N_t(j) - u^N_t(i)\right) \\ &\quad -\gamma_x Y^N_{N^kT-t}(i)\left(u^N_t(i)\right)^2 \quad (i\in\Omega_{N,L}) \quad . \end{aligned} \tag{9.19}$$

We introduce the following shorthand notation for (9.19)

$$\frac{\partial}{\partial t} u^N_t = \clubsuit_N u^N_t - \gamma_x Y^N_{N^kT-t}\left(u^N_t\right)^2 \quad . \tag{9.20}$$

Applying this to our situation (we are interested in convergence of empirical averages), we now and henceforth take the following case into consideration:

$$\varphi_\ell(i) := \varphi^N_\ell(i) := \begin{cases} \frac{1}{N^k} a_\ell & \text{if} \quad i\in B_k(0) \\ 0 & \text{else} \end{cases} \tag{9.21}$$

where $(a_\ell)_{\ell\in\{1,...,n\}} \in \mathbb{R}^n_+$ and

$$B_k(0) := \{i\in\Omega_{N,L} : d(0,i)\leq k\} \quad . \tag{9.22}$$

Moreover, we assume:

$$\sum_{\ell=1}^{n} a_\ell \leq 1 \quad , \tag{9.23}$$

making use of the property of Laplace transforms that they are uniquely determined by their evaluations on a dense set. Furthermore, let

$$\bar{u}^{N,k} := \frac{1}{N^k}\sum_{i\in B_k(0)} u^N(i) \quad . \tag{9.24}$$

We want to compare the dual processes $\left\{(u_t^N)_{t\geq 0}\right\}_{N\in\mathbb{N}}$ with a sequence of processes, $\left\{(q_t^N)_{t\geq 0}\right\}_{N\in\mathbb{N}}$. The latter sequence is the sequence of dual processes of $\sum_{\ell=1}^n a_\ell \hat{X}_{t_\ell}^{\infty,k}$. To be more precise, we want $(q_t^N)_{t\geq 0}$ to solve the following equation for any $N\in\mathbb{N}$:

$$\mathbf{E}_{reac}^Y\left[\exp\left(-\sum_{\ell=1}^n a_\ell \hat{X}_{t_\ell}^{\infty,k}\right)\right] = \exp\left(-\sum_{i\in\Omega_{N,L}} q_{N^kT}^N(i)\,\theta_0^x\right) \quad . \tag{9.25}$$

The dual processes $(q_t^N)_{t\geq 0}$ $(N\in\mathbb{N})$ can be described as solution of the following equation:

- initial condition:
$$q_0^N \equiv 0 \quad , \tag{9.26}$$
- evolution at non-jump times: for $N^kT - t \notin \left\{N^k t_1, ..., N^k t_n\right\}$:
$$\frac{\partial}{\partial t} q_t^N = \clubsuit_N q_t^N - \gamma_x \bar{Y}_{N^kT-t}^{\infty,k}\left(q_t^N\right)^2 \quad , \tag{9.27}$$
- jumps:
$$\lim_{t'\to N^k(T-t_\ell)+} q_{N^k(T-t_\ell)}^N = \lim_{t'\to N^k(T-t_\ell)-} q_{t'}^N + \varphi_{t_\ell} \quad . \tag{9.28}$$

As an immediate consequence of these equations, we get

$$\forall i\in\Omega_{N,L} : \bar{q}_t^{N,k}(i) = q_t^N(i) \quad \left(t\in\left[0, N^kT\right]\right) \quad , \tag{9.29}$$

where $\bar{q}_t^{N,k}$ is defined as a k-level average (see (9.24)). Equation (9.29) is an immediate consequence of the fact that $q_t^N(i)$ only depends on k-level averages of the catalyst, not on its microscopic structure.

We obtain that the following is equivalent to our claim (9.14):

$$\lim_{N\to\infty} \mathbf{E}_{cat}\left[\left|\exp\left(-\sum_{i\in\Omega_{N,L}} q_{N^kT}^N(i)\,\theta_0^x\right)\right.\right.$$
$$\left.\left. -\mathbf{E}_{reac}^Y\left[\exp\left(-\left\langle u_{N^kT}^N, X_0^N\right\rangle\right)\right]\right|\right] = 0 \quad . \tag{9.30}$$

Before we give a proof of this limit property, we want to provide a list of tools which we are going to use:

- Skorohod representation: This enables us to use almost surely convergent catalyst processes.
- Feynman-Kac representation: It is possible to describe dual processes in terms of Feynman-Kac representations.
- Coupling of random walks: To the Feynman-Kac representation, there correspond continuous time random walks. We compare processes characterized by Feynman-Kac representations by defining the corresponding random walks on a common probability space. This enables us to couple the random walks such that

– the random walks evolve together after they have met,
– the random time at which the random walks meet is "small" in expectation.

We have already carried out this program to prove Theorem 1. In fact, the main difference between the two proofs consists of the estimation of the coupling time of the considered continuous time random walks.

The proof of (9.30) consists of two steps:

1.

$$\lim_{N\to\infty} \mathbf{E}_{cat}\left[\left|\exp\left(-\sum_{i\in\Omega_{N,L}} q^N_{N^kT}(i)\cdot\theta^x_0\right) - \exp\left(-\sum_{i\in\Omega_{N,L}} u^N_{N^kT}(i)\cdot\theta^x_0\right)\right|\right] = 0 \quad . \tag{9.31}$$

It is sufficient to compare the arguments of the exponentials. We prove

$$\lim_{N\to\infty} \mathbf{E}_{cat}\left[\left|\sum_{i\in\Omega_{N,L}} \left(q^N_{N^kT}(i) - u^N_{N^kT}(i)\right)\right| \wedge 1\right] = 0 \quad . \tag{9.32}$$

2.

$$\lim_{N\to\infty} \mathbf{E}_{cat}\left[\left|\exp\left(-\sum_{i\in\Omega_{N,L}} u^N_{N^kT}(i)\cdot\theta^x_0\right) - \exp\left(-\sum_{i\in\Omega_{N,L}} u^N_{N^kT}(i)\, X^N_0(i)\right)\right|\right] = 0 \quad . \tag{9.33}$$

We only have to verify

$$\lim_{N\to\infty} \mathbf{E}\left[\left|\sum_{i\in\Omega_{N,L}} u^N_{N^kT}(i)\cdot\left(X^N_0(i) - \theta^x_0\right)\right|\right] = 0 \quad . \tag{9.34}$$

The proof of these steps is contained in the next two subsections.
Equation (9.32) is proven in 9.1.5, equation (9.34) is proven in 9.1.6 (p. 199pp).

9.1.5 Convergence of dual processes – Proof of (9.32)

First, we recall the differential equations we consider, namely (9.20) and (9.27):

$$\frac{\partial}{\partial t}u^N_t = \clubsuit_N u^N_t - \gamma_x Y^N_{N^kT-t}\left(u^N_t\right)^2 \quad , \tag{9.35}$$

$$\frac{\partial}{\partial t}q^N_t = \clubsuit_N q^N_t - \gamma_x \bar{Y}^{\infty,k}_{N^kT-t}\left(q^N_t\right)^2 \quad . \tag{9.36}$$

$$\left(N^k(T-t) \notin \{t_1,...,t_n\}\right)$$

We define

$$\mathbf{\Delta}^N := u^N - q^N \quad .$$

Note that the discontinuities of $\left(u_t^N\right)_{t\geq 0}$ and $\left(q_t^N\right)_{t\geq 0}$ coincide. Therefore, $\mathbf{\Delta}^N$ is continuous and piecewise differentiable. For $N^k(T-t) \notin \{t_1, ..., t_n\}$ we have

$$\begin{aligned}\frac{\partial}{\partial t}\left(\sum_{i\in B_k(0)} \mathbf{\Delta}_t^N(i)\right) &= \sum_{i\in B_k(0)} \clubsuit_N \mathbf{\Delta}_t^N(i) \\ &\quad -\gamma_x \sum_{i\in B_k(0)} Y_{N^kT-t}^N(i)\left(u_t^N(i)\right)^2 \\ &\quad +\gamma_x \sum_{i\in B_k(0)} \bar{Y}_{N^kT-t}^{\infty,k}\left(q_t^N(i)\right)^2 \end{aligned} \tag{9.37}$$

By a telescopic expansion, we get:

$$\begin{aligned}&\frac{\partial}{\partial t}\left(\sum_{i\in B_k(0)} \mathbf{\Delta}_t^N(i)\right) \\ &= \sum_{i\in B_k(0)} \clubsuit_N \mathbf{\Delta}_t^N(i) \\ &\quad + \gamma_x \sum_{i\in B_k(0)} \left(-Y_{N^kT-t}^N(i)\left(\bar{u}_t^{N,k}(i)\right)^2 \quad + Y_{N^kT-t}^N(i)\left[\left(\bar{u}_t^{N,k}(i)\right)^2 - \left(u_t^N(i)\right)^2\right]\right) \\ &\quad + \gamma_x \sum_{i\in B_k(0)} \left(\bar{Y}_{N^kT-t}^{N,k}\left(q_t^N(i)\right)^2 \quad + \left(\bar{Y}_{N^kT-t}^{N,\infty} - \bar{Y}_{N^kT-t}^{N,k}\right)\left(q_t^N(i)\right)^2\right) \end{aligned} \tag{9.38}$$

Rearranging terms, we get:

$$\begin{aligned}&= \sum_{i\in B_k(0)} \clubsuit_N \mathbf{\Delta}_t^N(i) \\ &\quad - \gamma_x \bar{Y}_{N^kT-t}^{N,k} N^k\left(\left(\bar{u}_t^{N,k}(0)\right)^2 - \left(\bar{q}_t^{N,k}(0)\right)^2\right) \\ &\quad + \gamma_x \sum_{i\in B_k(0)} Y_{N^kT-t}^N(i)\left[\left(\bar{u}_t^{N,k}(0)\right)^2 - \left(u_t^N(i)\right)^2\right] \\ &\quad + \gamma_x \sum_{i\in B_k(0)} \left(\bar{Y}_{N^kT-t}^{N,\infty} - \bar{Y}_{N^kT-t}^{N,k}\right)\left(q_t^N(i)\right)^2 \end{aligned} \tag{9.39}$$

Next, we apply "$x^2 - y^2 = (x+y)(x-y)$"

$$\begin{aligned}&= \sum_{i\in B_k(0)} \clubsuit_N \mathbf{\Delta}_t^N(i) \\ &\quad -\gamma_x \bar{Y}_{N^kT-t}^{N,k} N^k\left(\left(\bar{u}_t^{N,k}(0)\right) + \left(\bar{q}_t^{N,k}(0)\right)\right)\left(\left(\bar{u}_t^{N,k}(0)\right) - \left(\bar{q}_t^{N,k}(0)\right)\right) \\ &\quad +\gamma_x \sum_{i\in B_k(0)} Y_{N^kT-t}^N(i)\left[\left(\bar{u}_t^{N,k}(0)\right)^2 - \left(u_t^N(i)\right)^2\right] \\ &\quad +\gamma_x \sum_{i\in B_k(0)} \left(\bar{Y}_{N^kT-t}^{N,\infty} - \bar{Y}_{N^kT-t}^{N,k}\right)\left(q_t^N(i)\right)^2 \end{aligned} \tag{9.40}$$

This way, another "$\mathbf{\Delta}$" appears on the right hand side

$$
\begin{aligned}
= & \sum_{i \in B_k(0)} \clubsuit_N \mathbf{\Delta}_t^N(i) \\
& -\gamma_x \bar{Y}^{N,k}_{N^kT-t} \left(\left(\bar{u}_t^{N,k}(0) \right) + \left(\bar{q}_t^{N,k}(0) \right) \right) \left(\sum_{i \in B_k(0)} \mathbf{\Delta}_t^N(i) \right) \\
& +\gamma_x \sum_{i \in B_k(0)} Y^N_{N^kT-t}(i) \left[\left(\bar{u}_t^{N,k} \right)^2 - \left(u_t^N(i) \right)^2 \right] \\
& +\gamma_x \sum_{i \in B_k(0)} \left(\bar{Y}^{N,\infty}_{N^kT-t} - \bar{Y}^{N,k}_{N^kT-t} \right) \left(q_t^N(i) \right)^2 \quad .
\end{aligned}
\tag{9.41}
$$

By the triangular inequality, the following limits are sufficient for (9.32):

- $$\lim_{N \to \infty} \mathbf{E}_{cat} \left[\left| \sum_{i \in B_k(0)} \mathbf{\Delta}^N_{N^kT}(i) \right| \wedge 1 \right] = 0 \quad . \tag{9.42}$$

- $$\lim_{N \to \infty} \mathbf{E}_{cat} \left[\left| \sum_{i \in \Omega_{N,L} \setminus B_k(0)} \mathbf{\Delta}^N_{N^kT}(i) \right| \wedge 1 \right] = 0 \quad . \tag{9.43}$$

Similar to the calculations carried out in 9.2.8, we approximate $|\cdot|$ by the sequence $(g_n)_{n \in \mathbb{N}}$ given by

$$g_n : x \longmapsto \frac{1}{n} \ln\left(\cosh\left(nx\right)\right) \quad . \tag{9.44}$$

This sequence has the properties (compare page 93)

- $g_n'(x) = \tanh(x)$.
- $g_n(x) \leq x g_n'(x) \leq |x| \leq \frac{1}{n} \ln 2 + g_n(x)$.

We get (recall the definition of $\clubsuit_N$):

$$\begin{aligned}
&\frac{\partial}{\partial t}\left(g_n\left(\sum_{i\in B_k(0)}\boldsymbol{\Delta}_t^N(i)\right)\right)\\
=\;& g_n'\left(\sum_{i\in B_k(0)}\boldsymbol{\Delta}_t^N(i)\right)\\
&\cdot\left[\underbrace{\sum_{i\in B_k(0)}\sum_{\ell=1}^{k}\frac{c_{\ell-1}}{N^{\ell-1}}\left[\bar{\boldsymbol{\Delta}}_t^{N,\ell}(i)-\boldsymbol{\Delta}_t^N(i)\right]}_{=0}+\sum_{i\in B_k(0)}\sum_{\ell=k+1}^{L}\frac{c_{\ell-1}}{N^{\ell-1}}\left[\bar{\boldsymbol{\Delta}}_t^{N,\ell}(i)\underline{\underline{-\boldsymbol{\Delta}_t^N(i)}}\right]\right]\\
&\underline{\underline{-g_n'\left(\sum_{i\in B_k(0)}\boldsymbol{\Delta}_t^N(i)\right)\left(\sum_{i\in B_k(0)}\boldsymbol{\Delta}_t^N(i)\right)\gamma_x\bar{Y}_{N^kT-t}^{N,k}\left(\left(\bar{u}_t^{N,k}(0)\right)+\left(\bar{q}_t^{N,k}(0)\right)\right)}}\\
&+g_n'\left(\sum_{i\in B_k(0)}\boldsymbol{\Delta}_t^N(i)\right)\cdot\gamma_x\sum_{i\in B_k(0)}Y_{N^kT-t}^N(i)\left[\left(\bar{u}_t^{N,k}\right)^2-\left(u_t^N(i)\right)^2\right]\\
&+g_n'\left(\sum_{i\in B_k(0)}\boldsymbol{\Delta}_t^N(i)\right)\cdot\gamma_x\sum_{i\in B_k(0)}\left(\bar{Y}_{N^kT-t}^{N,\infty}-\bar{Y}_{N^kT-t}^{N,k}\right)\left(q_t^N(i)\right)^2 \quad ,
\end{aligned}\tag{9.45}$$

where $\bar{\boldsymbol{\Delta}}_t^{N,k}(i)$ is defined by $\bar{\boldsymbol{\Delta}}_t^{N,\ell}(i) := \frac{1}{N^\ell}\sum_{\varkappa\in B_\ell(i)}\boldsymbol{\Delta}_t^N(\varkappa)$.

Note that the underbraced term is zero.

Applying

- $g_n'(\cdot)\leq 1$,
- $\forall x\in\mathbb{R}:\quad xg_n'(x)\geq 0$,

we get an upper bound by leaving out the underlined terms in (9.45):

$$\frac{\partial}{\partial t}\left(g_n\left(\sum_{i\in B_k(0)}\boldsymbol{\Delta}_t^N(i)\right)\right)$$

$$\leq \sum_{i\in B_k(0)}\sum_{\ell=k+1}^{L}\frac{c_{\ell-1}}{N^{\ell-1}}\left[\bar{\boldsymbol{\Delta}}_t^{N,\ell}(i)\right] \tag{9.46}$$

$$+\gamma_x\left|\sum_{i\in B_k(0)}Y_{N^kT-t}^N(i)\left[\left(\bar{u}_t^{N,k}\right)^2-\left(u_t^N(i)\right)^2\right]\right| \tag{9.47}$$

$$+\gamma_x\left|\sum_{i\in B_k(0)}\left(\bar{Y}_{N^kT-t}^{N,\infty}-\bar{Y}_{N^kT-t}^{N,k}\right)\left(q_t^N(i)\right)^2\right| \quad . \tag{9.48}$$

To obtain (9.42) we have to prove:

α)

$$\lim_{N\to\infty} \mathbf{E}_{cat}\left[\int_0^{N^kT}\left|\sum_{i\in B_k(0)}\sum_{\ell=k+1}^{L}\frac{c_{\ell-1}}{N^{\ell-1}}\left[\bar{\mathbf{\Delta}}_t^{N,\ell}(i)\right]\right|dt\right]=0 \quad . \tag{9.49}$$

β)

$$\lim_{N\to\infty} \mathbf{E}_{cat}\left[\int_0^{N^kT}\left|\sum_{i\in B_k(0)} Y^N_{N^kT-t}(i)\left[\left(\bar{u}_t^{N,k}\right)^2-\left(u_t^N(i)\right)^2\right]\right|dt\right]=0 \quad . \tag{9.50}$$

γ)

$$\lim_{N\to\infty} \mathbf{E}_{cat}\left[\int_0^{N^kT}\sum_{i\in B_k(0)}\left|\bar{Y}^{N,\infty}_{N^kT-t}-\bar{Y}^{N,k}_{N^kT-t}\right|\left(q_t^N(i)\right)^2 dt\wedge 1\right]=0 \quad . \tag{9.51}$$

In addition to α, β, and γ we have to prove (9.43), which is given by:

δ)

$$\lim_{N\to\infty} \mathbf{E}_{cat}\left[\left|\sum_{i\in\Omega_{N,L}\setminus B_k(0)}\left(q^N_{N^kT}(i)-u^N_{N^kT}(i)\right)\right|\wedge 1\right]=0 \quad . \tag{9.52}$$

The proofs of these four equations are carried out in the following order: $\alpha,\delta,\gamma,\beta$. [first the short proofs, then the longer ones]

Proof of (9.49)

We have to prove

$$\limsup_{N\to\infty} \mathbf{E}_{cat}\left[\int_0^{N^kT}\left|\sum_{i\in B_k(0)}\sum_{\ell=k+1}^{L}\frac{c_{\ell-1}}{N^{\ell-1}}\left[\bar{\mathbf{\Delta}}_t^{N,\ell}(i)\right]\right|dt\right]=0 \quad . \tag{9.53}$$

It is an immediate consequence of (9.23) and the equations defining u^N and q^N that we have for $\ell\in\{1,...,L\}$

$$\bar{\mathbf{\Delta}}_t^{N,\ell}(i)\le\frac{1}{N^\ell} \quad . \tag{9.54}$$

We hence get, continuing (9.53):

$$\begin{aligned}
&\le \limsup_{N\to\infty}\int_0^{N^kT}\left|\sum_{i\in B_k(0)}\sum_{\ell=k+1}^{L}\frac{c_{\ell-1}}{N^{\ell-1}}\left[\frac{1}{N^\ell}\right]\right|dt\\
&\le \limsup_{N\to\infty} N^kT\cdot N^k\sum_{\ell=k+1}^{L}\frac{c_{\ell-1}}{N^{2\ell-1}}\\
&= \limsup_{N\to\infty} T\cdot\sum_{\ell=1}^{L-k}\frac{c_{\ell-1}}{N^{2\ell-1}}=0 \quad .
\end{aligned} \tag{9.55}$$

Proof of (9.52)

Next, we prove (δ), namely

$$\lim_{N\to\infty}\mathbf{E}_{cat}\left[\left|\sum_{i\in\Omega_{N,L}\setminus B_k(0)}\left(q^N_{N^kT}(i)-u^N_{N^kT}(i)\right)\right|\wedge 1\right]=0 \quad . \tag{9.56}$$

By calculations similar to those carried out deriving (6.161), p. 95 it turns out that

$$\lim_{N\to\infty}\mathbf{E}_{cat}\left[\int_0^{N^kT}\sum_{i\in\Omega_{N,L}\setminus B_k(0)}\left|Y^N_{N^kT-t}(i)-\bar{Y}^{\infty,k}_{N^kT-t}\right|\left(\left(u^N_t(i)\right)^2+\left(q^N_t(i)\right)^2\right)dt\wedge 1\right]=0 \quad . \tag{9.57}$$

is sufficient for (9.52).

By (9.172) and the evolution equations of u^N and q^N we have for all $i\in B_{m+1}(0)\setminus B_m(0)$ with $m\in\mathbb{N}$

$$\sup_{t\in[0,T]}u^N_{N^kt}(i)\le\frac{1}{N^{m+1}}\text{ and }\sup_{t\in[0,T]}q^N_{N^kt}(i)\le\frac{1}{N^{m+1}} \tag{9.58}$$

Applying this to (9.57), we get:

$$\begin{aligned}
&\lim_{N\to\infty}\mathbf{E}_{cat}\left[\int_0^{N^kT}\sum_{i\in\Omega_{N,L}\setminus B_k(0)}\left|Y^N_{N^kT-t}(i)-\bar{Y}^{\infty,k}_{N^kT-t}\right|\left(\left(u^N_t(i)\right)^2+\left(q^N_t(i)\right)^2\right)dt\wedge 1\right]\\
&\le\lim_{N\to\infty}\mathbf{E}_{cat}\left[\int_0^{N^kT}\sum_{i\in\Omega_{N,L}\setminus B_k(0)}\left(Y^N_{N^kT-t}(i)+\bar{Y}^{\infty,k}_{N^kT-t}\right)\left(\left(u^N_t(i)\right)^2+\left(q^N_t(i)\right)^2\right)dt\wedge 1\right]\\
&\le\lim_{N\to\infty}\mathbf{E}_{cat}\left[\int_0^{N^kT}\sum_{m=k}^{\infty}\sum_{i\in B_{m+1}(0)\setminus B_m(0)}\left(Y^N_{N^kT-t}(i)+\bar{Y}^{\infty,k}_{N^kT-t}\right)\cdot 2\cdot\left(\frac{1}{N^{m+1}}\right)^2dt\wedge 1\right]\\
&\le\lim_{N\to\infty}\sum_{m=k}^{\infty}N^{m+1}N^kT\left(2\theta_y\right)\cdot 2\cdot\left(\frac{1}{N^{m+1}}\right)^2\\
&\le\lim_{N\to\infty}4T\theta_y\sum_{m=k}^{\infty}\frac{N^k}{N^{m+1}}\\
&\le\lim_{N\to\infty}\frac{4T\theta_y}{N-1}=0 \quad .
\end{aligned}\tag{9.59}$$

Proof of (9.51)

We have to verify equation (9.51), namely

$$\lim_{N\to\infty}\mathbf{E}_{cat}\left[\int_0^{N^kT}\sum_{i\in B_k(0)}\left|\bar{Y}^{N,\infty}_{N^kT-t}-\bar{Y}^{N,k}_{N^kT-t}\right|\left(q^N_t(i)\right)^2dt\wedge 1\right]=0 \quad . \tag{9.60}$$

With

$$q_t^N(i) = \bar{q}_t^{N,k}(i) \tag{9.61}$$

and (compare (9.23))

$$\bar{q}_t^{N,k}(0) \leq \frac{1}{N^k} \tag{9.62}$$

we get:

$$\begin{aligned}
&\limsup_{N\to\infty} \mathbf{E}_{cat}\left[\int_0^{N^kT} \sum_{i\in B_k(0)} \left|\bar{Y}^{N,\infty}_{N^kT-t} - \bar{Y}^{N,k}_{N^kT-t}\right| \left(q_t^N(i)\right)^2 dt \wedge 1\right] \\
= \ &\limsup_{N\to\infty} \mathbf{E}_{cat}\left[\int_0^{N^kT} \sum_{i\in B_k(0)} \left|\bar{Y}^{N,\infty}_{N^kT-t} - \bar{Y}^{N,k}_{N^kT-t}\right| \left(\bar{q}_t^{N,k}(0)\right)^2 dt \wedge 1\right] \\
\leq \ &\limsup_{N\to\infty} \mathbf{E}_{cat}\left[\int_0^{N^kT} N^k \left|\bar{Y}^{N,\infty}_{N^kT-t} - \bar{Y}^{N,k}_{N^kT-t}\right| \left(\frac{1}{N^k}\right)^2 dt \wedge 1\right] \qquad (9.63) \\
= \ &\limsup_{N\to\infty} \mathbf{E}_{cat}\left[\frac{1}{N^k}\int_0^{N^kT} \left|\bar{Y}^{N,\infty}_{N^kT-t} - \bar{Y}^{N,k}_{N^kT-t}\right| dt \wedge 1\right] \qquad (9.64)
\end{aligned}$$

By assumption (9.8) which is based on Skorohod's representation, we have

$$\lim_{N\to\infty} \sup_{t\in[0,N^kT]} \left|\bar{Y}_t^{N,\infty} - \bar{Y}_t^{N,k}\right| = 0 \quad P_{cat}\text{-a.s.} \tag{9.65}$$

With dominated convergence (that's why we have to use the "$\wedge 1$") we get

$$\limsup_{N\to\infty} \mathbf{E}_{cat}\left[\frac{1}{N^k}\int_0^{N^kT} \left|\bar{Y}^{N,\infty}_{N^kT-t} - \bar{Y}^{N,k}_{N^kT-t}\right| dt \wedge 1\right] = 0 \quad . \tag{9.66}$$

Proof of (9.50)

It remains to prove (9.50), namely

$$\lim_{N\to\infty} \mathbf{E}_{cat}\left[\int_0^{N^kT} \left| \sum_{i\in B_k(0)} Y^N_{N^kT-t}(i) \left[\left(\bar{u}_t^{N,k}\right)^2 - \left(u_t^N(i)\right)^2\right]\right| dt\right] = 0 \quad . \tag{9.67}$$

We write the difference of squares as a product – our claim becomes

$$\lim_{N\to\infty} \mathbf{E}_{cat}\left[\int_0^{N^kT} \left| \sum_{i\in B_k(0)} Y^N_{N^kT-t}(i) \left(u_t^N(i) - \bar{u}_t^{N,k}\right)\left(u_t^N(i) + \bar{u}_t^{N,k}\right)\right| dt\right] = 0 \quad . \tag{9.68}$$

With (9.23) we have

$$u_t^N(i) + \bar{u}_t^{N,k} \le \frac{2}{N^k} \quad (i \in B_k(0)). \tag{9.69}$$

Hence, there only remains to prove:

$$\lim_{N\to\infty} \mathbf{E}_{cat}\left[\frac{1}{N^k}\int_0^{N^kT}\left|\sum_{i\in B_k(0)} Y^N_{N^kT-t}(i)\left(u_t^N(i) - \bar{u}_t^{N,k}\right)\right| dt\right] = 0 \quad . \tag{9.70}$$

It suffices to show

$$\lim_{N\to\infty} \sup_{t\in[0,N^kT]} \mathbf{E}_{cat}\left[\left|\sum_{i\in B_k(0)} Y^N_{N^kT-t}(i)\left(u_t^N(i) - \bar{u}_t^{N,k}\right)\right|\right] = 0 \quad . \tag{9.71}$$

The first step is to express $u_t^N(i)$ and $\bar{u}_t^{N,k}$ by means of a Feynman-Kac representation. A key ingredient of such formulas is a Markov process (in this case: a continuous time random walk). The main tool used to compare $u_t^N(i)$ and $\bar{u}_t^{N,k}$ is to use a coupling of the random walks occuring in the corresponding Feynman-Kac representations [3].

Ingredients of the Feynman-Kac representation:

Auxiliary random walk:

- Let
$$\left(\tilde{\Omega}, \tilde{\mathcal{F}}, \mathbb{P}\right)$$
be a probability space with filtration $\left(\tilde{\mathcal{F}}_t\right)_{t\ge 0}$.
- Let
$$\left(\tilde{\zeta}_t^{0,N}\right)_{t\ge 0}$$
be a random walk attaining values in $\Omega_{L,N}$ and being adapted to $\left(\tilde{\mathcal{F}}_t\right)_{t\ge 0}$. The initial condition of this random walk is given by $\tilde{\zeta}_0^{0,N} = 0$, its jump rates are given by $\clubsuit_N$.

Stopping time: first jump over a distance exceeding $k-2$

- Let ϑ be the first time the process $\tilde{\zeta}^{i,N}$ leaves $B_{k-1}(i)$:

$$\vartheta := \inf\left\{t \ge 0 : \tilde{\zeta}_t^{0,N} \notin B_{k-1}(0)\right\} \quad . \tag{9.72}$$

[3]The derivation of the Feynman-Kac representation interrupts the thread of our argumentation. The discussion of (9.71) is continued on page 196.

The two random walks we need On the probability space $\left(\tilde{\Omega}, \tilde{\mathcal{F}}, \mathbb{P}\right)$ we define another two random walks. These random walks are denoted by $\zeta^{i,N}$ and $\bar{\zeta}^{N}$.

- Let $\zeta^{i,N}$ be defined by

$$\zeta_t^{i,N} = \begin{cases} \tilde{\zeta}_t^{0,N} + i & \text{for } t < \vartheta \\ \tilde{\zeta}_t^{0,N} & \text{for } t \geq \vartheta \end{cases} \quad (i \in B_k(0)) \quad . \tag{9.73}$$

  Note that the transition rates of $\zeta^{i,N}$ are given by $\clubsuit_N$.

- For the definition of $\bar{\zeta}^{N}$ we need a random permutation. Namely, let $\pi_{N,k} \in (\Omega_{L,N})^{\Omega_{L,N}}$ be a random permutation of the hierarchical group $\Omega_{L,N}$. We assume that the law $\mathbb{P}\circ(\pi_{N,k})^{-1}$ has the following properties:

  - The random variables $\pi_{N,k}$ and $\left(\tilde{\zeta}_t^{0,N}\right)_{t\geq 0}$ are independent.
  - Let W be the set of permutations of $(\Omega_{L,N})^{\Omega_{L,N}}$ under which $\Omega_{L,N}\backslash B_k(0)$ contains only fixed points:

$$W := \left\{\pi \in (\Omega_{L,N})^{\Omega_{L,N}} : \pi \text{ bijective and } \forall j \in \Omega_{L,N}\backslash B_k(0) : \pi(j) = j\right\} \quad . \tag{9.74}$$

    Let $\mathbb{P}\circ(\pi_{N,k})^{-1}$ be the uniform distribution on W. (In particular, every permutation contained in W has probability $\left(N^k!\right)^{-1}$)

  Let $\bar{\zeta}^{N}$ be defined by

$$\bar{\zeta}_t^{N} = \begin{cases} \pi_{N,k}\left(\tilde{\zeta}_t^{0,N}\right) & \text{for } t < \vartheta \\ \tilde{\zeta}_t^{0,N} & \text{for } t \geq \vartheta \end{cases} \quad . \tag{9.75}$$

Summary: Things probably get clearer if we express in words what we have done so far: We have introduced random walks $\zeta^{i,N}$ $(i \in B_k(0))$, $\bar{\zeta}_t^{N}$, all with generator $\clubsuit_N$, but with different initial conditions. These random walks live on a common probability space and they are coupled in such a way that they jump all at the same times, and meet and stay together after the first jump over a distance of $k-1$ – this is typically as soon as possible. The rate of such jumps is well-known, namely $\frac{c_{k-1}}{N^{k-1}} \cdot \left(1 - \frac{1}{N}\right)$.

Lemma 12 *The following formulas of Feynman-Kac type hold for $\tilde{t} \in [0, T]$:*

$$\begin{aligned} & u_{N^k\tilde{t}}^N(i) \\ &= \mathbb{E}\left[\sum_{\ell=1}^{n} \frac{a_\ell}{N^k} \mathbb{I}_{\left\{\zeta^{i,N}_{N^k(\tilde{t}-(T-t_\ell))} \in B_k(0)\right\}} \mathbb{I}_{\{\tilde{t}>(T-t_\ell)\}}\right. \\ &\quad \left. \cdot \exp\left(-\int_0^{N^k(\tilde{t}-(T-t_\ell))} u_{N^k\tilde{t}-r}^N\left(\zeta_r^{i,N}\right) Y_{N^k(T-\tilde{t})+r}^N\left(\zeta_r^{i,N}\right) dr\right)\right] \end{aligned} \tag{9.76}$$

and

$$\begin{aligned}
&\bar{u}^{N,k}_{N^k\bar{t}}(0)\\
&= \mathbb{E}\left[\sum_{\ell=1}^{n} \frac{a_\ell}{N^k} \mathbb{I}_{\left\{\bar{\zeta}^N_{N^k(\bar{t}-(T-t_\ell))} \in B_k(0)\right\}} \mathbb{I}_{\{\bar{t}>(T-t_\ell)\}} \right.\\
&\quad \left. \cdot \exp\left(-\int_0^{N^k(\bar{t}-(T-t_\ell))} u^N_{N^k\bar{t}-r}\left(\bar{\zeta}^N_r\right) Y^N_{N^k(T-\bar{t})+r}\left(\bar{\zeta}^N_r\right) dr\right)\right] \quad . \qquad (9.77)
\end{aligned}$$

Remark 19 *Note that this Feynman-Kac representation is consistent with the Feynman-Kac representation we have made use of when proving the finite system scheme (compare Lemma 7, p. 117).*

Proof. We only give a proof of (9.76). The second formula (9.77) can be obtained by similar considerations.

Define for $i \in \Omega_{L,N}$

$$\begin{aligned}
\tilde{u}^N_t(i) := \mathbb{E}&\left[\sum_{\ell=1}^{n} \frac{a_\ell}{N^k} \mathbb{I}_{\left\{\zeta^{i,N}_{(t-N^k(T-t_\ell))} \in B_k(0)\right\}} \mathbb{I}_{\{t>N^k(T-t_\ell)\}} \right.\\
&\left. \cdot \exp\left(-\int_0^{t-N^k(T-t_\ell)} u^N_{t-r}\left(\zeta^{i,N}_r\right) Y^N_{N^kT-t+r}\left(\zeta^{i,N}_r\right) dr\right)\right] \quad . \qquad (9.78)
\end{aligned}$$

We want to obtain $u^N_t = \tilde{u}^N_t$. For this purpose, we have to verify:

1. $u^N_0 = \tilde{u}^N_0$.
2. for $t \in \left\{N^k(T-t_1), ..., N^k(T-t_n)\right\}$:

$$\lim_{\varsigma \searrow t} \tilde{u}^N_\varsigma - \lim_{\varsigma \nearrow t} \tilde{u}^N_\varsigma = \lim_{\varsigma \searrow t} u^N_\varsigma - \lim_{\varsigma \nearrow t} u^N_\varsigma \quad . \qquad (9.79)$$

3. for $t \notin \left\{N^k(T-t_1), ..., N^k(T-t_n)\right\}$:

$$\frac{\partial}{\partial t}\tilde{u}^N_t = \frac{\partial}{\partial t}u^N_t \quad . \qquad (9.80)$$

Since the right hand side is well-defined, this suffices.

Proof of 1. By assumption, we have $t_\ell < T$ for all $\ell \in \{1, ..., n\}$. Therefore $\mathbb{I}_{\{0>N^k(T-t_\ell)\}}$ vanishes for all $\ell \in \{1, ..., n\}$. This yields (compare (9.78))

$$u^N_0 = \tilde{u}^N_0 = 0 \quad . \qquad (9.81)$$

Proof of 2. For $t = N^k\left(T - t_{\tilde{\ell}}\right)$ with $\tilde{\ell} \in \{1, ..., n\}$ we have

$$\begin{aligned}
&\lim_{\varsigma\searrow t} \tilde{u}_t^N - \lim_{\varsigma\nearrow t} \tilde{u}_\varsigma^N \\
&= \lim_{\varsigma\searrow t} \mathbb{E}\left[\sum_{\ell=1}^{n} \frac{a_\ell}{N^k} \mathbb{I}_{\left\{\zeta^{i,N}_{(t-N^k(T-t_\ell))} \in B_k(0)\right\}} \mathbb{I}_{\{t>N^k(T-t_\ell)\}}\right. \\
&\quad \left.\cdot \exp\left(-\int_0^{t-N^k(T-t_\ell)} u^N_{t-r}\left(\zeta^{i,N}_r\right) Y^N_{N^kT-\tilde{t}+r}\left(\zeta^{i,N}_r\right) dr\right)\right] \\
&\quad - \lim_{\varsigma\nearrow t} \mathbb{E}\left[\sum_{\ell=1}^{n} \frac{a_\ell}{N^k} \mathbb{I}_{\left\{\zeta^{i,N}_{(\varsigma-N^k(T-t_\ell))} \in B_k(0)\right\}} \mathbb{I}_{\{\varsigma>N^k(T-t_\ell)\}}\right. \\
&\quad \left.\cdot \exp\left(-\int_0^{\varsigma-N^k(T-t_\ell)} u^N_{\varsigma-r}\left(\zeta^{i,N}_r\right) Y^N_{N^kT-\varsigma+r}\left(\zeta^{i,N}_r\right) dr\right)\right] \\
&= \mathbb{E}\left[\frac{a_{\tilde{\ell}}}{N^k} \mathbb{I}_{\left\{\zeta^{i,N}_{(t-N^k(T-t_\ell))} \in B_k(0)\right\}}\right] \\
&= \mathbb{E}\left[\frac{a_{\tilde{\ell}}}{N^k} \mathbb{I}_{\left\{\zeta^{i,N}_0 \in B_k(0)\right\}}\right] = \frac{a_{\tilde{\ell}}}{N^k} \mathbb{I}_{\{i\in B_k(0)\}}(i) \quad .
\end{aligned} \tag{9.82}$$

This coincides with $\lim_{\varsigma\searrow t} u_t^N - \lim_{\varsigma\nearrow t} u_\varsigma^N$.

Proof of 3. Let $t \in \left[0, N^kT\right] \setminus \left\{N^k\left(T - t_1\right), ..., N^k\left(T - t_n\right)\right\}$.
We have to consider

$$\begin{aligned}
&\frac{\partial}{\partial t} \tilde{u}_t^N(i) \\
&= \frac{\partial}{\partial t} \mathbb{E}\left[\sum_{\ell=1}^{n} \frac{a_\ell}{N^k} \mathbb{I}_{\left\{\zeta^{i,N}_{(t-N^k(T-t_\ell))} \in B_k(0)\right\}} \mathbb{I}_{\{t>N^k(T-t_\ell)\}}\right. \\
&\quad \left.\cdot \exp\left(-\int_0^{t-N^k(T-t_\ell)} u^N_{t-r}\left(\zeta^{i,N}_r\right) Y^N_{N^kT-t+r}\left(\zeta^{i,N}_r\right) dr\right)\right] \quad .
\end{aligned} \tag{9.83}$$

For $i, j \in \Omega_{N,L}, \ell \in \{1, ..., n\}$, we introduce conditioned random walks $\zeta^{i\to j,N,\ell}_{s\in[0,t-N^k(T-t_\ell)]}$ by conditioning on the final state

$$\begin{aligned}
&\mathcal{L}\left\{\left(\zeta^{i\to j,N,\ell}_s\right)_{s\in[0,t-N^k(T-t_\ell)]}\right\} \\
&:= \mathcal{L}\left\{\left(\zeta^{i,N}_s\right)_{s\in[0,t-N^k(T-t_\ell)]} \middle| \zeta^{i,N}_{t-N^k(T-t_\ell)} = j\right\} \quad .
\end{aligned} \tag{9.84}$$

Using this processes, we can write:

$$\begin{aligned}
&\tilde{u}_t^N(i) \\
&= \mathbb{E}\left[\sum_{\ell=1}^{n} \frac{a_\ell}{N^k} \sum_{j\in B_k(0)} \mathbb{P}\left[\zeta^{i,N}_{(t-N^k(T-t_\ell))} = j\right] \mathbb{I}_{\{t>N^k(T-t_\ell)\}}\right. \\
&\quad \left.\cdot \exp\left(-\int_0^{t-N^k(T-t_\ell)} u^N_{t-r}\left(\zeta^{i\to j,N,\ell}_r\right) Y^N_{N^kT-t+r}\left(\zeta^{i\to j,N,\ell}_r\right) dr\right)\right] \quad .
\end{aligned} \tag{9.85}$$

Since the transition kernel of $\zeta^{i,N}_{s\in[0,t-N^k(T-t_\ell)]}$ is symmetric (for all i), we have for $t \geq N^k(T-t_\ell)$

$$\mathcal{L}\left\{\left(\zeta_s^{i\to j,N,\ell}\right)_{s\in[0,t-N^k(T-t_\ell)]}\right\} = \mathcal{L}\left\{\left(\zeta^{j\to i,N,\ell}_{t-N^k(T-t_\ell)-s}\right)_{s\in[0,t-N^k(T-t_\ell)]}\right\} \quad . \tag{9.86}$$

Hence we can write in continuation of (9.85)

$$= \mathbb{E}\left[\sum_{\ell=1}^n \frac{a_\ell}{N^k} \sum_{j\in B_k(0)} \mathbb{P}\left[\zeta^{i,N}_{(t-N^k(T-t_\ell))} = j\right] \mathbb{1}_{\{t>N^k(T-t_\ell)\}}\right.$$
$$\left. \cdot \exp\left(-\int_0^{t-N^k(T-t_\ell)} u^N_{t-r}\left(\zeta^{j\to i,N,\ell}_{t-N^k(T-t_\ell)-r}\right) Y^N_{N^kT-t+r}\left(\zeta^{j\to i,N,\ell}_{t-N^k(T-t_\ell)-r}\right) dr\right)\right] \quad . \tag{9.87}$$

Substituting $r \longmapsto t - N^k(T-t_\ell) - r$, we obtain

$$= \mathbb{E}\left[\sum_{\ell=1}^n \frac{a_\ell}{N^k} \sum_{j\in B_k(0)} \mathbb{P}\left[\zeta^{i,N}_{(t-N^k(T-t_\ell))} = j\right] \mathbb{1}_{\{t>N^k(T-t_\ell)\}}\right.$$
$$\left. \cdot \exp\left(-\int_0^{t-N^k(T-t_\ell)} u^N_{N^kT-N^kt_\ell+r}\left(\zeta_r^{j\to i,N,\ell}\right) Y^N_{N^kt_\ell-r}\left(\zeta_r^{j\to i,N,\ell}\right) dr\right)\right]$$
$$= \mathbb{E}\left[\sum_{\ell=1}^n \frac{a_\ell}{N^k} \sum_{j\in B_k(0)} \mathbb{P}\left[\zeta^{i,N}_{(t-N^k(T-t_\ell))} = j\right] \mathbb{1}_{\{t>N^k(T-t_\ell)\}}\right.$$
$$\cdot \frac{\mathbb{1}_{\left\{\zeta^{j,N}_{(t-N^k(T-t_\ell))}=i\right\}}}{\mathbb{P}\left[\zeta^{j,N}_{(t-N^k(T-t_\ell))} = i\right]}$$
$$\left. \cdot \exp\left(-\int_0^{t-N^k(T-t_\ell)} u^N_{N^k(T-t_\ell)+r}\left(\zeta_r^{j,N}\right) Y^N_{N^kt_\ell-r}\left(\zeta_r^{j,N}\right) dr\right)\right] \tag{9.88}$$

With

$$\forall i,j \in B_k(0) : \mathbb{P}\left[\zeta^{j,N}_{(t-N^k(T-t_\ell))} = i\right] = \mathbb{P}\left[\zeta^{i,N}_{(t-N^k(T-t_\ell))} = j\right] \tag{9.89}$$

this equals

$$\mathbb{E}\left[\sum_{\ell=1}^n \frac{a_\ell}{N^k} \sum_{j\in B_k(0)} \mathbb{1}_{\{t\geq N^k(T-t_\ell)\}}\right. \quad .$$
$$\left. \cdot \exp\left(-\int_0^{t-N^k(T-t_\ell)} u^N_{N^k(T-t_\ell)+r}\left(\zeta_r^{j,N}\right) Y^N_{N^kt_\ell-r}\left(\zeta_r^{j,N}\right) dr\right) \mathbb{1}_{\left\{\zeta^{j,N}_{(t-N^k(T-t_\ell))}=i\right\}}\right] \tag{9.90}$$

We have to differentiate with respect to t. We proceed as in the proof of Lemma 7 (p. 117) and get:

$$\begin{aligned}
&\frac{\partial}{\partial t}\mathbb{E}\left[\sum_{\ell=1}^{n}\frac{a_\ell}{N^k}\sum_{j\in B_k(0)}\mathbb{1}_{\{t>N^k(T-t_\ell)\}}\right.\\
&\quad\left.\cdot\exp\left(-\int_0^{t-N^k(T-t_\ell)}u^N_{N^k(T-t_\ell)+r}\left(\zeta^{j,N}_r\right)Y^N_{N^k t_\ell-r}\left(\zeta^{j,N}_r\right)dr\right)\mathbb{1}_{\left\{\zeta^{j,N}_{(t-N^k(T-t_\ell))}=i\right\}}\right]\\
&=\sum_{\ell=1}^{n}\frac{a_\ell}{N^k}\sum_{j\in B_k(0)}\mathbb{1}_{\{t>N^k(T-t_\ell)\}}\\
&\quad\cdot\mathbb{E}\Bigg[-u^N_t\left(\zeta^{j,N}_{t-N^k(T-t_\ell)}\right)Y^N_{N^kT-t}\left(\zeta^{j,N}_{t-N^k(T-t_\ell)}\right)\\
&\qquad\qquad e^{-\int_0^{t-N^k(T-t_\ell)}u^N_{N^k(T-t_\ell)+r}\left(\zeta^{j,N,\ell}_r\right)Y^N_{N^k t_\ell-r}\left(\zeta^{j,N,\ell}_r\right)dr}\\
&\qquad\qquad\mathbb{1}_{\left\{\zeta^{j,N}_{(t-N^k(T-t_\ell))}=i\right\}}\Bigg]\\
&+\sum_{\ell=1}^{n}\frac{a_\ell}{N^k}\sum_{j\in B_k(0)}\mathbb{1}_{\{t\geq N^k(T-t_\ell)\}}\\
&\quad\cdot\mathbb{E}\Bigg[\exp\left(-\int_0^{t-N^k(T-t_\ell)}u^N_{N^k(T-t_\ell)+r}\left(\zeta^{j,N}_r\right)Y^N_{N^k t_\ell-r}\left(\zeta^{j,N}_r\right)dr\right)\\
&\qquad\qquad\cdot\sum_{\tilde{\imath}\in\Omega_{L,N}}\left(a^N(i,\tilde{\imath})\left[\mathbb{1}_{\left\{\zeta^{j,N}_{(t-N^k(T-t_\ell))}=\tilde{\imath}\right\}}-\mathbb{1}_{\left\{\zeta^{j,N}_{(t-N^k(T-t_\ell))}=i\right\}}\right]\right)\Bigg]
\end{aligned}$$

$$
\begin{aligned}
&= \sum_{\ell=1}^{n} \frac{a_\ell}{N^k} \sum_{j\in B_k(0)} \mathbb{1}_{\{t>N^k(T-t_\ell)\}} \\
&\quad \cdot \mathbb{E}\Bigg[-u_t^N(i)\, Y_{N^kT-t}^N(i) \\
&\qquad \cdot e^{-\int_0^{t-N^k(T-t_\ell)} u_{N^k(T-t_\ell)+r}^N\left(\zeta_r^{j,N,\ell}\right) Y_{N^kt_\ell-r}^N\left(\zeta_r^{j,N,\ell}\right) dr} \\
&\qquad \cdot \mathbb{1}_{\left\{\zeta_{(t-N^k(T-t_\ell))}^{j,N}=i\right\}} \Bigg] \\
&+ \sum_{\ell=1}^{n} \frac{a_\ell}{N^k} \sum_{j\in B_k(0)} \mathbb{1}_{\{t>N^k(T-t_\ell)\}} \\
&\quad \cdot \exp\left(-\int_0^{t-N^k(T-t_\ell)} u_{N^k(T-t_\ell)+r}^N\left(\zeta_r^{j,N}\right) Y_{N^kt_\ell-r}^N\left(\zeta_r^{j,N}\right) dr \right) \\
&\quad \cdot \sum_{\tilde{\imath}\in\not\in_{L,N}} \left(a^N(i,\tilde{\imath}) \left[\mathbb{1}_{\left\{\zeta_{(t-N^k(T-t_\ell))}^{j,N}=\tilde{\imath}\right\}} - \mathbb{1}_{\left\{\zeta_{(t-N^k(T-t_\ell))}^{j,N}=i\right\}} \right] \right) \\
&= -\tilde{u}_t^N(i)\, u_t^N(i)\, Y_{N^kT-t}^N(i) + \sum_{\tilde{\imath}\in\Omega_{L,N}} \left(a^N(i,\tilde{\imath}) \left[\tilde{u}_t^N(\tilde{\imath}) - \tilde{u}_t^N(i) \right] \right) \quad . \qquad (9.91)
\end{aligned}
$$

This completes the proof of the lemma.

■

We draw further the line of argumentation interrupted on page 190 by the Feynman-Kac lemma and its proof.

We have to prove (9.71), namely

$$
\lim_{N\to\infty} \sup_{t\in[0,N^kT]} \mathbf{E}_{cat}\left[\left| \sum_{i\in B_k(0)} Y_{N^kT-t}^N(i) \left(u_t^N(i) - \bar{u}_t^{N,k} \right) \right| \right] = 0 \quad . \qquad (9.92)
$$

Applying the Feynman-Kac representation[4], we get:

$$\sum_{i\in B_k(0)} Y^N_{N^k(T-\bar{t})}(i)\left|u^N_{N^k\bar{t}}(i)-\bar{u}^{N,k}_{N^k\bar{t}}(0)\right|$$

$$= \sum_{i\in B_k(0)} Y^N_{N^k(T-\bar{t})}(i)\left|\mathbb{E}\left[\sum_{\ell=1}^{n}\frac{a_\ell}{N^k}\mathbb{1}_{\left\{\bar{\zeta}^N_{N^k(\bar{t}-(T-t_\ell))}\in B_k(0)\right\}}\mathbb{1}_{\{\bar{t}>(T-t_\ell)\}}\right.\right.$$

$$\cdot\left(\exp\left(-\int_0^{N^k(\bar{t}-(T-t_\ell))} u^N_{N^k\bar{t}-r}\left(\zeta^{i,N}_r\right)Y^N_{N^k(T-\bar{t})+r}\left(\zeta^{i,N}_r\right)dr\right)\right.$$

$$\left.\left.\left.-\exp\left(-\int_0^{N^k(\bar{t}-(T-t_\ell))} u^N_{N^k\bar{t}-r}\left(\bar{\zeta}^N_r\right)Y^N_{N^k(T-\bar{t})+r}\left(\bar{\zeta}^N_r\right)dr\right)\right)\right]\right|$$

We apply

- Jensen's inequality: $|\mathbb{E}[\cdot]|\le\mathbb{E}|\cdot|$.
- the inequality $|e^{-x}-e^{-y}|\le|x-y|\quad(x,y\ge 0)$

and get:

$$\le\sum_{\ell=1}^{n}\frac{a_\ell}{N^k}\sum_{i\in B_k(0)}Y^N_{N^k(T-\bar{t})}(i)$$

$$\cdot\mathbb{1}_{\{\bar{t}>(T-t_\ell)\}}\mathbb{E}\left|\mathbb{1}_{\left\{\bar{\zeta}^N_{N^k(\bar{t}-(T-t_\ell))}\in B_k(0)\right\}}\right.$$

$$\cdot\left(\int_0^{N^k(\bar{t}-(T-t_\ell))} u^N_{N^k\bar{t}-r}\left(\zeta^{i,N}_r\right)Y^N_{N^k(T-\bar{t})+r}\left(\zeta^{i,N}_r\right)dr\right. \tag{9.93}$$

$$\left.\left.-\int_0^{N^k(\bar{t}-(T-t_\ell))} u^N_{N^k\bar{t}-r}\left(\bar{\zeta}^N_r\right)Y^N_{N^k(T-\bar{t})+r}\left(\bar{\zeta}^N_r\right)dr\right)\right|$$

The arguments of the two integrals coincide for $r\ge\vartheta$, since the random walks used for the Feynman-Kac representation coincide after time ϑ.

Hence, we can write:

$$\le\sum_{\ell=1}^{n}\frac{a_\ell}{N^k}\sum_{i\in B_k(0)}Y^N_{N^k(T-\bar{t})}(i)$$

$$\cdot\mathbb{E}\left|\int_0^{\vartheta}\left(u^N_{N^k\bar{t}-r}(i)\,Y^N_{N^k(T-\bar{t})+r}(i)-u^N_{N^k\bar{t}-r}(\pi_{N,k}(i))\,Y^N_{N^k(T-\bar{t})+r}(\pi_{N,k}(i))\right)dr\right| \tag{9.94}$$

[4]Note that by the assumption contained in (9.73) and (9.75), the following events are equal:

$$\left\{\zeta^{i,N}_{N^k(\bar{t}-(T-t_\ell))}\in B_k(0)\right\}=\left\{\bar{\zeta}^N_{N^k(\bar{t}-(T-t_\ell))}\in B_k(0)\right\}\quad(\ell\in\{1,\dots,n\}).$$

We clearly get an upper bound replacing "-" by "+":

$$\leq \sum_{\ell=1}^{n} \frac{a_\ell}{N^k} \sum_{i \in B_k(0)} Y^N_{N^k(T-\tilde{t})}(i)$$
$$\cdot \mathbb{E}\left| \int_0^{\vartheta} \left(u^N_{N^k\tilde{t}-r}\left(\zeta_r^{i,N}\right) Y^N_{N^k(T-\tilde{t})+r}\left(\zeta_r^{i,N}\right) + u^N_{N^k\tilde{t}-r}\left(\pi_{N,k}\left(\zeta_r^{i,N}\right)\right) Y^N_{N^k(T-\tilde{t})+r}\left(\pi_{N,k}\left(\zeta_r^{i,N}\right)\right)\right) dr \right| \quad . \tag{9.95}$$

With

$$u^N_{N^k\tilde{t}-r}(i) \leq \frac{1}{N^k} \quad ,$$

we finally get applying (9.23)

$$\begin{aligned}
& \left| \sum_{i \in B_k(0)} Y^N_{N^kT-t}(i) \left(u^N_t(i) - \bar{u}^{N,k}_t(0)\right)\right| \\
\leq & \sum_{i \in B_k(0)} Y^N_{N^k(T-\tilde{t})}(i) \left| u^N_{N^k\tilde{t}}(i) - \bar{u}^{N,k}_{N^k\tilde{t}}(0)\right| \\
\leq & \frac{1}{N^k} \sum_{i \in B_k(0)} Y^N_{N^k(T-\tilde{t})}(i) \\
& \cdot \mathbb{E}\left[\int_0^{\vartheta} \left(\frac{1}{N^k} \left(Y^N_{N^k(T-\tilde{t})+r}\left(\zeta_r^{i,N}\right) + Y^N_{N^k(T-\tilde{t})+r}\left(\pi_{N,k}\left(\zeta_r^{i,N}\right)\right)\right)\right) dr \right] \\
= & \frac{1}{N^{2k}} \sum_{i \in B_k(0)} Y^N_{N^k(T-\tilde{t})}(i) \\
& \cdot \mathbb{E}\left[\int_0^{\vartheta} \left(\left(Y^N_{N^k(T-\tilde{t})+r}\left(\zeta_r^{i,N}\right) + Y^N_{N^k(T-\tilde{t})+r}\left(\pi_{N,k}\left(\zeta_r^{i,N}\right)\right)\right)\right) dr \right] \quad .
\end{aligned} \tag{9.96}$$

According to (9.71), we have to show:

$$\lim_{N \to \infty} \sup_{t \in [0,T]} \mathbf{E}_{cat}\left[\left| \sum_{i \in B_k(0)} Y^N_{N^k(T-t)}(i) \left(u^N_{N^kt}(i) - \bar{u}^{N,k}_{N^kt}(0)\right)\right| dt \right] = 0 \quad . \tag{9.97}$$

This follows immediately from (9.96):

$$\begin{aligned}
&\mathbf{E}_{cat}\left[\sum_{i\in B_k(0)}\left|Y^N_{N^k(T-t)}(i)\left(u^N_{N^kt}(i)-\bar{u}^{N,k}_{N^kt}(0)\right)\right|dt\right]\\
=\ &\mathbf{E}_{cat}\left[\frac{1}{N^{2k}}\sum_{i\in B_k(0)}Y^N_{N^k(T-\tilde{t})}(i)\,\mathbb{E}\left[\int_0^{\vartheta}\left(\left(Y^N_{N^k(T-\tilde{t})+r}\left(\zeta^{i,N}_r\right)+Y^N_{N^k(T-\tilde{t})+r}\left(\pi_{N,k}\left(\zeta^{i,N}_r\right)\right)\right)\right)dr\right]\right]\\
=\ &\frac{1}{N^{2k}}\sum_{i\in B_k(0)}\mathbb{E}\otimes\mathbf{E}_{cat}\left[\int_0^{\vartheta}Y^N_{N^k(T-\tilde{t})}(i)\left(Y^N_{N^k(T-\tilde{t})+r}\left(\zeta^{i,N}_r\right)+Y^N_{N^k(T-\tilde{t})+r}\left(\pi_{N,k}\left(\zeta^{i,N}_r\right)\right)\right)dr\right]\\
=\ &\frac{1}{N^{2k}}\sum_{i\in B_k(0)}\mathbb{E}\left[\int_0^{\vartheta}\mathbf{E}_{cat}\left[\left(Y^N_{N^k(T-\tilde{t})}(i)\right)^2+\frac{\left(Y^N_{N^k(T-\tilde{t})+r}\left(\zeta^{i,N}_r\right)\right)^2}{2}\right.\right.\\
&\qquad\left.\left.+\frac{\left(Y^N_{N^k(T-\tilde{t})+r}\left(\pi_{N,k}\left(\zeta^{i,N}_r\right)\right)\right)^2}{2}\right]dr\right]\\
\leq\ &\frac{1}{N^{2k}}N^k\cdot\mathbb{E}[\vartheta]\cdot\sup_{t\in[0,T]}\mathbf{E}_{cat}\left[2\left(Y^N_{N^kt}(i)\right)^2\right]\quad.
\end{aligned}\tag{9.98}$$

Since the rate of jumps over distance $k-1$ is $\frac{c_{k-1}}{N^{k-1}}\cdot\left(1-\frac{1}{N}\right)$ we get

$$\mathbb{E}[\vartheta]=\frac{N^{k-1}}{c_{k-1}}\left(1-\frac{1}{N}\right)^{-1}\quad. \tag{9.99}$$

This yields (compare (9.98)):

$$\begin{aligned}
&\mathbf{E}_{cat}\left[\sum_{i\in B_k(0)}\left|Y^N_{N^k(T-t)}(i)\left(u^N_{N^kt}(i)-\bar{u}^{N,k}_{N^kt}\right)\right|dt\right]\\
\leq\ &\frac{1}{N\cdot c_{k-1}}\left(1-\frac{1}{N}\right)^{-1}\sup_{t\in[0,T]}\mathbf{E}_{cat}\left[2\left(Y^N_{N^kt}(i)\right)^2\right]\quad.
\end{aligned}\tag{9.100}$$

We know by Lemma 14 (p. 247) that second moments are uniformly bounded even in rescaled time. Hence we have

$$\sup_{t\in[0,T]}\mathbf{E}_{cat}\left[2\left(Y^N_{N^kt}(i)\right)^2\right]<\infty\quad. \tag{9.101}$$

Therefore, we get (9.97) and complete the proof of (9.71).

9.1.6 Independence of the initial condition – Proof of (9.34)

In this subsection, we prove

$$\lim_{N\to\infty}\mathbf{E}\left[\left|\sum_{i\in\Omega_{N,L}}u^N_{N^kT}(i)\cdot\left(X^N_0(i)-\theta^x_0\right)\right|\right]=0\quad. \tag{9.102}$$

Since

$$\bar{q}^{N,k}(i) = q^N(i) \tag{9.103}$$

for all $i \in \Omega_{N,L}$, we get

$$\lim_{N\to\infty} \mathbf{E}\left[\left|\sum_{i\in\Omega_{N,L}} q^N_{N^kT}(i) \cdot \left(X^N_0(i) - \theta^x_0\right)\right|\right] = 0 \tag{9.104}$$

applying the law of large numbers ($(X^N_0(i))_{i\in\Omega_{N,L}}$ is by assumption an iid family). Therefore, it is sufficient to verify

$$\lim_{N\to\infty} \mathbf{E}\left[\left|\sum_{i\in\Omega_{N,L}} \left(q^N_{N^kT}(i) - u^N_{N^kT}(i)\right) \cdot \left(X^N_0(i) - \theta^x_0\right)\right|\right] = 0 \quad . \tag{9.105}$$

Because of the first part (Subsection 9.1.5) of the proof, there only remains to prove

$$\lim_{N\to\infty} \mathbf{E}\left[\left|\sum_{i\in\Omega_{N,L}} \left(q^N_{N^kT}(i) - u^N_{N^kT}(i)\right) \cdot X^N_0(i)\right|\right] = 0 \quad . \tag{9.106}$$

Clearly[5], it is sufficient to prove

$$\lim_{N\to\infty} \sum_{i\in\Omega_{N,L}} \mathbf{E}\left[\left|q^N_{N^kT}(i) - u^N_{N^kT}(i)\right| \cdot X^N_0(i)\right] = 0 \quad . \tag{9.107}$$

Since – for arbitrary N – the difference $q^N_{N^kT}(i) - u^N_{N^kT}(i)$ is a function of the catalyst, therefore measurable with respect to the σ-algebra generated by the catalyst, this is equivalent to:

$$\lim_{N\to\infty} \mathbf{E}_{cat}\left[\sum_{i\in\Omega_{N,L}} \left|q^N_{N^kT}(i) - u^N_{N^kT}(i)\right| \mathbf{E}^Y_{reac}\left[X^N_0(i)\right]\right] = 0 \quad . \tag{9.108}$$

But we have

$$\mathbf{E}^Y_{reac}\left[X^N_0(i)\right] = \theta_x \quad , \tag{9.109}$$

and

$$\lim_{N\to\infty} \mathbf{E}_{cat}\left[\sum_{i\in\Omega_{N,L}} \left|q^N_{N^kT}(i) - u^N_{N^kT}(i)\right| \theta_x\right] = 0 \tag{9.110}$$

has already been proven in Subsection 9.1.5. This completes the proof of (9.34).

9.1.7 Tightness

We want to prove that for any $m \in \{1, ..., L\}$ the sequence of laws of the processes of time-transformed empirical averages

$$\left(\mathcal{L}\left\{\left(\bar{X}^{N,m}_{N^mt}(0), \bar{Y}^{N,m}_{N^mt}(0)\right)_{t\geq 0}\right\}\right)_{N\in\mathbb{N}} \tag{9.111}$$

[5] Expectation and summation can be interchanged due to (9.23).

is tight.

Note that the proof we give here is quite similar to the tightness proof provided in the context of the Finite System Scheme (compare page 106).

The evolution of this processes can be expressed as follows:

$$d\bar{X}^{N,m}_{N^m t}(0) = N^m \sum_{k=m+1}^{L} \frac{c_{k-1}}{N^{k-1}} \left(\bar{X}^{N,k}_{N^m t}(0) - \bar{X}^{N,m}_{N^m t}(0)\right) dt + \sum_{i\in B_m(0)} \sqrt{\frac{2\gamma_x X^N_{N^m t}(i)\, Y^N_{N^m t}(i)}{N^m}} dB^x_t(i) \quad , \tag{9.112}$$

$$d\bar{Y}^{N,m}_{N^m t}(0) = N^m \sum_{k=m+1}^{L} \frac{c_{k-1}}{N^{k-1}} \left(\bar{Y}^{N,k}_{N^m t}(0) - \bar{Y}^{N,m}_{N^m t}(0)\right) dt + \sum_{i\in B_m(0)} \sqrt{\frac{2\gamma_y Y^N_{N^m t}(i)}{N^m}} dB^y_t(i) \quad . \tag{9.113}$$

Applying the Kolmogorov criterion (see [EK86], Proposition 3.6.3), we have to prove that there exist $C < \infty, \tilde{p} > 2$ such that for $u, t \in \mathbb{R}_+$ with $u + 1 > t > u > 0$ we have

$$\limsup_{N\to\infty} \mathbf{E}\left[\left|\bar{X}^{N,m}_{N^m t}(0) - \bar{X}^{N,m}_{N^m u}(0)\right|^{\tilde{p}}\right] \le C\,|t-u|^{\frac{\tilde{p}}{2}} \quad , \tag{9.114}$$

$$\limsup_{N\to\infty} \mathbf{E}\left[\left|\bar{Y}^{N,m}_{N^m t}(0) - \bar{Y}^{N,m}_{N^m u}(0)\right|^{\tilde{p}}\right] \le C\,|t-u|^{\frac{\tilde{p}}{2}} \quad . \tag{9.115}$$

We only provide a prove of (9.114); Eq. (9.115) can be verified by similar, but simpler calculations.

Applying Lemma B.11 (p. 249), we have to prove that there exists $C < \infty, \tilde{p} > 2$ with

$$\limsup_{N\to\infty} \mathbf{E}\left[\left|\int_u^t N^m \sum_{k=m+1}^{L} \frac{c_{k-1}}{N^{k-1}} \left(\bar{X}^{N,k}_{N^m \varrho}(0) - \bar{X}^{N,m}_{N^m \varrho}(0)\right) d\varrho\right|^{\tilde{p}}\right] \le C\,|t-u|^{\frac{\tilde{p}}{2}} \tag{9.116}$$

and

$$\limsup_{N\to\infty} \mathbf{E}\left[\left|\sum_{i\in B_m(0)} \int_u^t \sqrt{\frac{2\gamma_x X^N_{N^m \varrho}(i)\, Y^N_{N^m \varrho}(i)}{N^m}} dB^x_\varrho(i)\right|^{\tilde{p}}\right] \le C\,|t-u|^{\frac{\tilde{p}}{2}} \quad . \tag{9.117}$$

The limit (9.116) can be proven as follows:

$$\begin{aligned}
&\limsup_{N\to\infty} \mathbf{E}\left[\left|\int_u^t N^m \sum_{k=m+1}^{L} \frac{c_{k-1}}{N^{k-1}} \left(\bar{X}^{N,k}_{N^m \varrho}(0) - \bar{X}^{N,m}_{N^m \varrho}(0)\right) d\varrho\right|^{\tilde{p}}\right] \\
= \; &\limsup_{N\to\infty} (t-u)^{\tilde{p}}\, \mathbf{E}\left[\left|\frac{1}{t-u}\int_u^t N^m \sum_{k=m+1}^{L} \frac{c_{k-1}}{N^{k-1}} \left(\bar{X}^{N,k}_{N^m \varrho}(0) - \bar{X}^{N,m}_{N^m \varrho}(0)\, e\right) d\varrho\right|^{\tilde{p}}\right] \quad ;
\end{aligned}$$

we apply Jensen's inequality:

$$\le \limsup_{N\to\infty} (t-u)^{\tilde p} N^m \sum_{k=m+1}^{L} \frac{c_{k-1}}{N^{k-1}} \frac{1}{t-u} \int_u^t \mathbf{E}\left[\left|\bar X^{N,k}_{N^m\varrho}(0) - \bar X^{N,m}_{N^m\varrho}(0)\right|^{\tilde p}\right] d\varrho$$

$$\le \limsup_{N\to\infty} (t-u)^{\tilde p} N^m \sum_{k=m+1}^{L} \frac{c_{k-1}}{N^{k-1}} \frac{1}{t-u} \int_u^t \mathbf{E}\left[\left|\bar X^{N,k}_{N^m\varrho}(0) + \bar X^{N,m}_{N^m\varrho}(0)\right|^{\tilde p}\right] d\varrho \quad .$$

There clearly exists $K > 0$ with

$$\begin{aligned}
&\le \limsup_{N\to\infty} (t-u)^{\tilde p} N^m \sum_{k=m+1}^{L} \frac{c_{k-1}}{N^{k-1}} K \\
&\quad \cdot \left(\sup_{\varrho\in[u,t]} \mathbf{E}\left[\left|\bar X^{N,k}_{N^m\varrho}(0)\right|^{\tilde p}\right] + \sup_{\varrho\in[u,t]} \mathbf{E}\left[\left|\bar X^{N,m}_{N^m\varrho}(0)\right|^{\tilde p}\right]\right) \\
&= \limsup_{N\to\infty} (t-u)^{\tilde p} N^m \sum_{k=m+1}^{L} \frac{c_{k-1}}{N^{k-1}} K \\
&\quad \cdot \left(\sup_{\varrho\in[u,t]} \mathbf{E}\left[\left|\frac{1}{N^k} \sum_{j\in B_k(0)} X^N_{N^m\varrho}(j)\right|^{\tilde p}\right] \right. \\
&\quad \left. + \sup_{\varrho\in[u,t]} \mathbf{E}\left[\left|\frac{1}{N^m} \sum_{j\in B_m(0)} X^N_{N^m\varrho}(j)\right|^{\tilde p}\right]\right) \quad .
\end{aligned} \tag{9.118}$$

Again using Jensen's inequality, we get:

$$\begin{aligned}
&\le \limsup_{N\to\infty} (t-u)^{\tilde p} N^m \sum_{k=m+1}^{L} \frac{c_{k-1}}{N^{k-1}} K \\
&\quad \cdot \left(\sup_{\varrho\in[u,t]} \frac{1}{N^k} \sum_{j\in B_k(0)} \mathbf{E}\left[\left|X^N_{N^m\varrho}(j)\right|^{\tilde p}\right] \right. \\
&\quad \left. + \sup_{\varrho\in[u,t]} \frac{1}{N^m} \sum_{j\in B_m(0)} \mathbf{E}\left[\left|X^N_{N^m\varrho}(j)\right|^{\tilde p}\right]\right) \quad .
\end{aligned} \tag{9.119}$$

With translation invariance, this yields:

$$\begin{aligned}
&\le \limsup_{N\to\infty} (t-u)^{\tilde p} N^m \sum_{k=m+1}^{L} \frac{c_{k-1}}{N^{k-1}} K \cdot 2 \sup_{\varrho\in[u,t]} \mathbf{E}\left[\left|X^N_{N^m\varrho}(0)\right|^{\tilde p}\right] \\
&= \limsup_{N\to\infty} (t-u)^{\tilde p} \sum_{k=1}^{L} \frac{c_{k-1}}{N^{k-1}} K \cdot 2 \sup_{\varrho\in[u,t]} \mathbf{E}\left[\left|X^N_{N^m\varrho}(0)\right|^{\tilde p}\right] \quad .
\end{aligned} \tag{9.120}$$

By Lemma 14 (in particular, equation B.3) we have

$$\limsup_{N\to\infty} \sup_{\varrho\in[u,t]} \mathbf{E}\left[\left|X^N_{N^m\varrho}(0)\right|^{\tilde p}\right] < \infty \quad . \tag{9.121}$$

Together with

$$\lim_{N\to\infty}\sum_{k=1}^{L}\frac{c_{k-1}}{N^{k-1}} = c_0 \quad , \tag{9.122}$$

This completes the proof of (9.116).

Next, we have to prove (9.117), namely

$$\limsup_{N\to\infty}\mathbf{E}\left[\left|\sum_{i\in B_m(0)}\int_u^t\sqrt{\frac{2\gamma_x X^N_{N^m\varrho}(i)\,Y^N_{N^m\varrho}(i)}{N^m}}dB^x_\varrho(i)\right|^{\tilde{p}}\right] \le C\,|t-u|^{\frac{\tilde{p}}{2}} \quad . \tag{9.123}$$

Similar to the argument carried out on page 152 we can apply Burkholder's inequality

$$\mathbf{E}\left[\left|\sum_{i\in B_m(0)}\int_u^t\sqrt{\frac{2\gamma_x X^N_{N^m\varrho}(i)\,Y^N_{N^m\varrho}(i)}{N^m}}dB^x_\varrho(i)\right|^{\tilde{p}}\right]$$
$$\le \quad (\tilde{p}-1)^{\tilde{p}}\,\mathbf{E}\left[\left(\sum_{i\in B_m(0)}\int_u^t\frac{2\gamma_x X^N_{N^m\varrho}(i)\,Y^N_{N^m\varrho}(i)}{N^m}d\varrho\right)^{\frac{\tilde{p}}{2}}\right] \quad . \tag{9.124}$$

By Jensen's inequality[6], this yields:

$$\begin{aligned}
&\le (\tilde{p}-1)^{\tilde{p}}\,\frac{1}{N^m}\sum_{i\in B_m(0)}|t-u|^{\frac{\tilde{p}}{2}}\,\mathbf{E}\left[\left(\frac{1}{t-u}\int_u^t 2\gamma_x X^N_{N^m\varrho}(i)\,Y^N_{N^m\varrho}(i)\,d\varrho\right)^{\frac{\tilde{p}}{2}}\right]\\
&\le (\tilde{p}-1)^{\tilde{p}}\,\frac{1}{N^m}\sum_{i\in B_m(0)}|t-u|^{\frac{\tilde{p}}{2}}\,\frac{1}{t-u}\int_u^t\mathbf{E}\left[\left(2\gamma_x X^N_{N^m\varrho}(i)\,Y^N_{N^m\varrho}(i)\,d\varrho\right)^{\frac{\tilde{p}}{2}}\right]\\
&\le (\tilde{p}-1)^{\tilde{p}}\,(2\gamma_x)^{\frac{\tilde{p}}{2}}\,\frac{1}{N^m}\sum_{i\in B_m(0)}|t-u|^{\frac{\tilde{p}}{2}}\sup_{\varrho\in[u,t]}\mathbf{E}\left[\left(X^N_{N^m\varrho}(i)\right)^{\tilde{p}}+\left(Y^N_{N^m\varrho}(i)\right)^{\tilde{p}}\right]\\
&= (\tilde{p}-1)^{\tilde{p}}\,(2\gamma_x)^{\frac{\tilde{p}}{2}}\,|t-u|^{\frac{\tilde{p}}{2}}\sup_{\varrho\in[u,t]}\mathbf{E}\left[\left(X^N_{N^m\varrho}(0)\right)^{\tilde{p}}+\left(Y^N_{N^m\varrho}(0)\right)^{\tilde{p}}\right]
\end{aligned} \tag{9.125}$$

By Lemma 14 (in particular, equation B.3 and the uniform boundedness of $\tilde{p}$th moments of the catalyst provided by Cox, Greven and Shiga [CGS95]) we have

$$\limsup_{N\to\infty}\sup_{\varrho\in[u,t]}\mathbf{E}\left[\left(X^N_{N^m\varrho}(0)\right)^{\tilde{p}}+\left(Y^N_{N^m\varrho}(0)\right)^{\tilde{p}}\right] < \infty \quad . \tag{9.126}$$

This completes the proof of (9.117).

[6] "$\left(\frac{1}{N^m}\sum(\bullet)\right)^{\frac{\tilde{p}}{2}} \le \frac{1}{N^m}\sum(\bullet)^{\frac{\tilde{p}}{2}}$"

9.2 Second part: local point of view

The proof of the second part of Theorem 3 is similar to that of the second part of Theorem 1 (see Section 7.2).

9.2.1 The claim

Let $k \leq j$, $s \in \mathbb{R}_+^{\mathbb{N}}$ with $\lim_{N\to\infty} s(N) = \infty$, $\lim_{N\to\infty} \frac{s(N)}{N} = 0$. We have to prove

$$\mathcal{L}\left\{\left(\bar{X}^{N,k}_{N^j s(N)+N^k t}(0), \bar{Y}^{N,k}_{N^j s(N)+N^k t}(0)\right)_{t\geq 0}\right\} \overset{N\to\infty}{\Longrightarrow} \int \hat{\Psi}^{\theta^*}_{c_k} \quad \mu^{j,k+1}_{\theta_0}(d\theta^*) \quad . \tag{9.127}$$

where the interaction chain has been defined by (4.27): For $k \leq j$ we have

$$\mu^{j,k}_{\theta}(\cdot) = \int_{\mathbb{R}_+\times\mathbb{R}_+} \cdots \int_{\mathbb{R}_+\times\mathbb{R}_+} \Gamma^{\theta_{j-k}}_{c_k}(\cdot) \cdots \Gamma^{\theta_1}_{c_{j-1}}(d\theta_2)\,\Gamma^{\theta}_{c_j}(d\theta_1) \quad , \tag{9.128}$$

and

$$\mu^{j,j+1}_{\theta}(\cdot) = \delta_{\{\theta\}} \quad .$$

Once again, recall the considered system of stochastic differential equations:

$$\begin{aligned} dX^N_t(i) &= \sum_{k=1}^{L} \frac{c_{k-1}}{N^{k-1}} \left(\frac{1}{N^k} \sum_{j\in\Omega_{N,L}:d(i,j)\leq k} \left[X^N_t(j) - X^N_t(i)\right] \right) dt \\ &\quad + \sqrt{2\gamma_x X^N_t(i)\, Y^N_t(i)}\, dB^x_t(i) \quad ; \end{aligned} \tag{9.129}$$

$$\begin{aligned} dY^N_t(i) &= \sum_{k=1}^{L} \frac{c_{k-1}}{N^{k-1}} \left(\frac{1}{N^k} \sum_{j\in\Omega_{N,L}:d(i,j)\leq k} \left[Y^N_t(j) - Y^N_t(i)\right] \right) dt \\ &\quad + \sqrt{2\gamma_y Y^N_t(i)}\, dB^y_i(i), \quad (i \in \Omega_{N,L}) \quad . \end{aligned} \tag{9.130}$$

The initial condition is assumed to be a product measure with finite pth moment for a $p > 2$; the intensity vector of reactant and catalyst ($\theta = (\theta_x, \theta_y)$) is assumed to be finite.

9.2.2 Central ideas of the proof

1. *Induction scheme*
   The proof follows an induction scheme leading us succesively from averages over large blocks to averages over small blocks. The starting point of induction will be the first part of Theorem 3 (namely: fluctuations of empirical averages), which we have proven above.

2. *Catalytic setting – Skorohod representation*
   We can use results by Dawson and Greven ([DG96]). They consider the process used as catalyst here. Their weak convergence results can be applied as results on almost sure convergence by defining all the considered processes on a common probability space. It is once again Skorohod's representation that allows us to do so.

9.2.3 Induction scheme

Induction will proceed from $k = j+1$ to $k = 0$.
initialization of the induction scheme:

We have

$$\mathcal{L}\left\{\left(\bar{X}^{N,j+1}_{N^j s(N)}(0), \bar{Y}^{N,j+1}_{N^j s(N)}(0)\right)_{t\geq 0}\right\} \overset{N\to\infty}{\Longrightarrow} \delta_{\{\theta\}} \quad . \tag{9.131}$$

This follows immediately from the first part of Theorem 3. In particular, note that

$$\delta_{\{\theta\}} = \mu_\theta^{j,j+1} \quad .$$

This serves as starting point of induction.

inductive step: $k+1 \longmapsto k$

Assume that we know

$$\mathcal{L}\left\{\left(\bar{X}^{N,k+1}_{N^j s(N)}(0), \bar{Y}^{N,k+1}_{N^j s(N)}(0)\right)\right\} \overset{N\to\infty}{\Longrightarrow} \mu_\theta^{j,k+1} \quad . \tag{9.132}$$

Then we claim to be able to conclude:

$$\mathcal{L}\left\{\left(\bar{X}^{N,k}_{N^j s(N)+N^k t}(0), \bar{Y}^{N,k}_{N^j s(N)+N^k t}(0)\right)_{t\geq 0}\right\} \overset{N\to\infty}{\Longrightarrow} \int \hat{\Psi}^{\theta^*}_{c_k} \quad \mu_\theta^{j,k+1}(d\theta^*) \quad . \tag{9.133}$$

In particular, (9.133) implies

$$\mathcal{L}\left\{\left(\bar{X}^{N,k}_{N^j s(N)}(0), \bar{Y}^{N,k}_{N^j s(N)}(0)\right)\right\} \overset{N\to\infty}{\Longrightarrow} \mu_\theta^{j,k} \quad . \tag{9.134}$$

9.2.4 Formulation of the claim (9.133) with respect to finite dimensional distributions

Using the Prohorov metric ϱ describing weak convergence on $\mathcal{M}^1(\mathcal{C}(\mathbb{R}_+, \mathbb{R}_+ \times \mathbb{R}_+))$, the claim (9.133) can be written as follows:

$$\lim_{N\to\infty} \varrho\left(\mathcal{L}\left\{\left(\bar{X}^{N,k}_{N^j s(N)+N^k t}(0), \bar{Y}^{N,k}_{N^j s(N)+N^k t}(0)\right)_{t\geq 0}\right\}, \int \hat{\Psi}^{\theta^*}_{c_k} \quad \mu_\theta^{j,k+1}(d\theta^*)\right) = 0 \quad . \tag{9.135}$$

Convergence of probability measures on the path space (here: $\mathcal{C}(\mathbb{R}_+, \mathbb{R}_+ \times \mathbb{R}_+)$) can be proven carrying out a two-step program. One has to show:

1. *Convergence of finite dimensional distributions:*
   First of all, we define a process that represents the limit object:
   Let $(\mathfrak{x}_t, \mathfrak{y}_t)_{t\geq 0}$ be a process with values in $\mathcal{C}(\mathbb{R}_+, \mathbb{R}_+ \times \mathbb{R}_+)$ and law

   $$\mathcal{L}\left\{(\mathfrak{x}_t, \mathfrak{y}_t)_{t\geq 0}\right\} = \int \hat{\Psi}^{\theta^*}_{c_k} \quad \mu_\theta^{j,k+1}(d\theta^*) \quad . \tag{9.136}$$

   We claim that for every finite set $\Upsilon = \{t_1, t_2, ..., t_n\} \subset \mathbb{R}_+$ the following weak convergence property is true:

   $$\mathcal{L}\left\{\left(\bar{X}^{N,k}_{N^j s(N)+N^k t}(0), \bar{Y}^{N,k}_{N^j s(N)+N^k t}(0)\right)_{t\in\Upsilon}\right\} \overset{N\to\infty}{\Longrightarrow} \mathcal{L}\left\{(\mathfrak{x}_t, \mathfrak{y}_t)_{t\in\Upsilon}\right\} \quad . \tag{9.137}$$

Notation 8 *In a certain abuse of notation, we often write*

$$\lim_{N\to\infty} \varrho^{\Upsilon}\left(\mathcal{L}\left\{\left(\bar{X}^{N,k}_{N^j s(N)+N^k t}(0), \bar{Y}^{N,k}_{N^j s(N)+N^k t}(0)\right)_{t\geq 0}\right\}, \int \hat{\Psi}^{\theta^*}_{c_k} \quad \mu^{j,k+1}_{\theta}(d\theta^*)\right) = 0 \tag{9.138}$$

instead of (9.137). This way, we do not have to introduce an auxiliary process (like $(\mathfrak{x}_t, \mathfrak{y}_t)_{t\geq 0}$) to be able to formulate convergence with respect to finite dimensional marginals.

2. *Tightness:*
It is of course not sufficient for pathwise convergence to have convergence with respect to finite dimensional marginals. The missing link is tightness. Namely, we have to prove that the sequence

$$\left(\mathcal{L}\left\{\left(\bar{X}^{N,k}_{N^j s(N)+N^k t}(0), \bar{Y}^{N,k}_{N^j s(N)+N^k t}(0)\right)_{t\geq 0}\right\}\right)_{N\in\mathbb{N}} \tag{9.139}$$

is a tight sequence of probability laws.
The proof of tightness is to be found in Subsection 9.2.11 on page 242.

We will first concentrate on the proof of convergence of finite dimensional distributions.

9.2.5 Survey of the proof of convergence of finite dimensional distributions

Before we proceed, we want to explain how the proof of (9.138) is organized and drop a note on the main ideas of the various steps of the proof. These steps are carried out in subsections 9.2.6 to 9.2.11.

Subsection 9.2.6: Skorohod's representation

We do not have to start at the very beginning, but we can use the following results. Namely, we know

- that the laws of the processes of $(k+1)$-level-averages at times $N^j s(N)$ converge weakly – this is the induction hypothesis.
- that the claim of the theorem holds for the catalyst. In fact, Dawson and Greven [DG96] have already carried out the Multiple Space Time Scale Analysis for the catalyst process.

We apply Skorohod's representation to convert these weak convergence properties into almost sure pathwise convergence, defining all the processes on a common probability space.

Subsection 9.2.7: Approximating processes

We split up the proof of (9.138) into three steps. The key idea is to introduce probability laws that are close to

$$\int \hat{\Psi}^{\theta^*}_{c_k} \quad \mu^{j,k+1}_{\theta}(d\theta^*) \tag{9.140}$$

in the limit $N \to \infty$. Namely, the following steps of approximation are carried out:

Steps of approximation

1. We will define random variables ϑ_{k+1} with $\mathcal{L}\{\vartheta_{k+1}\} = \mu_\theta^{j,k+1}$. This gives the following formulation of (9.140):

$$\int \hat{\Psi}_{c_k}^{\theta^*} \quad \mu_\theta^{j,k+1}(d\theta^*) = \int \hat{\Psi}_{c_k}^{\theta^*} \quad \mathcal{L}\{\vartheta_{k+1}\}(d\theta^*) \quad . \tag{9.141}$$

By the induction hypothesis the laws of $(k+1)$-level averages at time $N^j(s(N))$ converge weakly with limit $\mathcal{L}\{\vartheta_{k+1}\}$. We introduce random variables $\left(\vartheta_{k+1}^N\right)_{N\in\mathbb{N}}$ that have the same laws as the $(k+1)$-level averages at time $N^j(s(N))$. We claim that the laws $\int \hat{\Psi}_{c_k}^{\theta^*} \quad \mathcal{L}\{\vartheta_{k+1}^N\}(d\theta^*)$ $(N \in \mathbb{N})$ approximate (9.140), namely:

$$\lim_{N\to\infty} \varrho^\Upsilon \left(\int \hat{\Psi}_{c_k}^{\theta^*} \quad \mathcal{L}\{\vartheta_{k+1}\}(d\theta^*) \, , \int \hat{\Psi}_{c_k}^{\theta^*} \quad \mathcal{L}\{\vartheta_{k+1}^N\}(d\theta^*) \right) = 0 \quad . \tag{9.142}$$

2. The law

$$\int \hat{\Psi}_{c_k}^{\theta^*} \quad \mathcal{L}\{\vartheta_{k+1}^N\}(d\theta^*) \tag{9.143}$$

is by definition a mixture of laws of stationary processes. We approximate this mixture of stationary processes by mixtures of non-stationary processes. These non-stationary processes are assumed to be processes that have already been evolving for a certain amount of time $\tilde{\ell}_N$ after being started in an atomar law. With $\tilde{\ell}_N \to \infty$ as $N \to \infty$, we hope that the processes we obtain get closer and closer to (9.143) as $N \to \infty$. Since we are only giving a survey at the moment, we do not provide the precise notation needed for this step of approximation. Nevertheless, we cite the second step of approximation to give a taste of how things look like:

$$\lim_{N\to\infty} \varrho^\Upsilon \left(\mathcal{L} \int \hat{\Psi}_{c_k}^{\theta^*} \quad \mathcal{L}\{\vartheta_{k+1}^N\}(d\theta^*) \, , \mathcal{L}\left\{ \left(\bar{\mathfrak{X}}_{N^k(t+\tilde{\ell}_N)}^{N,\infty,k}, \bar{\mathfrak{Y}}_{N^k(t+\tilde{\ell}_N)}^{N,\infty,k} \right)_{t\geq 0} \right\} \right) = 0 \quad . \tag{9.144}$$

3. As third step, we prove that the law $\mathcal{L}\left\{ \left(\bar{\mathfrak{X}}_{N^k(t+\tilde{\ell}_N)}^{N,\infty,k}, \bar{\mathfrak{Y}}_{N^k(t+\tilde{\ell}_N)}^{N,\infty,k} \right)_{t\geq 0} \right\}$ – which is still undefined and somehow mysterious at the moment – gets close to

$$\mathcal{L}\left\{ \left(\bar{X}_{N^j s(N)+N^k t}^{N,k}(0) \, , \bar{Y}_{N^j s(N)+N^k t}^{N,k}(0) \right)_{t\geq 0} \right\} \quad . \tag{9.145}$$

as N grows to infinity. Defining $\left(\bar{\mathfrak{X}}_{N^k t}^{N,k}, \bar{\mathfrak{Y}}_{N^k t}^{N,k} \right)_{t\geq 0}$ s.t.

$$\mathcal{L}\left\{ \left(\bar{\mathfrak{X}}_{N^k t}^{N,k}, \bar{\mathfrak{Y}}_{N^k t}^{N,k} \right)_{t\geq 0} \right\} = \mathcal{L}\left\{ \left(\bar{X}_{N^j s(N)+N^k t}^{N,k}(0) \, , \bar{Y}_{N^j s(N)+N^k t}^{N,k}(0) \right)_{t\geq 0} \right\} \quad ,$$

the corresponding formula will have the following shape

$$\lim_{N\to\infty} \varrho^\Upsilon \left(\mathcal{L}\left\{ \left(\bar{\mathfrak{X}}_{N^k t}^{N,k}, \bar{\mathfrak{Y}}_{N^k t}^{N,k} \right)_{t\geq 0} \right\} , \mathcal{L}\left\{ \left(\bar{\mathfrak{X}}_{N^k(t+\tilde{\ell}_N)}^{N,\infty,k}, \bar{\mathfrak{Y}}_{N^k(t+\tilde{\ell}_N)}^{N,\infty,k} \right)_{t\geq 0} \right\} \right) = 0 \quad . \tag{9.146}$$

Subsection 9.2.8

In this subsection (p. 210pp), we prove (9.146). This is the most technical part of the proof. In particular, we apply duality techniques to get the result.

Subsection 9.2.9

In this subsection (p. 241), we prove (9.144). The key point is that we have to verify that the laws of the stationary processes depend continuously of the intensity parameters.

Subsection 9.2.10

In this subsection (p. 242), we prove (9.142). The most important point is that we have to verify that there is uniform convergence of the considered processes to their stationary laws.

Subsection 9.2.11

This subsection (p. 242pp) contains the proof of tightness.

9.2.6 Skorohod's representation

The situation at time $N^j s(N)$

Notation 9 *Let*

$$\left(\mathfrak{X}_t^N, \mathfrak{Y}_t^N\right)_{t\geq 0} := \left(X^N_{N^j s(N)+t}, Y^N_{N^j s(N)+t}\right)_{t\geq 0} \quad . \tag{9.147}$$

We are interested in the properties of the processes at time $N^j s(N)$. This time will play the role of a time of initialization in the following. At this time(s) the processes have certain weak convergence properties (given by the induction hypothesis and the results on the catalyst provided by Dawson and Greven [DG96]). We convert these weak convergence properties into properties of convergence of paths by defining the processes on a common probability space applying Skorohod's representation.

Notation 10 *We introduce the following notation for* $(k+1)$*-level averages at time* $N^j s(N)$:

$$\left(\vartheta^{N,\mathfrak{X}}_{k+1}, \vartheta^{N,\mathfrak{Y}}_{k+1}\right) := \left(\bar{X}^{N,k+1}_{N^j s(N)}, \bar{Y}^{N,k+1}_{N^j s(N)}\right) \quad . \tag{9.148}$$

Weak limit property I: We know from the induction hypothesis (9.132) that there is a well-defined limit law:

$$\mathcal{L}\left\{\left(\vartheta^{N,\mathfrak{X}}_{k+1}, \vartheta^{N,\mathfrak{Y}}_{k+1}\right)\right\} \overset{N\to\infty}{\Longrightarrow} \mu_\theta^{j,k+1} \quad . \tag{9.149}$$

Convergence of the catalyst

As we have already mentioned, the Multiple Space Time Scale Analysis has already been carried out for the catalyst process by Dawson and Greven [DG96]. They prove:

Weak limit property II: The limit of the local behavior of the catalyst is given by

$$\mathcal{L}\left\{\left(\bar{Y}^{N,k}_{N^j s(N)+N^k t}(i)\right)_{t\geq 0}\right\} \overset{N\to\infty}{\Longrightarrow} \int \hat{\Psi}^{\theta^*}_{c_k} \; \mu^{j,k+1}_{\theta_0}(d\theta^*) \quad . \tag{9.150}$$

Common probability space and pathwise convergence properties

By Skorohod's representation, there exists a probability space $\left(\tilde{\Omega}, \tilde{\mathfrak{A}}, \tilde{P}\right)$ and $\Omega_{N,L}$-valued stochastic processes $\left(V^N, W^N\right)$ – all defined on $\left(\tilde{\Omega}, \tilde{\mathfrak{A}}, \tilde{P}\right)$ – with

$$\tilde{P} \circ \left(V^N, W^N\right)^{-1} = P \circ \left(X^N, Y^N\right)^{-1} \tag{9.151}$$

such that after substituting

$$\begin{array}{ccc} X & \longmapsto & V \\ Y & \longmapsto & W \end{array} \tag{9.152}$$

we get $\tilde{P}$-a.s. convergence in (9.149) and (9.150). Not wishing to change all the notation in the spirit of (9.152), we assume that the processes $\left(X^N, Y^N\right)_{N\in\mathbb{N}}$ live all on a probability space $(\Omega, \mathcal{A}, P)$ s. t.:

- There exists a $\mathbb{R}_+ \times \mathbb{R}_+$-valued random variable $\vartheta_{k+1} = \left(\vartheta_{k+1}^{\mathfrak{X}}, \vartheta_{k+1}^{\mathfrak{Y}}\right)$ for which we have
$$\lim_{N\to\infty} \left(\left(\bar{X}_{N^j s(N)}^{N,k+1}(0), \bar{Y}_{N^j s(N)}^{N,k+1}(0)\right) - \vartheta_{k+1}\right) = 0 \quad P-a.s. \tag{9.153}$$
In particular we have
$$\mathcal{L}\left\{\vartheta_{k+1}\right\} = \mu_\theta^{j,k+1} \quad . \tag{9.154}$$

- There exists a $\mathcal{M}\left(\mathcal{C}\left(\mathbb{R}_+, \mathbb{R}_+\right)\right)$-valued process $\left(\bar{\mathfrak{Y}}_t^{\infty,k}\right)_{t\geq 0}$ with
$$\forall T>0: \lim_{N\to\infty} \sup_{t\in[0,T]} \left|\bar{Y}_{N^j s(N)+N^k t}^{N,k}(0) - \bar{\mathfrak{Y}}_{N^k t}^{\infty,k}\right| = 0 \quad P-\text{a.s.} \tag{9.155}$$
The law of $\left(\bar{\mathfrak{Y}}_{N^k t}^{\infty,k}\right)_{t\geq 0}$ is given by
$$\begin{aligned} \mathcal{L}\left\{\left(\bar{\mathfrak{Y}}_{N^k t}^{\infty,k}\right)_{t\geq 0}\right\} &= \int \hat{\Psi}_{c_k}^{\theta^*} \quad \mu_{\theta_y}^{j,k+1}(d\theta^*) \\ &= \int \hat{\Psi}_{c_k}^{\theta^*} \quad \mathcal{L}\left\{\vartheta_{k+1}^{\mathfrak{Y}}\right\}(d\theta^*) \quad . \end{aligned} \tag{9.156}$$

9.2.7 Approximating processes

The process $\left(\bar{\mathfrak{X}}_{N^k t}^{\infty,k}, \bar{\mathfrak{Y}}_{N^k t}^{\infty,k}\right)_{t\geq 0}$ with law[7]

$$\begin{aligned} \mathcal{L}\left\{\left(\bar{\mathfrak{X}}_t^{\infty,k}, \bar{\mathfrak{Y}}_t^{\infty,k}\right)_{t\geq 0}\right\} &= \int \hat{\Psi}_{c_k}^{\theta^*} \quad \mu_\theta^{j,k+1}(d\theta^*) \\ &= \int \hat{\Psi}_{c_k}^{\theta^*} \quad \mathcal{L}\left\{\vartheta_{k+1}\right\}(d\theta^*) \end{aligned} \tag{9.157}$$

is the process we claim to be the weak limit process of

$$\left(\bar{\mathfrak{X}}_{N^k t}^{N,k}, \bar{\mathfrak{Y}}_{N^k t}^{N,k}\right)_{t\geq 0} := \left(\bar{X}_{N^j s(N)+N^k t}^{N,k}, \bar{Y}_{N^j s(N)+N^k t}^{N,k}\right)_{t\geq 0} \tag{9.158}$$

[7] Note that this is consistent with (9.156).

as we let $N \to \infty$. As an intermediate step of approximation, we introduce processes $\left(\bar{\mathfrak{X}}_t^{N,\infty,k}, \bar{\mathfrak{Y}}_t^{N,\infty,k}\right)_{t\geq 0}$ $(N \in \mathbb{N})$ that have laws given by[8]

$$\mathcal{L}\left\{\left(\bar{\mathfrak{X}}_{N^k t}^{N,\infty,k}, \bar{\mathfrak{Y}}_{N^k t}^{N,\infty,k}\right)_{t\geq 0}\right\} := \int \Psi_{c_k}^{\theta^*} \quad \mathcal{L}\left\{\vartheta_{k+1}^N\right\}(d\theta^*) \quad . \tag{9.159}$$

Namely, we assume that the processes $\left(\bar{\mathfrak{X}}_{N^k t}^{N,\infty,k}, \bar{\mathfrak{Y}}_{N^k t}^{N,\infty,k}\right)_{t\geq 0}$ $(N \in \mathbb{N})$ live on the same common probability space all the other processes are already defined on. In particular, we assume that

$$\mathcal{L}\left\{\left(\bar{\mathfrak{X}}_{N^k t}^{N,\infty,k}, \bar{\mathfrak{Y}}_{N^k t}^{N,\infty,k}\right)_{t\geq 0} \cdot \mathbb{1}_{\{\vartheta_{k+1}^N \in A\}}\right\} = \int_A \Psi_{c_k}^{\theta^*} \quad \mathcal{L}\left\{\vartheta_{k+1}^N\right\}(d\theta^*) \tag{9.160}$$

holds for every set A for which $\left\{\vartheta_{k+1}^N \in A\right\}$ is measurable.

Using this intermediate step, the following is sufficient for the claim (9.127):

There exists an increasing sequence $\left(\tilde{\ell}_N\right)_{N\in\mathbb{N}} \in \mathbb{R}_+^N$ with

$$\lim_{N\to\infty} \tilde{\ell}_N = \infty \quad , \tag{9.161}$$

such that

1.

$$\lim_{N\to\infty} \varrho^\Upsilon \left(\mathcal{L}\left\{\left(\bar{\mathfrak{X}}_{N^k t}^{N,k}, \bar{\mathfrak{Y}}_{N^k t}^{N,k}\right)_{t\geq 0}\right\} \quad , \quad \mathcal{L}\left\{\left(\bar{\mathfrak{X}}_{N^k(t+\tilde{\ell}_N)}^{N,\infty,k}, \bar{\mathfrak{Y}}_{N^k(t+\tilde{\ell}_N)}^{N,\infty,k}\right)_{t\geq 0}\right\}\right) = 0 \quad . \tag{9.162}$$

2.

$$\lim_{N\to\infty} \varrho^\Upsilon \left(\mathcal{L}\left\{\left(\bar{\mathfrak{X}}_{N^k(t+\tilde{\ell}_N)}^{N,\infty,k}, \bar{\mathfrak{Y}}_{N^k(t+\tilde{\ell}_N)}^{N,\infty,k}\right)_{t\geq 0}\right\} \quad , \quad \int \hat{\Psi}_{c_k}^{\theta^*} \quad \mathcal{L}\left\{\vartheta_{k+1}^N\right\}(d\theta^*)\right) = 0 \quad . \tag{9.163}$$

3.

$$\lim_{N\to\infty} \varrho^\Upsilon \left(\int \hat{\Psi}_{c_k}^{\theta^*} \quad \mathcal{L}\left\{\vartheta_{k+1}^N\right\}(d\theta^*) \quad , \quad \int \hat{\Psi}_{c_k}^{\theta^*} \quad \mathcal{L}\left\{\vartheta_{k+1}\right\}(d\theta^*)\right) = 0 \quad . \tag{9.164}$$

9.2.8 Proof of (9.162)

It suffices to prove (9.162) for given "pseudo-initial intensities" $\left(\vartheta_{k+1}^N\right)_{N\in\mathbb{N}}$. We claim that for almost all realizations of $\left(\vartheta_{k+1}^N\right)_{N\in\mathbb{N}}$ we have[9]

$$\lim_{N\to\infty} \varrho\Bigg(\mathcal{L}\left\{\left(\bar{\mathfrak{X}}_{N^k t}^{N,k}, \bar{\mathfrak{Y}}_{N^k t}^{N,k}\right)_{t\geq 0} \Bigg| \left(\vartheta_{k+1}^N\right)_{N\in\mathbb{N}}\right\} , \tag{9.165}$$

$$\mathcal{L}\left\{\left(\bar{\mathfrak{X}}_{N^k(t+\tilde{\ell}_N)}^{N,\infty,k}, \bar{\mathfrak{Y}}_{N^k(t+\tilde{\ell}_N)}^{N,\infty,k}\right)_{t\geq 0} \Bigg| \left(\vartheta_{k+1}^N\right)_{N\in\mathbb{N}}\right\}\Bigg) = 0 \quad . \tag{9.166}$$

[8]Note that Ψ denotes a non-stationary process, whereas $\hat{\Psi}$ denotes a stationary process.

[9]In fact, expressions (9.165) and (9.166) are Radon-Nikodyn derivatives. Their existence (namely the corresponding statement of absolute continouity) is implied by the definition of the sequence $\left(\vartheta_{k+1}^N\right)_{N\in\mathbb{N}}$ and (9.160).

This modification of the claim is justified by the proof of Lemma 2, p. 44, where a similar property is proven.[10] In the following, we use the following notation

$$\mathcal{L}^{(\vartheta)}\{\bullet\} := \mathcal{L}\left\{\bullet \,\middle|\, (\vartheta^N_{k+1})_{N\in\mathbb{N}}\right\} \tag{9.167}$$

We have introduced the time shift $\left(\tilde{\ell}_N\right)$. This gives the motivation for the following generalization of (9.155):

For arguments similar to those nedded for (7.126), there exists an increasing sequence $(\ell_N)_{N\in\mathbb{N}} \in \mathbb{R}_+^{\mathbb{N}}$ with $\lim\limits_{N\to\infty} \ell_N = \infty$ such that the following holds:

$$\lim_{N\to\infty} \sup_{t\in[0,T+\ell_N]} \left|\bar{\mathfrak{Y}}^{N,k}_{N^k t} - \bar{\mathfrak{Y}}^{N,\infty,k}_{N^k t}\right| = 0 \quad P\text{-a.s.} \quad . \tag{9.168}$$

As a by-product this provides us with an upper bound of $\left(\tilde{\ell}_N\right)$, namely (for N large):

$$\tilde{\ell}_N \leq \ell_N \quad .$$

Quenched point of view

We only have to compare the laws of the reactant for given catalyst. In the language of random media, this corresponds to the quenched point of view. In particular we assume that the catalyst converges pathwise as described in (9.168). Taking such a catalyst realization for given, we compare the laws of the reactant conditioned on the catalyst realizations.

In particular, we can decompose $\mathcal{L}^{(\vartheta)}$ as follows[11]:

$$\mathcal{L}^{(\vartheta)} = \int \mathcal{L}^{(\vartheta),\mathfrak{Y}}_{reac} \quad P^{(\vartheta)}_{cat}(d\mathfrak{Y})$$

Our claim is that for any sequence $\tilde{\ell}_N$ with

$$\lim_{N\to\infty} \tilde{\ell}_N = \infty \quad . \tag{9.169}$$

and

$$\tilde{\ell}_N \leq \ell_N \quad (N \in \mathbb{N}) \tag{9.170}$$

we have for P-almost all ϑ

$$\lim_{t\to\infty} \varrho^{\Upsilon}\left(\mathcal{L}^{(\vartheta),\mathfrak{Y}}_{reac}\left\{\left(\bar{\mathfrak{X}}^{N,k}_{N^k(t+\tilde{\ell}_N)}\right)_{t\geq 0}\right\}, \mathcal{L}^{(\vartheta),\mathfrak{Y}}_{reac}\left\{\left(\bar{\mathfrak{X}}^{N,\infty,k}_{t+\tilde{\ell}_N}\right)_{t\geq 0}\right\}\right) = 0 \quad P^{(\vartheta)}_{cat} - a.s. \tag{9.171}$$

Again, the proof relies on duality..

[10]Very briefly: The Prohorov distance of mixtures of probability laws converges to zero if the expectation (resp. the mixing law) of the Prohorov distances of the mixed laws converges to zero.

[11]The measurability of the mapping $\mathfrak{Y} \longmapsto \mathcal{L}^{(\vartheta),\mathfrak{Y}}_{reac}$ follows similar to Lemma 6 (p. 109).

Duality

We introduce a sequence $(\psi_\alpha)_{\alpha\in\{1,...,n\}} \in \mathbb{R}^n_+$ of "observations" corresponding to the elements of the set Υ, which contains the "observation times". We assume

$$\sum_{\alpha=1}^{n} \psi_\alpha \leq 1 \quad . \tag{9.172}$$

Then $\mathbf{E}^{(\vartheta),\mathfrak{Y}}_{reac}\left[\exp\left(-\sum_{\alpha=1}^{n}\psi_\alpha \bar{\mathfrak{X}}^{N,k}_{N^k t_\alpha}(0)\right)\right]$ describes (Cramér-Wold device) the quenched distrubution of $\left(\bar{\mathfrak{X}}^{N,k}_{N^k t}(0)\right)_{t\in\Upsilon}$.

We introduce catalyst-dependent processes $\left(\mathfrak{u}^N_t\right)_{t\geq 0} \in \left(\mathcal{M}\left(\Omega_{N,L}\right)\right)^{\mathbb{R}_+}$, for which the following equation of duality holds

$$\mathbf{E}^{(\vartheta),\mathfrak{Y}}_{reac}\left[\exp\left(-\left\langle \mathfrak{u}^N_{N^k(T+\bar{\ell}_N)}, \mathfrak{X}^N_0\right\rangle\right)\right] = \mathbf{E}^{(\vartheta),\mathfrak{Y}}_{reac}\left[\exp\left(-\sum_{\alpha=1}^{n}\psi_\alpha \bar{\mathfrak{X}}^{N,k}_{N^k t_\alpha}(0)\right)\right] \quad . \tag{9.173}$$

Note that the situation is similar to the lattice case discussed in the context of Proposition 1 (p. 49). This justifies that $\left(\mathfrak{u}^N_t\right)_{t\geq 0}$ has to be defined the following way:

initial condition:

$$\mathfrak{u}^N_0 \equiv 0 \quad . \tag{9.174}$$

evolution at non-jump times: For $N^k(T-t) \notin \{t_1, ..., t_n\}$ the evolution equation is

$$\begin{aligned}\frac{\partial}{\partial t}\mathfrak{u}^N_t(i) &= \sum_{\ell=1}^{L}\frac{c_{\ell-1}}{N^{\ell-1}}\left(\frac{1}{N^\ell}\sum_{\substack{j\in\Omega_{N,L}:\\ d(i,j)\leq \ell}}\mathfrak{u}^N_t(j) - \mathfrak{u}^N_t(i)\right)\\ &\quad - \gamma_x \mathfrak{Y}^N_{N^k(T+\bar{\ell}_N)-t}(i)\left(\mathfrak{u}^N_t(i)\right)^2 \quad (i\in\Omega_{N,L}).\end{aligned} \tag{9.175}$$

jumps: At times t with $N^k(T-t) \in \{t_1, ..., t_n\}$ there are jumps:

$$\begin{aligned}\lim_{s\to N^k(T-t_\alpha)+}\mathfrak{u}^N_{N^k(T-t_\alpha)}(i) &= \lim_{s\to N^k(T-t_\alpha)-}\mathfrak{u}^N_s(i)\\ &\quad + \frac{\psi_\alpha}{N^k}\mathbb{1}_{B_k(0)}(i); \quad (\alpha\in\{1,...,n\}) \quad .\end{aligned} \tag{9.176}$$

We want to compare this process with the dual process

$$\left(\mathfrak{v}^N_t\right)_{t\geq 0} \in \left(\mathcal{M}\left(\{0,\Delta\}\right)\right)^{\mathbb{R}_+}$$

of $\left\{\bar{\mathfrak{X}}^{N,\infty,k}_{N^k t}(0)\right\}_{t\geq 0}$. Namely, the finite dimensional distributions of $\left\{\bar{\mathfrak{X}}^{N,\infty,k}_{N^k t}(0)\right\}_{t\geq 0}$ shall be characterized by $\left(\mathfrak{v}^N_t\right)_{t\geq 0}$ in the following way:

$$\begin{aligned}&\mathbf{E}^{(\vartheta),\mathfrak{Y}}_{reac}\left[\exp\left(-\mathfrak{v}^N_{(T+\bar{\ell}_N)}(0)\,\bar{\mathfrak{X}}^{N,\infty,k}_0(0) - \mathfrak{v}^N_{(T+\bar{\ell}_N)}(\Delta)\,\vartheta^{N,\mathfrak{X}}_{k+1}\right)\right]\\ =\ &\mathbf{E}^{(\vartheta),\mathfrak{Y}}_{reac}\left[\exp\left(-\sum_{\alpha=1}^{n}\psi_\alpha\bar{\mathfrak{X}}^{N,\infty,k}_{N^k(t_\alpha+\bar{\ell}_N)}(0)\right)\right]\end{aligned} \tag{9.177}$$

The equations defining $\left(\mathfrak{v}^N_t\right)_{t\geq 0}$ have to be chosen as follows (compare Chapter 8):

initial condition: Let

$$\mathfrak{v}_0^N(0) = \mathfrak{v}_0^N(\Delta) = 0 \tag{9.178}$$

evolution at non-jump times: For $T-t \notin \{t_1, ..., t_n\}$ we assume the following evolution:

$$\frac{\partial}{\partial t}\mathfrak{v}_t^N(0) = -c_k\mathfrak{v}_t^N(0) - \gamma\bar{\mathfrak{Y}}^{N,\infty,k}_{N^k(T+\tilde{\ell}_N-t)}(0)\left(\mathfrak{v}_t^N(0)\right)^2 \tag{9.179}$$

$$\frac{\partial}{\partial t}\mathfrak{v}_t^N(\Delta) = c_k\mathfrak{v}_t^N(0) \tag{9.180}$$

jumps: At times t with $(T-t) \in \{t_1, ..., t_n\}$ there are (left-continuous) jumps:

$$\lim_{s\to T-t_\alpha+}\mathfrak{v}_{T-t_\alpha}^N(0) = \lim_{s\to T-t_\alpha-}\mathfrak{v}_s^N(0) + \psi_\alpha, \tag{9.181}$$

$$\lim_{s\to T-t_\alpha+}\mathfrak{v}_{T-t_\alpha}^N(\Delta) = \lim_{s\to T-t_\alpha-}\mathfrak{v}_s^N(\Delta); \quad (\alpha \in \{1, ..., n\}) \tag{9.182}$$

It is our objective to prove that

$$\mathcal{L}_{reac}^{(\vartheta),\mathfrak{Y}}\left\{\left(\bar{\mathfrak{X}}^{N,k}_{N^k(t+\tilde{\ell}_N)}(0)\right)_{t\geq 0}\right\} \tag{9.183}$$

and

$$\mathcal{L}_{reac}^{(\vartheta),\mathfrak{Y}}\left\{\left(\bar{\mathfrak{X}}^{N,\infty,k}_{N^k(t+\tilde{\ell}_N)}(0)\right)_{t\geq 0}\right\} \tag{9.184}$$

coincide more and more in terms of finite dimensional distributions. This is equivalent to the claim that

$$\mathbf{E}_{reac}^{(\vartheta),\mathfrak{Y}}\left[\exp\left(-\sum_{\alpha=1}^{n}\psi_\alpha\bar{\mathfrak{X}}^{N,\infty,k}_{N^k(t_\alpha+\tilde{\ell}_N)}(0)\right)\right] - \mathbf{E}_{reac}^{(\vartheta),\mathfrak{Y}}\left[\exp\left(-\sum_{\alpha=1}^{n}\psi_\alpha\bar{\mathfrak{X}}^{N,k}_{N^k(t_\alpha+\tilde{\ell}_N)}(0)\right)\right] \tag{9.185}$$

converges to zero in P_{cat}-probability (Cramér-Wold device).

It suffices to prove this in L^1 – namely, we claim

$$\lim_{N\to\infty}\mathbf{E}_{cat}\left|\mathbf{E}_{reac}^{(\vartheta),\mathfrak{Y}}\left[\exp\left(-\sum_{\alpha=1}^{n}\psi_\alpha\bar{\mathfrak{X}}^{N,\infty,k}_{N^k(t_\alpha+\tilde{\ell}_N)}(0)\right)\right]\right. \tag{9.186}$$

$$\left. - \mathbf{E}_{reac}^{(\vartheta),\mathfrak{Y}}\left[\exp\left(-\sum_{\alpha=1}^{n}\psi_\alpha\bar{\mathfrak{X}}^{N,k}_{N^k(t_\alpha+\tilde{\ell}_N)}(0)\right)\right]\right| = 0 \quad . \tag{9.187}$$

The dual formulation of this claim is

$$\lim_{N\to\infty}\mathbf{E}_{cat}\left|\mathbf{E}_{reac}^{(\vartheta),\mathfrak{Y}}\left[\exp\left(-\mathfrak{v}^N_{(T+\tilde{\ell}_N)}(0)\,\bar{\mathfrak{X}}_0^{N,\infty,k}(0) - \mathfrak{v}^N_{(T+\tilde{\ell}_N)}(\Delta)\,\vartheta^{N,\mathfrak{X}}_{k+1}\right)\right]\right.$$
$$\left.-\mathbf{E}_{reac}^{(\vartheta),\mathfrak{Y}}\left[\exp\left(-\left\langle \mathfrak{u}^N_{N^k(T+\tilde{\ell}_N)}, \mathfrak{X}_0^N\right\rangle\right)\right]\right| = 0 \tag{9.188}$$

With $|e^{-x} - e^{-y}| \leq |x-y| \wedge 1$ for $x, y \geq 0$, it is sufficient to prove

$$\lim_{N\to\infty}\mathbf{E}\left[\left|\mathfrak{v}^N_{(T+\tilde{\ell}_N)}(0)\,\bar{\mathfrak{X}}_0^{N,\infty,k}(0) + \mathfrak{v}^N_{(T+\tilde{\ell}_N)}(\Delta)\,\vartheta^{N,\mathfrak{X}}_{k+1}\right.\right. \tag{9.189}$$

$$\left.\left. - \left\langle \mathfrak{u}^N_{N^k(T+\tilde{\ell}_N)}, \mathfrak{X}_0^N\right\rangle\right| \wedge 1\right] = 0 \tag{9.190}$$

We split up the sums to prove this limit statement. The following points have to be proven one by one:

1.
$$\lim_{N\to\infty} \mathbf{E}\left[\mathfrak{v}^N_{(T+\tilde{\ell}_N)}(0)\,\bar{\mathfrak{X}}^{N,\infty,k}_0(0)\wedge 1\right] = 0 \quad . \tag{9.191}$$

2.
$$\lim_{N\to\infty} \mathbf{E}\left[\sum_{i\in B_k(0)} \mathfrak{u}^N_{N^k(T+\tilde{\ell}_N)}(i)\,\mathfrak{X}^N_0(i)\wedge 1\right] = 0 \quad . \tag{9.192}$$

3.
$$\lim_{N\to\infty} \mathbf{E}\left[\left|\sum_{i\in\Omega_{N,L}\setminus B_k(0)} \mathfrak{u}^N_{N^k(T+\tilde{\ell}_N)}(i)\left(\mathfrak{X}^N_0(i)-\vartheta^{N,\mathfrak{X}}_{k+1}\right)\right|\wedge 1\right] = 0 \quad . \tag{9.193}$$

4.
$$\lim_{N\to\infty} \mathbf{E}\left[\left|\sum_{i\in\Omega_{N,L}\setminus B_k(0)} \mathfrak{u}^N_{N^k(T+\tilde{\ell}_N)}(i) - \mathfrak{v}^N_{T+\tilde{\ell}_N}(\Delta)\right|\wedge 1\right] = 0 \quad . \tag{9.194}$$

Proof of 1. We can conclude from the differential equations defining $\left(\mathfrak{v}^N_t\right)_{t\geq 0}$ (compare in particular (9.181) and (9.172))

$$\mathfrak{v}^N_T(0)\leq 1 \quad . \tag{9.195}$$

There are no jumps in the interval $\left(T, T+\tilde{\ell}_N\right]$.By (9.179), $\mathfrak{v}^N(0)$ is at least exponentially decaying within this interval. Therefore, we get

$$\mathfrak{v}^N_{T+\tilde{\ell}_N}(0)\leq \exp\left(-c_k\tilde{\ell}_N\right) \quad . \tag{9.196}$$

This tends to zero as N increases. With

$$\mathbf{E}\left[\bar{\mathfrak{X}}^{N,\infty,k}_0(0)\right] = \theta_x \quad , \tag{9.197}$$

we get (9.191).

Proof of 2. A similar argument can be applied: The evolution equation of $\mathfrak{u}^N$ implies that for $i\in B_k(0)$ we have the upper bound

$$\begin{aligned}
\mathfrak{u}^N_{N^k(T+\tilde{\ell}_N)}(i) &\leq \left(\frac{1}{N^k}+e^{-c_{k-1}N\tilde{\ell}_N}\right)\exp\left(-c_k\tilde{\ell}_N\right)+\frac{1}{N^{k+1}}\\
&= \frac{1}{N^k}\left(\left(1+e^{k\ln N - c_{k-1}N\tilde{\ell}_N}\right)\exp\left(-c_k\tilde{\ell}_N\right)+\frac{1}{N}\right)
\end{aligned} \tag{9.198}$$

We can conclude:

$$\begin{aligned}
&\lim_{N\to\infty} \mathbf{E}\left[\sum_{i\in B_k(0)} \mathfrak{u}^N_{N^k(T+\tilde{\ell}_N)}(i)\,\mathfrak{X}^N_0(i)\right]\\
\leq\; &\lim_{N\to\infty}\left(\left(1+\exp\left(k\ln N - c_{k-1}N\tilde{\ell}_N\right)\right)\exp\left(-c_k\tilde{\ell}_N\right)+\frac{1}{N}\right)\theta_x\\
\leq\; &\lim_{N\to\infty}\left(\left(1+\frac{N^k}{\exp\left(c_{k-1}N\tilde{\ell}_N\right)}\right)\exp\left(-c_k\tilde{\ell}_N\right)+\frac{1}{N}\right)\theta_x\\
=\; &0 \quad .
\end{aligned} \tag{9.199}$$

This yields (9.192).

Proof of 3. We have to prove (9.193), namely

$$\lim_{N\to\infty} \mathbf{E}\left[\left|\sum_{i\in\Omega_{N,L}\setminus B_k(0)} \mathfrak{u}^N_{N^k(T+\tilde{\ell}_N)}(i)\left(\mathfrak{x}^N_0(i)-\vartheta^{N,\mathfrak{Y}}_{k+1}\right)\right|\wedge 1\right]=0 \quad . \tag{9.200}$$

We use the following decomposition:

$$\begin{aligned}
&\mathbf{E}\left[\left|\sum_{i\in\Omega_{N,L}\setminus B_k(0)} \mathfrak{u}^N_{N^k(T+\tilde{\ell}_N)}(i)\left(\mathfrak{x}^N_0(i)-\vartheta^{N,\mathfrak{Y}}_{k+1}\right)\right|\wedge 1\right]\\
\le\ &\mathbf{E}\left[\left|\sum_{i\in B_{k+1}(0)\setminus B_k(0)} \mathfrak{u}^N_{N^k(T+\tilde{\ell}_N)}(i)\left(\mathfrak{x}^N_0(i)-\vartheta^{N,\mathfrak{Y}}_{k+1}\right)\right|\wedge 1\right] &(9.201)\\
&+\mathbf{E}\left[\left|\sum_{i\in\Omega_{N,L}\setminus B_{k+1}(0)} \mathfrak{u}^N_{N^k(T+\tilde{\ell}_N)}(i)\left(\mathfrak{x}^N_0(i)-\vartheta^{N,\mathfrak{Y}}_{k+1}\right)\right|\wedge 1\right] . &(9.202)
\end{aligned}$$

The second summand (9.202) vanishes in expectation in the limit $N\to\infty$:

$$\begin{aligned}
&\mathbf{E}\left[\left|\sum_{i\in\Omega_{N,L}\setminus B_{k+1}(0)} \mathfrak{u}^N_{N^k(T+\tilde{\ell}_N)}(i)\left(\mathfrak{x}^N_0(i)-\vartheta^{N,\mathfrak{Y}}_{k+1}\right)\right|\wedge 1\right]\\
\le\ &\mathbf{E}\left[\left|\sum_{i\in\Omega_{N,L}\setminus B_{k+1}(0)} \mathfrak{u}^N_{N^k(T+\tilde{\ell}_N)}(i)\left(\mathfrak{x}^N_0(i)+\vartheta^{N,\mathfrak{Y}}_{k+1}\right)\right|\wedge 1\right]\\
\le\ &2\theta_x\mathbf{E}\left[\sum_{i\in\Omega_{N,L}\setminus B_{k+1}(0)} \mathfrak{u}^N_{N^k(T+\tilde{\ell}_N)}(i)\right]\\
\le\ &2\theta_x\left(\frac{c_{k+1}}{N}+\frac{c_{k+2}}{N^2}+\ldots\right)\left(T+\tilde{\ell}_N\right)\overset{N\to\infty}{\longrightarrow}0 &(9.203)
\end{aligned}$$

We consider the first summand (9.201). First we make a telescopic expansion:

$$\begin{aligned}
&\mathbf{E}\left[\left|\sum_{i\in B_{k+1}(0)\setminus B_k(0)} \mathfrak{u}^N_{N^k(T+\tilde{\ell}_N)}(i)\left(\mathfrak{x}^N_0(i)-\vartheta^{N,\mathfrak{Y}}_{k+1}\right)\right|\wedge 1\right]\\
\le\ &\mathbf{E}\left[\left|\sum_{i\in B_{k+1}(0)\setminus B_k(0)} \left(\mathfrak{u}^N_{N^k(T+\tilde{\ell}_N)}(i)-\bar{\mathfrak{u}}^{N,k+1}_{N^k(T+\tilde{\ell}_N)}(0)\right)\mathfrak{x}^N_0(i)\right|\wedge 1\right] &(9.204)\\
&+\mathbf{E}\left[\left|\sum_{i\in B_{k+1}(0)\setminus B_k(0)} \bar{\mathfrak{u}}^{N,k+1}_{N^k(T+\tilde{\ell}_N)}(0)\left(\vartheta^{N,\mathfrak{Y}}_{k+1}-\mathfrak{x}^N_0(i)\right)\right|\wedge 1\right] &(9.205)\\
&+\mathbf{E}\left[\left|\sum_{i\in B_{k+1}(0)\setminus B_k(0)} \left(\bar{\mathfrak{u}}^{N,k+1}_{N^k(T+\tilde{\ell}_N)}(0)-\mathfrak{u}^N_{N^k(T+\tilde{\ell}_N)}(i)\right)\vartheta^{N,\mathfrak{Y}}_{k+1}\right|\wedge 1\right] . &(9.206)
\end{aligned}$$

Next, we prove that (9.204) (9.205) and (9.206) vanish in the limit $N \to \infty$.

First, we consider (9.205): A consideration similar to (9.192)[12] yields that

$$\lim_{N\to\infty} \mathbf{E}\left[\left|\sum_{B_k(0)} \bar{\mathrm{u}}^{N,k+1}_{N^k(T+\tilde{\ell}_N)}(0)\left(\vartheta^{N,\mathfrak{Y}}_{k+1} - \mathfrak{x}^N_0(i)\right)\right| \wedge 1\right] = 0 \quad . \tag{9.207}$$

Hence we get:

$$\begin{aligned}
&\limsup_{N\to\infty}\mathbf{E}\left[\left|\sum_{i\in B_{k+1}(0)\setminus B_k(0)} \bar{\mathrm{u}}^{N,k+1}_{N^k(T+\tilde{\ell}_N)}(0)\left(\vartheta^{N,\mathfrak{Y}}_{k+1} - \mathfrak{x}^N_0(i)\right)\right| \wedge 1\right] \\
= &\limsup_{N\to\infty}\mathbf{E}\left[\left|\sum_{i\in B_{k+1}(0)} \bar{\mathrm{u}}^{N,k+1}_{N^k(T+\tilde{\ell}_N)}(0)\left(\vartheta^{N,\mathfrak{Y}}_{k+1} - \mathfrak{x}^N_0(i)\right)\right| \wedge 1\right] \\
= &\limsup_{N\to\infty}\mathbf{E}\left[\left|\sum_{i\in B_{k+1}(0)} \bar{\mathrm{u}}^{N,k+1}_{N^k(T+\tilde{\ell}_N)}(0)\,\vartheta^{N,\mathfrak{Y}}_{k+1} - \sum_{i\in B_{k+1}(0)} \bar{\mathrm{u}}^{N,k+1}_{N^k(T+\tilde{\ell}_N)}(0)\,\mathfrak{x}^N_0(i)\right| \wedge 1\right] \\
= &\limsup_{N\to\infty}\mathbf{E}\left[\left|N^k\bar{\mathrm{u}}^{N,k+1}_{N^k(T+\tilde{\ell}_N)}(0)\,\vartheta^{N,\mathfrak{Y}}_{k+1} - \bar{\mathrm{u}}^{N,k+1}_{N^k(T+\tilde{\ell}_N)}(0)\,N^k\vartheta^{N,\mathfrak{Y}}_{k+1}\right| \wedge 1\right] \\
= &\ 0 \quad .
\end{aligned} \tag{9.208}$$

In particular, we have used $\vartheta^{N,\mathfrak{Y}}_{k+1} = \frac{1}{N^{k+1}}\sum_{i\in B_{k+1}(0)}\mathfrak{x}^N_0(i)$ (compare eq. (9.148)).

Next, we consider (9.206):

$$\begin{aligned}
&\lim_{N\to\infty}\mathbf{E}\left[\left|\sum_{i\in B_{k+1}(0)\setminus B_k(0)}\left(\bar{\mathrm{u}}^{N,k+1}_{N^k(T+\tilde{\ell}_N)}(0) - \mathrm{u}^N_{N^k(T+\tilde{\ell}_N)}(i)\right)\vartheta^{N,\mathfrak{Y}}_{k+1}\right| \wedge 1\right] \\
= &\lim_{N\to\infty}\mathbf{E}\left[\left|\left(\left(N^{k+1} - N^k\right)\frac{1}{N^{k+1}}\sum_{i\in B_{k+1}(0)} \mathrm{u}^N_{N^k(T+\tilde{\ell}_N)}(i)\right.\right.\right. \\
&\left.\left.\left. - \sum_{i\in B_{k+1}(0)\setminus B_k(0)} \mathrm{u}^N_{N^k(T+\tilde{\ell}_N)}(i)\right)\vartheta^{N,\mathfrak{Y}}_{k+1}\right| \wedge 1\right] \\
= &\lim_{N\to\infty}\mathbf{E}\left[\left|\left(\left(1 - \frac{1}{N}\right)\sum_{i\in B_{k+1}(0)} \mathrm{u}^N_{N^k(T+\tilde{\ell}_N)}(i)\right.\right.\right. \\
&\left.\left.\left. - \sum_{i\in B_{k+1}(0)\setminus B_k(0)} \mathrm{u}^N_{N^k(T+\tilde{\ell}_N)}(i)\right)\vartheta^{N,\mathfrak{Y}}_{k+1}\right| \wedge 1\right] \\
= &\lim_{N\to\infty}\mathbf{E}\left[\vartheta^{N,\mathfrak{Y}}_{k+1}\sum_{i\in B_k(0)} \mathrm{u}^N_{N^k(T+\tilde{\ell}_N)}(i) \wedge 1\right]
\end{aligned} \tag{9.209}$$

This limit equals zero as a consequence of the evolution equations (9.175), (9.176) and (9.172) – again compare(9.192).

[12] Approximately all "u-mass" has left $B_k(0)$ at time $N^k\left(T + \tilde{\ell}_N\right)$ in the limit $N \to \infty$.

The third summand that we have to consider is (9.204). We have to verify that (9.204) vanishes in the limit $N \to \infty$; namely, we claim

$$\lim_{N\to\infty} \mathbf{E}\left[\left|\sum_{i\in B_{k+1}(0)\setminus B_k(0)} \left(\mathrm{u}^N_{N^k(T+\tilde{\ell}_N)}(i) - \bar{\mathrm{u}}^{N,k+1}_{N^k(T+\tilde{\ell}_N)}(0)\right) \cdot \mathfrak{X}^N_0(i)\right| \wedge 1\right] = 0 \quad . \tag{9.210}$$

As first step, we get that for every $\tilde{\imath} \in B_{k+1}(0)\setminus B_k(0)$ we have

$$\begin{aligned} &\mathbf{E}\left[\left|\sum_{i\in B_{k+1}(0)\setminus B_k(0)} \left(\mathrm{u}^N_{N^k(T+\tilde{\ell}_N)}(i) - \bar{\mathrm{u}}^{N,k+1}_{N^k(T+\tilde{\ell}_N)}(0)\right) \mathfrak{X}^N_0(i)\right| \wedge 1\right] \\ \leq\ & \mathbf{E}\left[\left(N^{k+1} - N^k\right)\left|\mathrm{u}^N_{N^k(T+\tilde{\ell}_N)}(\tilde{\imath}) - \bar{\mathrm{u}}^{N,k+1}_{N^k(T+\tilde{\ell}_N)}(0)\right| \theta_x \wedge 1\right] \\ \leq\ & \mathbf{E}\left[N^{k+1}\left|\mathrm{u}^N_{N^k(T+\tilde{\ell}_N)}(\tilde{\imath}) - \bar{\mathrm{u}}^{N,k+1}_{N^k(T+\tilde{\ell}_N)}(0)\right| \theta_x \wedge 1\right] \quad . \end{aligned} \tag{9.211}$$

This way, to get (9.193) it only remains to prove that for arbitrary $i \in B_{k+1}(0)\setminus B_k(0)$ we have

$$\lim_{N\to\infty} \mathbf{E}\left[N^{k+1}\left|\mathrm{u}^N_{N^k(T+\tilde{\ell}_N)}(i) - \bar{\mathrm{u}}^{N,k+1}_{N^k(T+\tilde{\ell}_N)}(0)\right| \wedge 1\right] = 0 \quad . \tag{9.212}$$

We decompose the term

$$N^{k+1}\bar{\mathrm{u}}^{N,k+1}_{N^k(T+\tilde{\ell}_N)}(0) \quad , \tag{9.213}$$

as follows:

$$\begin{aligned} & N^{k+1}\bar{\mathrm{u}}^{N,k+1}_{N^k(T+\tilde{\ell}_N)}(0) \\ =\ & \sum_{\beta\in B_{k+1}(0)} \mathrm{u}^N_{N^k(T+\tilde{\ell}_N)}(\beta) \\ =\ & \sum_{\beta\in B_{k+1}(0)\setminus B_k(0)} \mathrm{u}^N_{N^k(T+\tilde{\ell}_N)}(\beta) + \sum_{\beta\in B_k(0)} \mathrm{u}^N_{N^k(T+\tilde{\ell}_N)}(\beta) \quad . \end{aligned} \tag{9.214}$$

In the limit $N \to \infty$, all mass has evaded[13] from $B_k(0)$ at time $N^k\left(T + \tilde{\ell}_N\right)$:

$$\lim_{N\to\infty} \sum_{\beta\in B_k(0)} \mathrm{u}^N_{N^k(T+\tilde{\ell}_N)}(\beta) = 0 \tag{9.215}$$

Hence the following statement (for $i \in B_{k+1}(0)\setminus B_k(0)$) is equivalent to (9.212):

$$\lim_{N\to\infty} \mathbf{E}\left[\left|N^{k+1}\left(\mathrm{u}^N_{N^k(T+\tilde{\ell}_N)}(i) - \frac{1}{N^{k+1} - N^k} \sum_{\beta\in B_{k+1}(0)\setminus B_k(0)} \mathrm{u}^N_{N^k(T+\tilde{\ell}_N)}(\beta)\right)\right| \wedge 1\right] = 0 \quad . \tag{9.216}$$

[13] again, compare (9.175) and (9.172).

Next, we focus on representing both

$$\mathrm{u}^N_{N^k(T+\tilde{\ell}_N)}(i) \tag{9.217}$$

and

$$\frac{1}{N^{k+1}-N^k} \sum_{\beta \in B_{k+1}(0)\setminus B_k(0)} \mathrm{u}^N_{N^k(T+\tilde{\ell}_N)}(\beta) \tag{9.218}$$

by means of Feynman-Kac representations.

Feynman-Kac representation

On a probability space $\left(\tilde{\Omega}, \mathfrak{A}, \mathbb{P}\right)$ we define n independent families

$$\left\{\left(\xi_s^{j,N,t_\alpha}\right)_{s \geq N^k(T-t_\alpha)}\right\}_{j \in B_k(0)}$$

– parametrized by $\alpha \in \{1, ..., n\}$ – of independent random walks:

As initial condition we choose

$$\xi^{j,N,t_\alpha}_{N^k(T-t_\alpha)} = j \quad ,$$

The transition rates are given by

$$a^N(j_1, j_2) = \sum_{k=d(j_1,j_2)}^{L} \frac{c_{k-1}}{N^{k-1}} \frac{1}{N^k} \quad (j_1, j_2 \in \Omega_{L,N}, j_1 \neq j_2) \quad . \tag{9.219}$$

To prevent a typographical mess, we have to introduce abbreviations:[14]

$\widehat{\mathfrak{Y}\xi}$	:	$-\int_{N^k(T-t_\alpha)}^{N^k(T+\tilde{\ell}_N)} \mathfrak{Y}^N_{N^k(T+\tilde{\ell}_N)-s}\left(\xi_s^{j,N,t_\alpha}\right) \cdot \mathrm{u}^N_s\left(\xi_s^{j,N,t_\alpha}\right) ds$
$\widehat{\mathfrak{Y}\tilde{\pi}\xi}$	:	$-\int_{N^k(T-t_\alpha)}^{N^k(T+\tilde{\ell}_N)} \mathfrak{Y}^N_{N^k(T+\tilde{\ell}_N)-s}\left(\tilde{\pi}\left(\xi_s^{j,N,t_\alpha}\right)\right) \cdot \mathrm{u}^N_s\left(\tilde{\pi}\left(\xi_s^{j,N,t_\alpha}\right)\right) ds$

Then we have the following Feynman-Kac representation[15]:

$$\begin{aligned} &\mathrm{u}^N_{N^k(T+\tilde{\ell}_N)}(i) \\ &= \sum_{\alpha=1}^{n} \psi_\alpha \frac{1}{N^k} \sum_{j \in B_k(0)} \mathbb{E}\left[\exp\left(\widehat{\mathfrak{Y}\xi}\right) \mathbb{I}_{\left\{\xi^{j,N,t_\alpha}_{N^k(T+\tilde{\ell}_N)}=i\right\}}\right] \end{aligned} \tag{9.220}$$

[14] $\tilde{\pi}$ is not defined at this point, but it will be defined a few lines below.

[15] The proof is similar to all the other Feynman-Kac proofs to be found in this thesis. We feel free to omit it.

Next, we construct a Feynman-Kac representation of the second term that occurs in (9.216), namely

$$\frac{1}{N^{k+1}-N^k}\sum_{\beta\in B_{k+1}(0)\backslash B_k(0)} \mathrm{u}^N_{N^k(T+\tilde{\ell}_N)}(\beta) \quad . \tag{9.221}$$

To this end, we introduce on $\left(\tilde{\Omega},\mathfrak{A},\mathbb{P}\right)$ a random permutation π of the elements of $B_{k+1}\backslash B_k$. We assume that π is independent of the families of random walks

$$\left\{\left(\xi_s^{j,N,t_\alpha}\right)_{s\geq N^k(T-t_\alpha)}\right\}_{j\in B_k(0)}$$

and that all permutations have the same probability $\left((N^{k+1}-N^k)!\right)^{-1}$. Let $\tilde{\pi}$ be the invertible (random) mapping defined by

$$\tilde{\pi}(i)=\begin{cases} i & \text{for } \; i\notin B_{k+1}\backslash B_k \\ \pi(i) & \text{for } \; i\in B_{k+1}\backslash B_k \end{cases} \quad . \tag{9.222}$$

For $i\in B_{k+1}(0)\backslash B_k(0)$ arbitrarily chosen, we obtain the following Feynman-Kac representation of (9.221):

$$\begin{aligned} &\frac{1}{N^{k+1}-N^k}\sum_{\beta\in B_{k+1}(0)\backslash B_k(0)} \mathrm{u}^N_{N^k(T+\tilde{\ell}_N)}(\beta) \\ &=\sum_{\alpha=1}^n \psi_\alpha \frac{1}{N^k}\sum_{j\in B_k(0)} \mathbb{E}\left[\exp\left(\widehat{\mathfrak{Y}\tilde{\pi}\xi}\right)\mathbb{I}_{\left\{\xi^{j,N,t_\alpha}_{N^k(T+\tilde{\ell}_N)}=i\right\}}\right] \quad . \end{aligned} \tag{9.223}$$

We have to compare (9.223) and (9.220). The crucial points are:

- the random walks ξ^{j,N,t_α} and $\tilde{\pi}\left(\xi^{j,N,t_\alpha}\right)$ coincide if they are inside $B_k(0)$ (due to (9.222)).
- For the region outside (namely $B_{k+1}(0)\backslash B_k(0)$), we have $\frac{1}{N^{k+1}}$ as an upper bound of the $\mathrm{u}^{\cdots}_{\cdots}$ processes (compare again (9.175), (9.176) and (9.172)).

We want to prove (9.216), namely

$$\begin{aligned} \lim_{N\to\infty}\mathbf{E}\Bigg[\Bigg|N^{k+1}\Bigg(&\mathrm{u}^N_{N^k(T+\tilde{\ell}_N)}(i) \\ &-\frac{1}{N^{k+1}-N^k}\sum_{\beta\in B_{k+1}(0)\backslash B_k(0)} \mathrm{u}^N_{N^k(T+\tilde{\ell}_N)}(\beta)\Bigg)\Bigg|\wedge 1\Bigg]=0 \quad . \end{aligned} \tag{9.224}$$

For this purpose, we focus on finding an upper bound of

$$\left|N^{k+1}\left(\mathrm{u}^N_{N^k(T+\tilde{\ell}_N)}(i)-\frac{1}{N^{k+1}-N^k}\sum_{\beta\in B_{k+1}(0)\backslash B_k(0)} \mathrm{u}^N_{N^k(T+\tilde{\ell}_N)}(\beta)\right)\right| \tag{9.225}$$

for $i \in B_{k+1}(0) \backslash B_k(0)$. We get:

$$\begin{aligned}
&\left| N^{k+1} \left(\mathrm{u}^N_{N^k(T+\tilde{\ell}_N)}(i) \right.\right. \\
&\qquad \left.\left. - \frac{1}{N^{k+1}-N^k} \sum_{\beta \in B_{k+1}(0)\backslash B_k(0)} \mathrm{u}^N_{N^k(T+\tilde{\ell}_N)}(\beta) \right) \right| \\
= \; & N^{k+1} \left| \sum_{\alpha=1}^{n} \psi_\alpha \frac{1}{N^k} \cdot \right. \qquad (9.226) \\
& \left. \cdot \sum_{j \in B_k(0)} \mathbb{E}\left[\mathbb{1}_{\left\{ \xi^{j,N,t_\alpha}_{N^k(T+\tilde{\ell}_N)} = i \right\}} \left(\exp\left(\widehat{\mathfrak{Y}\xi}\right) - \exp\left(\widehat{\mathfrak{Y}\tilde{\pi}\xi}\right) \right) \right] \right| . \qquad (9.227)
\end{aligned}$$

The value of the term

$$\mathbb{E}\left[\mathbb{1}_{\left\{ \xi^{j,N,t_\alpha}_{N^k(T+\tilde{\ell}_N)} = i \right\}} \left(\exp\left(\widehat{\mathfrak{Y}\xi}\right) - \exp\left(\widehat{\mathfrak{Y}\tilde{\pi}\xi}\right) \right) \right] \qquad (9.228)$$

does not depend on $j \in B_k(0)$. This is a simple consequence of the hierarchical stucture of the underlying group. Hence, we do not really have to carry out the summation $\sum_{j \in B_k(0)}$, but we can fix $j = 0$ and multiply with N^k. This factor is compensated by the factor $\frac{1}{N^k}$ in line (9.226).

Therefore, we get in contiuation of (9.227):

$$\begin{aligned}
= \; & N^{k+1} \left| \sum_{\alpha=1}^{n} \psi_\alpha \mathbb{E}\left[\mathbb{1}_{\left\{ \xi^{0,N,t_\alpha}_{N^k(T+\tilde{\ell}_N)} = i \right\}} \cdot \right.\right. \\
& \left.\left. \cdot \left(\exp\left(\widehat{\mathfrak{Y}\xi}\right) - \exp\left(\widehat{\mathfrak{Y}\tilde{\pi}\xi}\right) \right) \right] \right| \qquad (9.229)
\end{aligned}$$

With $\widehat{\mathfrak{Y}\xi}, \widehat{\mathfrak{Y}\tilde{\pi}\xi} \le 0$ we get:

$$\left| \exp\left(\widehat{\mathfrak{Y}\xi}\right) - \exp\left(\widehat{\mathfrak{Y}\tilde{\pi}\xi}\right) \right| \le \left| \left(\widehat{\mathfrak{Y}\xi} - \widehat{\mathfrak{Y}\tilde{\pi}\xi} \right) \right| \wedge 1 \qquad (9.230)$$

Since – we still assume $i \in B_{k+1}(0) \backslash B_k(0)$ – the $\mathbb{P}$-probability of $\left\{ \xi^{0,N,t_\alpha}_{N^k(T+\tilde{\ell}_N)} = i \right\}$ is smaller than $\frac{1}{N^{k+1}}$, we get in continuation of (9.229):

$$\begin{aligned}
&\leq N^{k+1}\sum_{\alpha=1}^{n}\psi_\alpha\frac{1}{N^{k+1}}\mathbb{E}\left[\left|\left(\widehat{\mathfrak{Y}\xi}-\widehat{\mathfrak{Y}\tilde{\pi}\xi}\right)\right|\right]\\
&\leq \sum_{\alpha=1}^{n}\psi_\alpha\mathbb{E}\left[\left|\left(\widehat{\mathfrak{Y}\xi}-\widehat{\mathfrak{Y}\tilde{\pi}\xi}\right)\right|\right]\\
&= \sum_{\alpha=1}^{n}\psi_\alpha\mathbb{E}\Bigg[\Bigg|\Bigg(\int_{N^k(T-t_\alpha)}^{N^k(T+\tilde{\ell}_N)}\mathfrak{Y}^N_{N^k(T+\tilde{\ell}_N)-s}\left(\xi_s^{j,N,t_\alpha}\right)\cdot\mathrm{u}_s^N\left(\xi_s^{j,N,t_\alpha}\right)ds\\
&\qquad -\int_{N^k(T-t_\alpha)}^{N^k(T+\tilde{\ell}_N)}\mathfrak{Y}^N_{N^k(T+\tilde{\ell}_N)-s}\left(\tilde{\pi}\left(\xi_s^{j,N,t_\alpha}\right)\right)\cdot\mathrm{u}_s^N\left(\tilde{\pi}\left(\xi_s^{j,N,t_\alpha}\right)\right)ds\Bigg)\Bigg|\Bigg]
\end{aligned}\tag{9.231}$$

By the definition of $\tilde{\pi}$ (compare (9.222)), we have

$$\begin{aligned}
&\mathfrak{Y}^N_{N^k(T+\tilde{\ell}_N)-s}\left(\xi_s^{j,N,t_\alpha}\right)\cdot\mathbb{1}_{\{\xi_s^{j,N,t_\alpha}\in B_k(0)\}}\\
=\ &\mathfrak{Y}^N_{N^k(T+\tilde{\ell}_N)-s}\left(\tilde{\pi}\left(\xi_s^{j,N,t_\alpha}\right)\right)\cdot\mathbb{1}_{\{\xi_s^{j,N,t_\alpha}\in B_k(0)\}}\quad .
\end{aligned}\tag{9.232}$$

We apply this to (9.231) and get (in continuation)

$$\begin{aligned}
=\sum_{\alpha=1}^{n}\psi_\alpha\mathbb{E}\Bigg[\mathbb{1}_{\{\xi_s^{0,N,t_\alpha}\notin B_k(0)\}}\Bigg|\Bigg(&\int_{N^k(T-t_\alpha)}^{N^k(T+\tilde{\ell}_N)}\mathfrak{Y}^N_{N^k(T+\tilde{\ell}_N)-s}\left(\xi_s^{j,N,t_\alpha}\right)\cdot\mathrm{u}_s^N\left(\xi_s^{j,N,t_\alpha}\right)ds\\
&-\int_{N^k(T-t_\alpha)}^{N^k(T+\tilde{\ell}_N)}\mathfrak{Y}^N_{N^k(T+\tilde{\ell}_N)-s}\left(\tilde{\pi}\left(\xi_s^{j,N,t_\alpha}\right)\right)\cdot\mathrm{u}_s^N\left(\tilde{\pi}\left(\xi_s^{j,N,t_\alpha}\right)\right)ds\Bigg)\Bigg|\Bigg]\quad .
\end{aligned}\tag{9.233}$$

Replacing the difference by a sum makes it larger

$$\begin{aligned}
&\leq\sum_{\alpha=1}^{n}\psi_\alpha\mathbb{E}\left[\mathbb{1}_{\{\xi_s^{0,N,t_\alpha}\notin B_k(0)\}}\cdot\int_{N^k(T-t_\alpha)}^{N^k(T+\tilde{\ell}_N)}\mathfrak{Y}^N_{N^k(T+\tilde{\ell}_N)-s}\left(\xi_s^{j,N,t_\alpha}\right)\cdot\mathrm{u}_s^N\left(\xi_s^{j,N,t_\alpha}\right)ds\right]\\
&\quad+\sum_{\alpha=1}^{n}\psi_\alpha\mathbb{E}\left[\mathbb{1}_{\{\xi_s^{0,N,t_\alpha}\notin B_k(0)\}}\cdot\int_{N^k(T-t_\alpha)}^{N^k(T+\tilde{\ell}_N)}\mathfrak{Y}^N_{N^k(T+\tilde{\ell}_N)-s}\left(\tilde{\pi}\left(\xi_s^{j,N,t_\alpha}\right)\right)\cdot\mathrm{u}_s^N\left(\tilde{\pi}\left(\xi_s^{j,N,t_\alpha}\right)\right)ds\right]\\
&=2\sum_{\alpha=1}^{n}\psi_\alpha\mathbb{E}\left[\mathbb{1}_{\{\xi_s^{0,N,t_\alpha}\notin B_k(0)\}}\cdot\int_{N^k(T-t_\alpha)}^{N^k(T+\tilde{\ell}_N)}\mathfrak{Y}^N_{N^k(T+\tilde{\ell}_N)-s}\left(\xi_s^{j,N,t_\alpha}\right)\cdot\mathrm{u}_s^N\left(\xi_s^{j,N,t_\alpha}\right)ds\right]\quad .
\end{aligned}\tag{9.234}$$

With $\mathfrak{u}_s^N(z) \le \frac{1}{N^{k+1}}$ for $z \notin B_k(0)$(by (9.172), (9.175) and (9.176)) we get

$$\left| N^{k+1} \left(\mathfrak{u}^N_{N^k(T+\tilde{\ell}_N)}(i) - \frac{1}{N^{k+1}-N^k} \sum_{\beta \in B_{k+1}(0)\setminus B_k(0)} \mathfrak{u}^N_{N^k(T+\tilde{\ell}_N)}(\beta) \right) \right|$$
$$\le \sum_{\alpha=1}^{n} \psi_\alpha \mathbb{E}\left[\left(2 \int\limits_{N^k(T-t_\alpha)}^{N^k(T+\tilde{\ell}_N)} \mathfrak{Y}^N_{N^k(T+\tilde{\ell}_N)-s}\left(\xi_s^{j,N,t_\alpha}\right) \cdot \frac{1}{N^{k+1}}\, ds \right)\right] \tag{9.235}$$

To obtain (9.216) we hence only have to prove that

$$\lim_{N\to\infty} \mathbf{E}\left[\sum_{\alpha=1}^{n} \psi_\alpha \mathbb{E}\left[\left(2 \int\limits_{N^k(T-t_\alpha)}^{N^k(T+\tilde{\ell}_N)} \mathfrak{Y}^N_{N^k(T+\tilde{\ell}_N)-s}\left(\xi_s^{j,N,t_\alpha}\right) \cdot \frac{1}{N^{k+1}}\, ds \right)\right]\right] = 0 \quad .$$

This tends to zero, as the following calculation shows:

$$\lim_{N\to\infty} \mathbf{E}\otimes\mathbb{E}\left[\sum_{\alpha=1}^{n} \psi_\alpha \left(2 \int\limits_{N^k(T-t_\alpha)}^{N^k(T+\tilde{\ell}_N)} \mathfrak{Y}^N_{N^k(T+\tilde{\ell}_N)-s}\left(\xi_s^{j,N,t_\alpha}\right) \cdot \frac{1}{N^{k+1}}\, ds \right)\right]$$
$$\le \lim_{N\to\infty} 2N^k\left(\tilde{\ell}_N + t_\alpha\right)\theta_y \cdot \frac{1}{N^{k+1}}$$
$$= \lim_{N\to\infty} \frac{2\left(\tilde{\ell}_N + t_\alpha\right)\theta_y}{N} = 0 \quad . \tag{9.236}$$

This completes the proof of 3.

Proof of 4. We have to prove that

$$\lim_{N\to\infty} \mathbf{E}\left[\left| \sum_{i\in\Omega_{N,L}\setminus B_k(0)} \mathfrak{u}^N_{N^k(T+\tilde{\ell}_N)}(i) - \mathfrak{v}^N_{T+\tilde{\ell}_N}(\Delta)\right| \wedge 1\right] = 0 \tag{9.237}$$

holds. We subdivide the proof into two steps:

Step I: It is sufficient to prove

$$\lim_{N\to\infty} \mathbf{E}\left[\left| \int\limits_0^{T+\tilde{\ell}_N} \left(\sum_{i\in B_k(0)} \mathfrak{u}^N_{N^k s}(i) - \mathfrak{v}^N_s(0)\right) ds \right| \wedge 1\right] = 0 \quad . \tag{9.238}$$

Step II: (9.238) is true.

Proof of Step I: By definition (compare (9.180)), we have

$$\mathfrak{v}^N_{T+\tilde{\ell}_N}(\Delta) = c_k \int\limits_0^{T+\tilde{\ell}_N} \mathfrak{v}^N_s(0)\, ds \quad . \tag{9.239}$$

Therefore we only have to verify

$$\lim_{N\to\infty} \mathbf{E}\left[\left|\sum_{i\in\Omega_{N,L}\setminus B_k(0)} \mathrm{u}^N_{N^k(T+\tilde{\ell}_N)}(i) - c_k \int_0^{T+\tilde{\ell}_N} \sum_{i\in B_k(0)} \mathrm{u}^N_{N^k s}(i)\right| \wedge 1\right] = 0 \quad . \tag{9.240}$$

Since jumps over distances exceeding k are not to be found at times of order $o\left(N^{k+1}\right)$ as $N\to\infty$, we can reformulate (9.240) as follows:

$$\lim_{N\to\infty} \mathbf{E}\left[\left|\sum_{i\in B_{k+1}(0)\setminus B_k(0)} \mathrm{u}^N_{N^k(T+\tilde{\ell}_N)}(i) - c_k \int_0^{T+\tilde{\ell}_N} \sum_{i\in B_k(0)} \mathrm{u}^N_{N^k s}(i)\right| \wedge 1\right] = 0 \quad . \tag{9.241}$$

With (9.175), the evolution of $\sum_{i\in B_{k+1}(0)\setminus B_k(0)} \mathrm{u}^N_{N^k t}(i)$ is given by

$$\begin{aligned}
&\frac{\partial}{\partial t} \sum_{i\in B_{k+1}(0)\setminus B_k(0)} \mathrm{u}^N_{N^k t}(i) \\
= \; & -N^k \sum_{i\in B_{k+1}(0)\setminus B_k(0)} \gamma \mathfrak{Y}^N_{N^k(T+\tilde{\ell}_N-t)}(i)\left(\mathrm{u}^N_{N^k t}(i)\right)^2 \\
& +c_k \sum_{i\in B_k(0)} \mathrm{u}^N_{N^k t}(i) \\
& -N^k \frac{c_{k+1}}{N^{k+1}} \sum_{i\in B_{k+1}(0)\setminus B_k(0)} \mathrm{u}^N_{N^k t}(i) \\
& +N^k \sum_{m=k+2}^{L} \frac{c_m}{N^m} \frac{1}{N^{m-(k+1)}} \sum_{i\in B_m(0)\setminus B_{m-1}(0)} \mathrm{u}^N_{N^k t}(i) \\
& -N^k \sum_{m=k+2}^{L} \frac{c_m}{N^m} \sum_{i\in B_{k+1}(0)\setminus B_k(0)} \mathrm{u}^N_{N^k t}(i) \quad .
\end{aligned} \tag{9.242}$$

The last four lines correspond to

- *immigration:* jumps from $B_k(0)$ to $B_{k+1}(0)\setminus B_k(0)$
- *emigration:* jumps from $B_{k+1}(0)\setminus B_k(0)$ to $B_k(0)$
- *immigration:* jumps from $B_m(0)\setminus B_{m-1}(0)$ to $B_{k+1}(0)\setminus B_k(0)$, $(k+2\le m\le L)$
- *emigration:* jumps from $B_{k+1}(0)\setminus B_k(0)$ to $B_m(0)\setminus B_{m-1}(0)$, $(k+2\le m\le L)$

Taking into account that by (9.172) we have

$$\sum_{i\in B_m(0)\setminus B_{m-1}(0)} \mathrm{u}^N_{N^k t}(i) < 1 \tag{9.243}$$

we realize that the last three lines are of order $O\left(\frac{1}{N}\right)$ or smaller as $N\to\infty$. They therefore can be ignored in the limit.

It is therefore left to prove that

$$\lim_{N\to\infty} \mathbf{E}\left[\int\limits_0^{N^k(T+\tilde{\ell}_N)} \sum_{i\in B_{k+1}(0)\setminus B_k(0)} \gamma\mathfrak{Y}^N_{N^k(T+\tilde{\ell}_N)-s}(i)\left(\mathfrak{u}^N_s(i)\right)^2 ds \wedge 1\right] = 0 \tag{9.244}$$

With

$$\mathfrak{u}^N_s(i) \le \frac{1}{N^{k+1}} \quad \forall i \in B_{k+1}(0)\setminus B_k(0) \tag{9.245}$$

we obtain (9.244) from

$$\begin{aligned}
&\lim_{N\to\infty} \mathbf{E}\left[\int\limits_0^{N^k(T+\tilde{\ell}_N)} \sum_{i\in B_{k+1}(0)\setminus B_k(0)} \gamma\mathfrak{Y}^N_{N^k(T+\tilde{\ell}_N)-s}(i)\left(\mathfrak{u}^N_s(i)\right)^2 ds \wedge 1\right] \\
\le\ &\lim_{N\to\infty} \int\limits_0^{N^k(T+\tilde{\ell}_N)} \sum_{i\in B_{k+1}(0)\setminus B_k(0)} \gamma\theta_y\left(\frac{1}{N^{k+1}}\right)^2 ds \\
\le\ &\lim_{N\to\infty} N^k\left(T+\tilde{\ell}_N\right) N^{k+1}\gamma\theta_y\left(\frac{1}{N^{k+1}}\right)^2 \\
\le\ &\lim_{N\to\infty} \frac{\left(T+\tilde{\ell}_N\right)\gamma\theta_y}{N} = 0 \quad .
\end{aligned} \tag{9.246}$$

Proof of Step II: In this step, we have to prove that (9.238) holds for $\left(\tilde{\ell}_N\right)_{N\in\mathbb{N}}$ increasing to infinity sufficiently slowly. We claim that for a suitable choice of $\left(\tilde{\ell}_N\right)_{N\in\mathbb{N}}$ we have

$$\lim_{N\to\infty} \mathbf{E}\left[\left|\int\limits_0^{T+\tilde{\ell}_N}\left(\sum_{i\in B_k(0)} \mathfrak{u}^N_{N^k s}(i) - \mathfrak{v}^N_s(0)\right) ds\right| \wedge 1\right] = 0 \quad . \tag{9.247}$$

It is clear that we have to compare the folowing catalyst-dependent processes:

- $$\left(N^k \bar{\mathfrak{u}}^{N,k}_{N^k t}(0)\right)_{t\in[0,T+\tilde{\ell}_N]} \quad , \tag{9.248}$$
- $$\left(\mathfrak{v}^N_t(0)\right)_{t\in[0,T+\tilde{\ell}_N]} \quad . \tag{9.249}$$

As notation for the difference process, we introduce.

$$\Delta^N_t := N^k \bar{\mathfrak{u}}^{N,k}_{N^k t}(0) - \mathfrak{v}^N_t(0) \quad . \tag{9.250}$$

By their defining equations (to be found on page 212), both processes are initialized at zero and have jumps at coinciding times. Hence the difference process $\left(\Delta^N_t\right)_{t\in[0,T+\tilde{\ell}_N]}$ is continuous and piecewise differentiable.

It is therefore possible to describe it by a differential equation.
For this purpose, we consider the derivatives of $\left(N^k \bar{\mathfrak{u}}^{N,k}_{N^k t}(0)\right)_{t\in[0,T+\tilde{\ell}_N]}$ and $\left(\mathfrak{v}^N_t(0)\right)_{t\in[0,T+\tilde{\ell}_N]}$ at the points of differentiability. In the following, let

$$t \in \{t \geq 0 : T-t \notin \{t_1,...,t_n\}\} \quad . \tag{9.251}$$

By the evolution equations (9.175) and (9.179), we have

$$\frac{\partial}{\partial t}\mathfrak{v}^N_t(0) = -c_k \mathfrak{v}^N_t(0) - \gamma \bar{\mathfrak{Y}}^{N,\infty,k}_{N^k(T+\tilde{\ell}_N-t)}\left(\mathfrak{v}^N_t(0)\right)^2 \tag{9.252}$$

and

$$\begin{aligned}
&\frac{\partial}{\partial t}\left(N^k \bar{\mathfrak{u}}^{N,k}_{N^k t}(0)\right) \\
= \;& N^{2k}\sum_{\ell=1}^{L}\frac{c_{k+\ell-1}}{N^{k+\ell-1}}\left(\bar{\mathfrak{u}}^{N,k+\ell}_{N^k t}(0) - \bar{\mathfrak{u}}^{N,k}_{N^k t}(0)\right) \\
& -\gamma N^k \sum_{j\in B_k(0)} \mathfrak{Y}^N_{N^k(T+\tilde{\ell}_N-t)}(j)\left(\mathfrak{u}^N_{N^k t}(j)\right)^2 \quad .
\end{aligned} \tag{9.253}$$

We split up the sum $\sum_{\ell=1}^{L}$ and obtain:

$$\begin{aligned}
&\frac{\partial}{\partial t}\left(N^k \bar{\mathfrak{u}}^{N,k}_{N^k t}(0)\right) \\
= \;& -c_k N^k \bar{\mathfrak{u}}^{N,k}_{N^k t}(0) \\
& + c_k N^k \bar{\mathfrak{u}}^{N,k+1}_{N^k t}(0) \\
& + N^{2k}\sum_{\ell=2}^{L}\frac{c_{k+\ell-1}}{N^{k+\ell-1}}\left(\bar{\mathfrak{u}}^{N,k+\ell}_{N^k t}(0) - \bar{\mathfrak{u}}^{N,k}_{N^k t}(0)\right) \\
& -\gamma N^k \sum_{j\in B_k(0)} \mathfrak{Y}^N_{N^k(T+\tilde{\ell}_N-t)}(j)\left(\mathfrak{u}^N_{N^k t}(j)\right)^2 \quad .
\end{aligned} \tag{9.254}$$

Substracting (9.254) from (9.252) we obtain the following differential equation (for $t \in \{t \geq 0 : T-t \notin \{t_1,...,t_n\}\}$)

$$\begin{aligned}
\frac{\partial}{\partial t}\Delta^N_t \;=\; & -c_k \Delta^N_t \\
& -\gamma N^k \sum_{j\in B_k(0)} \mathfrak{Y}^N_{N^k(T+\tilde{\ell}_N-t)}(j)\left(\mathfrak{u}^N_{N^k t}(j)\right)^2 \\
& +\gamma \bar{\mathfrak{Y}}^{N,\infty,k}_{N^k(T+\tilde{\ell}_N-t)}\left(\mathfrak{v}^N_t(0)\right)^2 \\
& +N^k c_k \bar{\mathfrak{u}}^{N,k+1}_{N^k t}(0) \\
& +N^{2k}\sum_{\ell=2}^{L}\frac{c_{k+\ell-1}}{N^{k+\ell-1}}\left(\bar{\mathfrak{u}}^{N,k+\ell}_{N^k t}(0) - \bar{\mathfrak{u}}^{N,k}_{N^k t}(0)\right)
\end{aligned} \tag{9.255}$$

The first three lines may remind us on some calculations carried out in Chapter 7 when considering the Finite System Scheme. A few lines later, we will apply similar methods. But first, we concentrate on the terms which are new for us, namely the two last lines. We introduce the following abbreviation:

$$K_t^N := N^k c_k \bar{\mathrm{u}}_{N^k t}^{N,k+1}(0) + N^{2k} \sum_{\ell=2}^{L} \frac{c_{k+\ell-1}}{N^{k+\ell-1}} \left(\bar{\mathrm{u}}_{N^k t}^{N,k+\ell}(0) - \bar{\mathrm{u}}_{N^k t}^{N,k}(0) \right) \quad . \tag{9.256}$$

In the next two paragraphs α) and β) we will show that for all $t \in \left[0, T + \tilde{\ell}_N\right]$ there exists $\hat{K} < \infty$ with

$$\limsup_{N\to\infty} \sup_{t\in\left[0,T+\tilde{\ell}_N\right]} N \cdot \left|K_t^N\right| < \hat{K} \quad . \tag{9.257}$$

α) For $t \in \left[0, T + \tilde{\ell}_N\right]$ we have

$$\begin{aligned} & \limsup_{N\to\infty} \left(N^k c_k \bar{\mathrm{u}}_{N^k t}^{N,k+1}(0) \right) \\ \leq \; & \limsup_{N\to\infty} \left(N \cdot \left(N^k c_k \bar{\mathrm{u}}_{N^k t}^{N,k+1}(0) \right) \right) \\ \leq \; & c_k \quad . \end{aligned} \tag{9.258}$$

This follows from

$$\bar{\mathrm{u}}_{N^k t}^{N,k+1}(0) \leq \frac{1}{N^{k+1}} \quad . \tag{9.259}$$

β) For $t \in \left[0, T + \tilde{\ell}_N\right]$:

$$\begin{aligned} & \limsup_{N\to\infty} \left(N \cdot \left(N^{2k} \sum_{\ell=2}^{L} \frac{c_{k+\ell-1}}{N^{k+\ell-1}} \left| \bar{\mathrm{u}}_{N^k t}^{N,k+\ell}(0) - \bar{\mathrm{u}}_{N^k t}^{N,k}(0) \right| \right) \right) \\ \leq \; & \limsup_{N\to\infty} \left(N \cdot \left(\sum_{\ell=2}^{L} \frac{c_{k+\ell-1}}{N^{\ell-1}} \right) \right) = c_{k+1} \end{aligned}$$

This follows from

$$\left| \bar{\mathrm{u}}_{N^k t}^{N,k+\ell}(0) - \bar{\mathrm{u}}_{N^k t}^{N,k}(0) \right| \leq \frac{1}{N^k} \quad . \tag{9.260}$$

and $\ell \geq 2$.

We have already mentioned (eq. (9.247)) that it is our objective to verify

$$\lim_{N\to\infty} \mathbf{E} \left[\left| \int_0^{T+\tilde{\ell}_N} \left(\sum_{i\in B_k(0)} \mathrm{u}_{N^k s}^N(i) - \mathfrak{v}_s^N(0) \right) ds \right| \wedge 1 \right] = 0 \quad . \tag{9.261}$$

Since $|\cdot|$ is not differentiable at zero, we have to approximate it. Namely, we approximate $|\cdot|$ by the sequence $(g_n)_{n\in\mathbb{N}}$ given by

$$g_n : x \longmapsto \frac{1}{n} \ln(\cosh(nx)) \quad . \tag{9.262}$$

This sequence has the following properties (compare page 93):

- $g_n'(x) = \tanh(x)$.
- $g_n(x) \leq x g_n'(x) \leq |x| \leq \frac{1}{n}\ln 2 + g_n(x)$.

Therefore, instead of (9.261) we can prove that for arbitrary $n \in \mathbb{N}$ the following is true:

$$\limsup_{N\to\infty} \mathbf{E}\left[g_n\left(\int_0^{T+\bar{\ell}_N} \left(\sum_{i\in B_k(0)} \mathfrak{u}^N_{N^k s}(i) - \mathfrak{v}^N_s(0) \right) ds \right) \wedge 1 \right] = 0 \quad . \tag{9.263}$$

Applying the product formula, we get

$$\begin{aligned}
& \frac{\partial}{\partial t} g_n\left(\Delta^N_t\right) \\
= \; & -c_k g_n'\left(\Delta^N_t\right)\Delta^N_t \\
& -\gamma g_n'\left(\Delta^N_t\right) N^k \sum_{j\in B_k(0)} \mathfrak{Y}^N_{N^k(T+\bar{\ell}_N-t)}(j)\left(\mathfrak{u}^N_{N^k t}(j)\right)^2 \\
& +\gamma g_n'\left(\Delta^N_t\right) \bar{\mathfrak{Y}}^{N,\infty,k}_{N^k(T+\bar{\ell}_N-t)}\left(\mathfrak{v}^N_t(0)\right)^2 \\
& +g_n'\left(\Delta^N_t\right) N^k c_k \bar{\mathfrak{u}}^{N,k+1}_{N^k t}(0) \\
& +g_n'\left(\Delta^N_t\right) N^{2k} \sum_{\ell=2}^{L} \frac{c_{k+\ell-1}}{N^{k+\ell-1}}\left(\bar{\mathfrak{u}}^{N,k+\ell}_{N^k t}(0) - \bar{\mathfrak{u}}^{N,k}_{N^k t}(0)\right) \\
\leq \; & -c_k g_n\left(\Delta^N_t\right) \\
& +K^N_t \\
& +g_n'\left(\Delta^N_t\right)\cdot\gamma\Bigg(\bar{\mathfrak{Y}}^{N,\infty,k}_{N^k(T+\bar{\ell}_N-t)}\left(\mathfrak{v}^N_t(0)\right)^2 \\
& \qquad -N^k \sum_{j\in B_k(0)} \mathfrak{Y}^N_{N^k(T+\bar{\ell}_N-t)}(j)\left(\mathfrak{u}^N_{N^k t}(j)\right)^2\Bigg)
\end{aligned} \tag{9.264}$$

Next, we discuss the difference term on the right hand side of (9.264), namely

$$\bar{\mathfrak{Y}}^{N,\infty,k}_{N^k(T+\bar{\ell}_N-t)}\left(\mathfrak{v}^N_t(0)\right)^2 - N^k \sum_{j\in B_k(0)} \mathfrak{Y}^N_{N^k(T+\bar{\ell}_N-t)}(j)\left(\mathfrak{u}^N_{N^k t}(j)\right)^2 \quad . \tag{9.265}$$

We can write it as a telescopic sum:

$$\begin{aligned}
& \bar{\mathfrak{Y}}^{N,\infty,k}_{N^k(T+\bar{\ell}_N-t)}\left(\mathfrak{v}^N_t(0)\right)^2 - N^k \sum_{\substack{j\in\Omega_{N,L} \\ j\in B_k(0)}} \mathfrak{Y}^N_{N^k(T+\bar{\ell}_N-t)}(j)\left(\mathfrak{u}^N_{N^k t}(j)\right)^2 \\
= \; & \bar{\mathfrak{Y}}^{N,\infty,k}_{N^k(T+\bar{\ell}_N-t)}\left(\left(\mathfrak{v}^N_t(0)\right)^2 - \left(N^k \bar{\mathfrak{u}}^{N,k}_{N^k t}(0)\right)^2\right) && (9.266) \\
& +\left(\bar{\mathfrak{Y}}^{N,\infty,k}_{N^k(T+\bar{\ell}_N-t)} - \bar{\mathfrak{Y}}^{N,k}_{N^k(T+\bar{\ell}_N-t)}(0)\right)\left(N^k \bar{\mathfrak{u}}^{N,k}_{N^k t}(0)\right)^2 && (9.267) \\
& +\bar{\mathfrak{Y}}^{N,k}_{N^k(T+\bar{\ell}_N-t)}(0)\left(N^k \bar{\mathfrak{u}}^{N,k}_{N^k t}(0)\right)^2 \\
& -N^k \sum_{j\in B_k(0)} \mathfrak{Y}^N_{N^k(T+\bar{\ell}_N-t)}(j)\left(\mathfrak{u}^N_{N^k t}(j)\right)^2 && (9.268)
\end{aligned}$$

Inserting this in (9.264) and carrying on the chain of inequalities interrupted at (9.264), we get:

$$\begin{aligned}
&\frac{\partial}{\partial t} g_n\left(\Delta_t^N\right)\\
\leq\ & -c_k g_n\left(\Delta_t^N\right) + K_t^N\\
&+g_n'\left(\Delta_t^N\right)\cdot \bar{\mathfrak{Y}}^{N,\infty,k}_{N^k(T+\tilde{\ell}_N-t)}\left(\left(\mathfrak{v}_t^N(0)\right)^2-\left(N^k\bar{\mathfrak{u}}^{N,k}_{N^kt}(0)\right)^2\right)\\
&+g_n'\left(\Delta_t^N\right)\cdot\left(\left(\bar{\mathfrak{Y}}^{N,\infty,k}_{N^k(T+\tilde{\ell}_N-t)}-\bar{\mathfrak{Y}}^{N,k}_{N^k(T+\tilde{\ell}_N-t)}(0)\right)\left(N^k\bar{\mathfrak{u}}^{N,k}_{N^kt}(0)\right)^2\right)\\
&+g_n'\left(\Delta_t^N\right)\cdot\gamma\left(\bar{\mathfrak{Y}}^{N,k}_{N^k(T+\tilde{\ell}_N-t)}(0)\left(N^k\bar{\mathfrak{u}}^{N,k}_{N^kt}(0)\right)^2\right)\\
&-g_n'\left(\Delta_t^N\right)\cdot\gamma\left(N^k\sum_{\substack{j\in\Omega_{N,L}\\ j\in B_k(0)}}\mathfrak{Y}^N_{N^k(T+\tilde{\ell}_N-t)}(j)\left(\mathfrak{u}^N_{N^kt}(j)\right)^2\right)
\end{aligned}$$

This is bounded above by

$$\begin{aligned}
\leq\ & K_t^N\\
&-g_n'\left(\Delta_t^N\right)\Delta_t^N\cdot\bar{\mathfrak{Y}}^{N,\infty,k}_{N^k(T+\tilde{\ell}_N-t)}\left(\mathfrak{v}_t^N(0)+N^k\bar{\mathfrak{u}}^{N,k}_{N^kt}(0)\right)\\
&+\left|\bar{\mathfrak{Y}}^{N,\infty,k}_{N^k(T+\tilde{\ell}_N-t)}-\bar{\mathfrak{Y}}^{N,k}_{N^k(T+\tilde{\ell}_N-t)}(0)\right|\left(N^k\bar{\mathfrak{u}}^{N,k}_{N^kt}(0)\right)^2\\
&+g_n'\left(\Delta_t^N\right)\cdot\gamma\left(\bar{\mathfrak{Y}}^{N,k}_{N^k(T+\tilde{\ell}_N-t)}(0)\left(N^k\bar{\mathfrak{u}}^{N,k}_{N^kt}(0)\right)^2\right)\\
&-g_n'\left(\Delta_t^N\right)\cdot\gamma\left(N^k\sum_{j\in B_k(0)}\mathfrak{Y}^N_{N^k(T+\tilde{\ell}_N-t)}(j)\left(\mathfrak{u}^N_{N^kt}(j)\right)^2\right)\\
\leq\ & K_t^N+\left|\bar{\mathfrak{Y}}^{N,\infty,k}_{N^k(T+\tilde{\ell}_N-t)}-\bar{\mathfrak{Y}}^{N,k}_{N^k(T+\tilde{\ell}_N-t)}(0)\right|\left(N^k\bar{\mathfrak{u}}^{N,k}_{N^kt}(0)\right)^2\\
&+\gamma\left|N^k\sum_{j\in B_k(0)}\mathfrak{Y}^N_{N^k(T+\tilde{\ell}_N-t)}(j)\left(\left(\bar{\mathfrak{u}}^{N,k}_{N^kt}(0)\right)^2-\left(\mathfrak{u}^N_{N^kt}(j)\right)^2\right)\right|\quad.
\end{aligned}\tag{9.269}$$

We recall our claim

$$\lim_{N\to\infty}\mathbf{E}\left[\left|\int_0^{T+\tilde{\ell}_N}\left(\sum_{i\in B_k(0)}\mathfrak{u}^N_{N^ks}(i)-\mathfrak{v}_s^N(0)\right)\right|ds\wedge 1\right]=0\quad.\tag{9.270}$$

It is sufficient to prove

$$\lim_{N\to\infty}\mathbf{E}\left[\int_0^{T+\tilde{\ell}_N}\left(\int_0^s\left|\Delta_t^N\right|dt\right)ds\wedge 1\right]=0\quad.\tag{9.271}$$

With (9.269), the following has to be verified:

i.

$$\lim_{N\to\infty} \mathbf{E}\left[\int_0^{T+\tilde{\ell}_N}\left(\int_0^s \left|K_t^N\right| dt\right) ds \wedge 1\right] = 0 \quad . \tag{9.272}$$

ii.

$$\lim_{N\to\infty} \mathbf{E}\Bigg[\int_0^{T+\tilde{\ell}_N}\int_0^s \left|\bar{\mathfrak{Y}}^{N,\infty,k}_{N^k(T+\tilde{\ell}_N-t)} - \bar{\mathfrak{Y}}^{N,k}_{N^k(T+\tilde{\ell}_N-t)}(0)\right| \cdot \left(N^k \bar{\mathrm{u}}^{N,k}_{N^k t}(0)\right)^2 dt\, ds \wedge 1\Bigg] = 0 \quad . \tag{9.273}$$

iii.

$$\lim_{N\to\infty} \mathbf{E}\Bigg[\int_0^{T+\tilde{\ell}_N}\int_0^s \Bigg|N^k \sum_{j\in B_k(0)} \mathfrak{Y}^N_{N^k(T+\tilde{\ell}_N-t)}(j) \cdot \left(\left(\bar{\mathrm{u}}^{N,k}_{N^k t}(0)\right)^2 - \left(\mathrm{u}^N_{N^k t}(j)\right)^2\right)\Bigg| dt\, ds \wedge 1\Bigg] = 0. \tag{9.274}$$

To be more precise, we claim that there exists $\left(\tilde{\ell}_N\right)_{N\in\mathbb{N}}$ with

$$\lim_{N\to\infty} \tilde{\ell}_N = \infty, \tag{9.275}$$

such that i., ii. and iii. hold.

<u>Proof of i.</u>

Equation (9.272) follows immediately from (9.257):

$$\begin{aligned}
&\limsup_{N\to\infty} \mathbf{E}\left[\int_0^{T+\tilde{\ell}_N}\int_0^s \left|K_t^N\right| dt\, ds \wedge 1\right] \\
\leq\ &\limsup_{N\to\infty} \mathbf{E}\left[\int_0^{T+\tilde{\ell}_N}\int_0^s \frac{\hat{K}}{N}\, dt\, ds \wedge 1\right] \\
=\ &\limsup_{N\to\infty} \mathbf{E}\left[\int_0^{T+\tilde{\ell}_N} s\frac{\hat{K}}{N}\, ds \wedge 1\right] \\
=\ &\limsup_{N\to\infty} \mathbf{E}\left[\frac{1}{2}\left(T+\tilde{\ell}_N\right)^2 \frac{\hat{K}}{N} \wedge 1\right] \quad .
\end{aligned} \tag{9.276}$$

The latter is zero if we choose $\left(\tilde{\ell}_N\right)_{N\in\mathbb{N}}$ grows sufficiently slowly.

Proof of ii.

We have to prove (9.273), namely

$$\lim_{N\to\infty} \mathbf{E}\left[\int_0^{T+\tilde{\ell}_N}\int_0^s \left|\bar{\mathfrak{Y}}^{N,\infty,k}_{N^k(T+\tilde{\ell}_N-t)} - \bar{\mathfrak{Y}}^{N,k}_{N^k(T+\tilde{\ell}_N-t)}(0)\right| \left(N^k \bar{\mathrm{u}}^{N,k}_{N^k t}(0)\right)^2 dt\, ds \wedge 1\right] = 0 \quad . \tag{9.277}$$

Since the cut-off "$\wedge 1$" enables us to apply dominated convergence, it is sufficient to prove

$$\lim_{N\to\infty} \int_0^{T+\tilde{\ell}_N}\int_0^s \left|\bar{\mathfrak{Y}}^{N,\infty,k}_{N^k(T+\tilde{\ell}_N-t)} - \bar{\mathfrak{Y}}^{N,k}_{N^k(T+\tilde{\ell}_N-t)}(0)\right| \left(N^k \bar{\mathrm{u}}^{N,k}_{N^k t}(0)\right)^2 dt\, ds = 0 \quad P_{cat}\text{-a.s.} \tag{9.278}$$

Using

$$N^k \bar{\mathrm{u}}^{N,k}_{N^k t}(0) \leq 1 \tag{9.279}$$

– this follows from (9.172), it is even sufficient to prove

$$\lim_{N\to\infty} \int_0^{T+\tilde{\ell}_N}\int_0^s \left|\bar{\mathfrak{Y}}^{N,\infty,k}_{N^k(T+\tilde{\ell}_N-t)} - \bar{\mathfrak{Y}}^{N,k}_{N^k(T+\tilde{\ell}_N-t)}(0)\right| dt\, ds = 0 \quad P_{cat}\text{-a.s.} \tag{9.280}$$

We get

$$\begin{aligned}
&\int_0^{T+\tilde{\ell}_N}\int_0^s \left|\bar{\mathfrak{Y}}^{N,\infty,k}_{N^k(T+\tilde{\ell}_N-t)} - \bar{\mathfrak{Y}}^{N,k}_{N^k(T+\tilde{\ell}_N-t)}(0)\right| dt\, ds \\
\leq\ &\int_0^{T+\tilde{\ell}_N} s \sup_{t\in[0,T+\ell_N]} \left|\bar{\mathfrak{Y}}^{N,k}_{N^k t} - \bar{\mathfrak{Y}}^{N,\infty,k}_{N^k t}\right| dt\, ds \\
=\ &\left(T+\tilde{\ell}_N\right)^2 \sup_{t\in[0,T+\ell_N]} \left|\bar{\mathfrak{Y}}^{N,k}_{N^k t} - \bar{\mathfrak{Y}}^{N,\infty,k}_{N^k t}\right| dt\, ds \quad .
\end{aligned} \tag{9.281}$$

By (9.168) we have

$$\lim_{N\to\infty} \sup_{t\in[0,T+\ell_N]} \left|\bar{\mathfrak{Y}}^{N,k}_{N^k t} - \bar{\mathfrak{Y}}^{N,\infty,k}_{N^k t}\right| = 0 \quad P_{cat} - a.s. \tag{9.282}$$

Therefore, (9.281) converges to zero P_{cat}-a.s., if $\tilde{\ell}_N$ grows to infinity sufficiently slowly as $N \to \infty$.

Proof of iii.

It remains to prove

$$\begin{aligned}
\lim_{N\to\infty} \mathbf{E}\Bigg[\int_0^{T+\tilde{\ell}_N}\int_0^s \Bigg| N^k \sum_{j\in B_k(0)} \mathfrak{Y}^N_{N^k(T+\tilde{\ell}_N-t)}(j) \\
\cdot\left(\left(\bar{\mathrm{u}}^{N,k}_{N^k t}(0)\right)^2 - \left(\mathrm{u}^N_{N^k t}(j)\right)^2\right)\Bigg| dt\, ds \wedge 1\Bigg] = 0 \quad .
\end{aligned} \tag{9.283}$$

We have

$$\begin{aligned}
&\int_0^{T+\tilde{\ell}_N}\int_0^{s}\left|N^k\sum_{j\in B_k(0)}\mathfrak{Y}^N_{N^k(T+\tilde{\ell}_N-t)}(j)\left(\left(\bar{\mathrm{u}}^{N,k}_{N^kt}(0)\right)^2-\left(\mathrm{u}^N_{N^kt}(j)\right)^2\right)\right|dt\,ds\\
\leq&\int_0^{T+\tilde{\ell}_N}\int_0^{T+\tilde{\ell}_N}\left|N^k\sum_{j\in B_k(0)}\mathfrak{Y}^N_{N^k(T+\tilde{\ell}_N-t)}(j)\left(\left(\bar{\mathrm{u}}^{N,k}_{N^kt}(0)\right)^2-\left(\mathrm{u}^N_{N^kt}(j)\right)^2\right)\right|dt\,ds\\
=&\left(T+\tilde{\ell}_N\right)\\
&\cdot\int_0^{T+\tilde{\ell}_N}\left|N^k\sum_{j\in B_k(0)}\mathfrak{Y}^N_{N^k(T+\tilde{\ell}_N-t)}(j)\left(\left(\bar{\mathrm{u}}^{N,k}_{N^kt}(0)\right)^2-\left(\mathrm{u}^N_{N^kt}(j)\right)^2\right)\right|dt\quad.
\end{aligned}\tag{9.284}$$

Being free to assume that $\left(\tilde{\ell}_N\right)_{N\in\mathbb{N}}$ grows to infinity very slowly[16], we only have to prove that there exists $\left(\tilde{\ell}_N\right)_{N\in\mathbb{N}}$ with

$$\lim_{N\to\infty}\mathbf{E}\left[\int_0^{T+\tilde{\ell}_N}\left|N^k\sum_{j\in B_k(0)}\mathfrak{Y}^N_{N^k(T+\tilde{\ell}_N-t)}(j)\left(\left(\bar{\mathrm{u}}^{N,k}_{N^kt}(0)\right)^2-\left(\mathrm{u}^N_{N^kt}(j)\right)^2\right)\right|\wedge 1\right]=0\quad.\tag{9.285}$$

We have

$$\begin{aligned}
&\left|N^k\sum_{j\in B_k(0)}\mathfrak{Y}^N_{N^k(T+\tilde{\ell}_N-t)}(j)\left(\left(\bar{\mathrm{u}}^{N,k}_{N^kt}(0)\right)^2-\left(\mathrm{u}^N_{N^kt}(j)\right)^2\right)\right|\\
=&\left|N^k\sum_{j\in B_k(0)}\mathfrak{Y}^N_{N^k(T+\tilde{\ell}_N-t)}(j)\left(\bar{\mathrm{u}}^{N,k}_{N^kt}(0)-\mathrm{u}^N_{N^kt}(j)\right)\left(\bar{\mathrm{u}}^{N,k}_{N^kt}(0)+\mathrm{u}^N_{N^kt}(j)\right)\right|\\
\leq&\ 2\left|\sum_{j\in B_k(0)}\mathfrak{Y}^N_{N^k(T+\tilde{\ell}_N-t)}(j)\left(\bar{\mathrm{u}}^{N,k}_{N^kt}(0)-\mathrm{u}^N_{N^kt}(j)\right)\right|\quad.
\end{aligned}\tag{9.286}$$

As to the last step, compare (9.279). It remains to prove that there exists a slowly increasing $\left(\tilde{\ell}_N\right)_{N\in\mathbb{N}}$ with

$$\lim_{N\to\infty}\int_0^{T+\tilde{\ell}_N}\mathbf{E}\left[\left|\sum_{j\in B_k(0)}\mathfrak{Y}^N_{N^k(T+\tilde{\ell}_N-s)}(j)\left(\bar{\mathrm{u}}^{N,k}_{N^ks}(0)-\mathrm{u}^N_{N^ks}(j)\right)\right|\right]ds=0\quad.\tag{9.287}$$

[16]We only claim that there exists an increasing sequence with $\left(\tilde{\ell}_N\right)_{N\in\mathbb{N}}$ for which the mentioned statements hold. Perhaps $\left(\tilde{\ell}_N\right)_{N\in\mathbb{N}}$ has to be chosen as small as $\log\log\log N$ – who knows.

Applying the Cauchy-Schwartz inequality, we get:

$$\begin{aligned}
&\int_0^{T+\tilde{\ell}_N} \mathbf{E}\left[\left|\sum_{j\in B_k(0)} \mathfrak{Y}^N_{N^k(T+\tilde{\ell}_N-s)}(j)\left(\bar{\mathrm{u}}^{N,k}_{N^k s}(0)-\mathrm{u}^N_{N^k s}(j)\right)\right|\right] ds \\
\leq &\int_0^{T+\tilde{\ell}_N} \mathbf{E}\left[\sqrt{\sum_{j\in B_k(0)}\left(\mathfrak{Y}^N_{N^k(T+\tilde{\ell}_N-s)}(j)\right)^2}\right] \\
&\qquad \cdot\mathbf{E}\left[\sqrt{\sum_{j\in B_k(0)}\left(\bar{\mathrm{u}}^{N,k}_{N^k s}(0)-\mathrm{u}^N_{N^k s}(j)\right)^2}\right] ds \\
\leq &\int_0^{T+\tilde{\ell}_N} \mathbf{E}\left[\sqrt{\sum_{j\in B_k(0)}\left(\mathfrak{Y}^N_{N^k(T+\tilde{\ell}_N-s)}(j)\right)^2}\right] \\
&\qquad \cdot\mathbf{E}\left[\sqrt{\frac{2}{N^k}\sum_{j\in B_k(0)}\left|\bar{\mathrm{u}}^{N,k}_{N^k s}(0)-\mathrm{u}^N_{N^k s}(j)\right|}\right] ds
\end{aligned} \tag{9.288}$$

The latter is a consequence of the finiteness provided by (9.172).
Next, we apply

- the fact that all sites $j \in B_k(0)$ are equivalent
- Jensen's inequality[17]

and get:

$$\begin{aligned}
\leq &\int_0^{T+\tilde{\ell}_N} \sqrt{\mathbf{E}\left[\sum_{j\in B_k(0)}\left(\mathfrak{Y}^N_{N^k(T+\tilde{\ell}_N-s)}(j)\right)^2\right]} \\
&\qquad \cdot\sqrt{\mathbf{E}\left[\frac{2}{N^k}\sum_{j\in B_k(0)}\left|\bar{\mathrm{u}}^{N,k}_{N^k s}(0)-\mathrm{u}^N_{N^k s}(j)\right|\right]} ds \\
= &\int_0^{T+\tilde{\ell}_N} \sqrt{\mathbf{E}\left[N^k\left(\mathfrak{Y}^N_{N^k(T+\tilde{\ell}_N-s)}(0)\right)^2\right]} \\
&\qquad \cdot\sqrt{2\mathbf{E}\left[\left|\bar{\mathrm{u}}^{N,k}_{N^k s}(0)-\mathrm{u}^N_{N^k s}(0)\right|\right]} ds \\
= &\int_0^{T+\tilde{\ell}_N} \sqrt{\mathbf{E}\left[\left(\mathfrak{Y}^N_{N^k(T+\tilde{\ell}_N-s)}(0)\right)^2\right]} \\
&\qquad \cdot\sqrt{2\mathbf{E}\left[N^k\left|\bar{\mathrm{u}}^{N,k}_{N^k s}(0)-\mathrm{u}^N_{N^k s}(0)\right|\right]} ds \quad .
\end{aligned} \tag{9.289}$$

Hence, the following is sufficient for (9.285):

[17] namely $\mathbf{E}\left[\sqrt{\bullet}\right] \leq \sqrt{\mathbf{E}\left[\bullet\right]}$.

- $$\limsup_{N\to\infty} \sup_{s\in[0,N^k(T+\tilde{\ell}_N)]} \mathbf{E}\left[\left(\mathfrak{Y}_s^N(0)\right)^2\right] < \infty \quad , \tag{9.290}$$

- $$\limsup_{N\to\infty} \left(T+\tilde{\ell}_N\right) \sup_{s\in[0,T+\tilde{\ell}_N]} \sqrt{\mathbf{E}\left[N^k \left|\bar{\mathrm{u}}_{N^k s}^{N,k}(0) - \mathrm{u}_{N^k s}^N(0)\right|\right]} = 0 \quad . \tag{9.291}$$

Remark 20 *It clearly is sufficient to prove that there exists an increasing sequence* $\left(\breve{\ell}_N\right)_{N\to\infty}$ *with* $\lim_{N\to\infty} \breve{\ell}_N = \infty$, *s.t. we have*

$$\limsup_{N\to\infty} \left(T+\breve{\ell}_N\right) \sup_{s\in[0,T+\breve{\ell}_N]} \sqrt{\mathbf{E}\left[N^k \left|\bar{\mathrm{u}}_{N^k s}^{N,k}(0) - \mathrm{u}_{N^k s}^N(0)\right|\right]} < \infty \quad . \tag{9.292}$$

Then we get (9.291) by choosing $\tilde{\ell}_N = \sqrt{\breve{\ell}_N}$, *for example.*

Proof of (9.290):

The Multiple Space Time Scale Analysis of Feller's branching diffusion has been discussed by Dawson and Greven in [DG96]. Due to their calculations, especially equation (2.3.), it suffices to prove

$$\sup_{N\in\mathbb{N}} \sup_{t\in[0,T+\tilde{\ell}_N]} \int_0^{N^k t} a_s^N(0,0)\,ds < \infty \tag{9.293}$$

to obtain (9.290). Clearly it suffices to prove

$$\sup_{N\in\mathbb{N}} \int_0^{N^k(T+\tilde{\ell}_N)} a_s^N(0,0)\,ds < \infty \quad . \tag{9.294}$$

Recall the migration rates

$$a^N(j_1,j_2) = \sum_{k=d(j_1,j_2)}^{L} \frac{c_{k-1}}{N^{k-1}} \frac{1}{N^k} \quad (j_1,j_2 \in \Omega_{L,N}, j_1 \neq j_2) \quad . \tag{9.295}$$

They have the following interpretation:
Consider random walks $\left(\zeta_t^N\right)_{t\geq 0}$ – defined on a common probability space – with transition rates given by (9.295). The jumps of this random walk can be classified as follows:

- with rate c_0 there is a *jump of the first kind,* that means if the jumping particle is at site i before the jump, its new site is sampled uniformly in $B_1(i)$. In particular, there is the possibility of degenerate jumps: with probability $\frac{1}{N}$ the particle stays at i.
- with rate $\frac{c_1}{N}$ there is a *jump of the second kind,* that means if the jumping particle is at site i before the jump, its new site is sampled uniformly in $B_2(i)$. In particular, there is the possibility of degenerate jumps: with probability $\frac{1}{N^2}$ the particle stays at i.

- with rate $\frac{c_\ell}{N^\ell}$ $(\ell \in \mathbb{N})$, there is a *jump of the ℓth kind,* that means if the jumping particle is at site i before the jump, its new site is sampled uniformly in $B_\ell(i)$. In particular, there is the possibility of degenerate jumps: with probability $\frac{1}{N^\ell}$ the particle stays at i.

Let

$$\vartheta_1^N, \vartheta_2^N, \ldots (N \in \mathbb{N})$$

be the following sequence of stopping times: ϑ_ℓ^N is the first time at which a jump of the Nth kind takes place. By construction, we can couple the random walks $\left(\zeta_t^N\right)_{t\geq 0}$ $(N \in \mathbb{N})$ by the following assumption:

Let

$$\vartheta_\ell^N = N^{\ell-1} \vartheta_\ell^1 \quad . \tag{9.296}$$

Being interested in the limit $N \to \infty$, it is no loss of generality to assume

$$\vartheta_1^N < \vartheta_2^N < \vartheta_3^N < \ldots \quad . \tag{9.297}$$

This construction yields:

$$P\left[\zeta_t^N = 0 \,\middle|\, t \in \left[\vartheta_\ell^N, \vartheta_{\ell+1}^N\right)\right] = \frac{1}{N^\ell} \quad . \tag{9.298}$$

We finally get:

$$\begin{aligned}
\int\limits_0^{N^k(T+\tilde{\ell}_N)} a_s^N(0,0)\, ds &= \int\limits_0^{N^k(T+\tilde{\ell}_N)} P\left(\zeta_t^N = 0\right) ds \\
&= \mathbf{E}\left[\sum_{\ell=1}^{k+1} \frac{1}{N^{\ell-1}} \left(\vartheta_\ell^N - \vartheta_{\ell-1}^N\right)\right] \\
&= \mathbf{E}\left[\sum_{\ell=1}^{k+1} \frac{1}{N^{\ell-1}} \left(\vartheta_\ell^N - \vartheta_{\ell-1}^N\right)\right] \\
&= \mathbf{E}\left[\sum_{\ell=1}^{k+1} \frac{1}{N^{\ell-1}} \left(N^{\ell-1}\vartheta_\ell^1 - N^{\ell-2}\vartheta_{\ell-1}^1\right)\right] \\
&\leq \mathbf{E}\left[\sum_{\ell=1}^{k+1} \vartheta_\ell^1\right] = \sum_{\ell=1}^{k+1} \frac{1}{c_{\ell-1}} < \infty \quad .
\end{aligned} \tag{9.299}$$

<u>Proof of (9.292):</u>

We have to prove:

$$\limsup_{N\to\infty} \left(T + \tilde{\ell}_N\right) \sup_{s\in\left[0,T+\tilde{\ell}_N\right]} \sqrt{\mathbf{E}\left[N^k \left|\bar{\mathrm{u}}_{N^k s}^{N,k}(0) - \mathrm{u}_{N^k s}^N(0)\right|\right]} < \infty \quad . \tag{9.300}$$

Clearly, the following is sufficient:

Claim 8 *There exists $K < \infty$ such that for all $h, i \in B_k(0), \tau \in \left[0, T + \tilde{\ell}_N\right]$ we have*

$$\limsup_{N\to\infty} N \cdot N^k \mathbf{E}\left[\mathrm{u}^N_{N^k\tau}(h) - \mathrm{u}^N_{N^k\tau}(i)\right] \le K \tag{9.301}$$

The proof of this claim relies on a Feynman-Kac representation similar to (9.220). Let

$$\mathsf{A} := \{\alpha \in \{1, ..., n\} : T - t_\alpha < \tau\} \tag{9.302}$$

For every $\alpha \in \mathsf{A}$ and $N \in \mathbb{N}$ we define a family

$$\left\{\xi^{j,N,t_\alpha}_{t\in\left[N^k(T-t_\alpha),N^k\tau\right]}\right\}_{j\in B_k(0)}$$

of random walks. These random walks shall have common probability space $(\mathbb{O}, \mathbb{A}, \mathbb{P})$. The evolution of the random walks is given by their initial conditions and their migration rates:

- initial condition:

$$\xi^{j,N,t_\alpha}_{N^k(T-t_\alpha)} = j \tag{9.303}$$

- migration rates

$$a^N(j_1, j_2) = \sum_{k=d(j_1,j_2)}^{L} \frac{c_{k-1}}{N^{k-1}} \frac{1}{N^k} \quad (j_1, j_2 \in \Omega_{L,N}, j_1 \neq j_2) \quad . \tag{9.304}$$

Definition 9 *Let*

$$\vartheta^{j,N,t_\alpha}$$

denote the last time at which the continuous time random walk ξ^{j,N,t_α} has a jump of the k-th kind[18] *(that means typically a jump over distance k) before time $N^k\tau$.*

Special case / convention: *If there is no such jump of the k-th kind within the interval $\left[N^k(T - t_\alpha), N^k\tau\right]$, let*

$$\vartheta^{j,N,t_\alpha} = N^k(T - t_\alpha) \quad . \tag{9.305}$$

Note that ϑ^{j,N,t_α} is not a stopping time, but a function of the whole path of ξ^{j,N,t_α}.

Definition 10 *Consider $i, h \in B_k(0)$ as in (9.301).*

1. Denote by $\Pi^{j,h,N}$ the set of permutations $p \in \Omega_{L,N}^{\Omega_{L,N}}$ with

$$p(i) = h \tag{9.306}$$

and

$$p(j) = j \text{ for } d(j, 0) \ge k + 1 \quad . \tag{9.307}$$

2. Denote by $\mathfrak{P} \in \mathcal{M}^1\left(\Pi^{j,h,N}\right)$ the uniform distribution on $\Pi^{j,h,N}$.

[18] the notion "jump of the k-th kind" was introduced on page 234. Note that a jump of the k-th kind happens with rate $\frac{N^{k-1}}{c_{k-1}}$.

3. *Let π^N be a $\Pi^{j,h,N}$-valued random variable defined on $(\mathbb{O}, \mathbb{A}, \mathbb{P})$ with the following two properties:*

   (a)
$$\mathbb{P} \circ \left(\pi^N\right)^{-1} = \mathfrak{P} \quad . \tag{9.308}$$

   (b) $\left\{\xi^{j,N,t_\alpha}_{t\in\left[N^k(T-t_\alpha),N^k\tau\right]}\right\}_{j\in B_k(0)}$ *and π^N are independent.*

Let $\left\{\tilde{\xi}^{j,N,t_\alpha}_{t\in\left[N^k(T-t_\alpha),N^k\tau\right]}\right\}_{j\in B_k(0)}$ be defined by

$$\tilde{\xi}^{j,N,t_\alpha}_s = \xi^{j,N,t_\alpha}_s \text{ for } s \in \left[N^k\left(T-t_\alpha\right), \vartheta^{j,N,t_\alpha}\right) \quad , \tag{9.309}$$
$$\tilde{\xi}^{j,N,t_\alpha}_s = \pi^N\left(\xi^{j,N,t_\alpha}_s\right) \text{ for } s \in \left[\vartheta^{j,N,t_\alpha}, N^k T\right] \quad . \tag{9.310}$$

This construction yields:

1. Whenever the final position of ξ^{j,N,t_α} is i, $\tilde{\xi}^{j,N,t_\alpha}$ has h as final position (this follows from (9.306))
$$\mathbb{I}_{\left\{\xi^{j,N,t_\alpha}_{N^k\tau}=i\right\}} = \mathbb{I}_{\left\{\tilde{\xi}^{j,N,t_\alpha}_{N^k\tau}=h\right\}} \tag{9.311}$$

2. The event $\left\{\xi^{j,N,t_\alpha}_{N^k\tau} = i\right\}$ is independent of the time of the last jump ϑ^{j,N,t_α}. This follows from property "3b" above)

We use a Feynman-Kac representation[19] to describe $\mathrm{u}^N_{N^k\tau}(h)$ and $\mathrm{u}^N_{N^k\tau}(i)$. We use the following abbreviations:

$\widehat{\mathcal{Y}\xi}$	:	$-\int_{N^k(T-t_\alpha)}^{N^k\tau} \mathfrak{Y}^N_{N^k(T+\tilde{\ell}_N)-s}\left(\xi^{j,N,t_\alpha}_s\right)\cdot \mathrm{u}^N_s\left(\xi^{j,N,t_\alpha}_s\right) ds$
$\widehat{\mathcal{Y}\tilde{\xi}}$	:	$-\int_{N^k(T-t_\alpha)}^{N^k\tau} \mathfrak{Y}^N_{N^k(T+\tilde{\ell}_N)-s}\left(\tilde{\xi}^{j,N,t_\alpha}_s\right)\cdot \mathrm{u}^N_s\left(\tilde{\xi}^{j,N,t_\alpha}_s\right) ds$
$\widehat{\mathcal{Y}h}$	:	$-\int_{\vartheta^{j,N,t_\alpha}}^{N^k\tau} \mathfrak{Y}^N_{N^k(T+\tilde{\ell}_N)-s}(h)\cdot \mathrm{u}^N_s(h)\, ds$
$\widehat{\mathcal{Y}i}$	:	$-\int_{\vartheta^{j,N,t_\alpha}}^{N^k\tau} \mathfrak{Y}^N_{N^k(T+\tilde{\ell}_N)-s}(i)\cdot \mathrm{u}^N_s(i)\, ds$
$\widehat{\widehat{\mathcal{Y}}}$	:	$-\int_{N^k(T-t_\alpha)}^{\vartheta^{j,N,t_\alpha}} \mathfrak{Y}^N_{N^k(T+\tilde{\ell}_N)-s}\left(\xi^{j,N,t_\alpha}_s\right)\cdot \mathrm{u}^N_s\left(\xi^{j,N,t_\alpha}_s\right) ds$

Remark 21 *Note that formally the following equations hold:*

$$\widehat{\widehat{\mathcal{Y}}} + \widehat{\mathcal{Y}i} = \widehat{\mathcal{Y}\xi} \quad , \tag{9.312}$$
$$\widehat{\widehat{\mathcal{Y}}} + \widehat{\mathcal{Y}h} = \widehat{\mathcal{Y}\tilde{\xi}} \quad . \tag{9.313}$$

[19]We skip the proof; it is similar to all the other Feynman-Kac arguments carried out in this thesis.

For $\mathrm{u}^N_{N^k\tau}(i)$ we have the Feynman-Kac representation

$$\mathrm{u}^N_{N^k\tau}(i) = \sum_{\alpha\in\mathsf{A}} \psi_\alpha \frac{1}{N^k} \sum_{j\in B_k(0)} \mathbb{E}\left[\exp\left(\widehat{\mathcal{Y}\xi}\right) \mathbb{1}_{\left\{\xi^{j,N,t_\alpha}_{N^k\tau}=i\right\}}\right] \quad . \tag{9.314}$$

(Note that $\widehat{\mathcal{Y}\xi}$ implicitely depends on j)

Similarly, $\mathrm{u}^N_{N^k\tau}(h)$ can be expressed by

$$\mathrm{u}^N_{N^k\tau}(h) = \sum_{\alpha\in\mathsf{A}} \psi_\alpha \frac{1}{N^k} \sum_{j\in B_k(0)} \mathbb{E}\left[\exp\left(\widehat{\mathcal{Y}\tilde{\xi}}\right) \mathbb{1}_{\left\{\tilde{\xi}^{j,N,t_\alpha}_{N^k\tau}=h\right\}}\right] \quad . \tag{9.315}$$

We get:

$$\begin{aligned}
& \left|\mathrm{u}^N_{N^k\tau}(i) - \mathrm{u}^N_{N^k\tau}(h)\right| \\
= & \sum_{\alpha\in\mathsf{A}} \psi_\alpha \frac{1}{N^k} \sum_{j\in B_k(0)} \mathbb{E}\left[e^{\hat{\hat{\mathcal{Y}}}} \left|e^{\widehat{\mathcal{Y}i}} - e^{\widehat{\mathcal{Y}h}}\right| \mathbb{1}_{\left\{\xi^{j,N,t_\alpha}_{N^k\tau}=i\right\}}\right] \\
\leq & \sum_{\alpha\in\mathsf{A}} \psi_\alpha \frac{1}{N^k} \sum_{j\in B_k(0)} \mathbb{E}\left[\left|\widehat{\mathcal{Y}i} - \widehat{\mathcal{Y}h}\right| \mathbb{1}_{\left\{\xi^{j,N,t_\alpha}_{N^k\tau}=i\right\}}\right] \\
\leq & \sum_{\alpha\in\mathsf{A}} \psi_\alpha \frac{1}{N^k} \sum_{j\in B_k(0)} \mathbb{E}\left[\left|\widehat{\mathcal{Y}i} + \widehat{\mathcal{Y}h}\right| \mathbb{1}_{\left\{\xi^{j,N,t_\alpha}_{N^k\tau}=i\right\}}\right] \\
\leq & \; 2\sum_{\alpha\in\mathsf{A}} \psi_\alpha \frac{1}{N^k} \sum_{j\in B_k(0)} \mathbb{E}\left[\left|\widehat{\mathcal{Y}i}\right| \mathbb{1}_{\left\{\xi^{j,N,t_\alpha}_{N^k\tau}=i\right\}}\right] \\
\leq & \; 2\sum_{\alpha\in\mathsf{A}} \psi_\alpha \\
& \cdot \frac{1}{N^k} \sum_{j\in B_k(0)} \mathbb{E}\left[\int_{\vartheta^{j,N,t_\alpha}}^{N^k\tau} \mathfrak{Y}^N_{N^k(T+\tilde{\ell}_N)-s}(i) \cdot \mathrm{u}^N_s(i)\, ds \right. \\
& \left. \cdot \mathbb{1}_{\left\{\xi^{j,N,t_\alpha}_{N^k\tau}=i\right\}} \right] \quad .
\end{aligned} \tag{9.316}$$

Remember that it is our objective to prove (9.301), namely

$$\exists K < \infty : \limsup_{N\to\infty} N \cdot N^k \mathbf{E}\left[\mathrm{u}^N_{N^k\tau}(h) - \mathrm{u}^N_{N^k\tau}(i)\right] \leq K \quad . \tag{9.317}$$

This can be obtained from (9.316) as follows:

$$\begin{aligned}
& \limsup_{N\to\infty} N \cdot N^k \mathbf{E}\left[\left|\mathrm{u}^N_{N^k\tau}(i) - \mathrm{u}^N_{N^k\tau}(h)\right|\right] \\
\leq & \limsup_{N\to\infty} N \cdot \mathbf{E}\left[2\sum_{\alpha\in\mathsf{A}} \psi_\alpha \right. \\
& \left. \cdot \sum_{j\in B_k(0)} \mathbb{E}\left[\int_{\vartheta^{j,N,t_\alpha}}^{N^k\tau} \mathfrak{Y}^N_{N^k(T+\tilde{\ell}_N)-s}(i) \cdot \mathrm{u}^N_s(i)\, ds\, \mathbb{1}_{\left\{\xi^{j,N,t_\alpha}_{N^k\tau}=i\right\}}\right]\right]
\end{aligned} \tag{9.318}$$

Since at time ϑ^{0,N,t_α} a jump of the k-th kind has taken place, we have for $s \geq \vartheta^{j,N,t_\alpha}$:

$$\mathrm{u}_s^N(i) \leq \frac{1}{N^k} \tag{9.319}$$

This yields for $N \in \mathbb{N}$:

$$\begin{aligned}
& N \cdot N^k \mathbf{E}\left[\left|\mathrm{u}_{N^k\tau}^N(i) - \mathrm{u}_{N^k\tau}^N(h)\right|\right] \\
\leq\ & N \cdot \mathbf{E}\left[2\sum_{\alpha\in\mathsf{A}} \psi_\alpha \sum_{j\in B_k(0)} \mathbb{E}\left[\frac{1}{N^k}\int_{\vartheta^{j,N,t_\alpha}}^{N^k\tau} \mathfrak{Y}^N_{N^k(T+\tilde{\ell}_N)-s}(i)\, ds\, \mathbb{1}_{\left\{\xi^{j,N,t_\alpha}_{N^k\tau}=i\right\}}\right]\right] \\
=\ & N\theta_y \cdot \mathbf{E}\left[2\sum_{\alpha\in\mathsf{A}} \psi_\alpha \sum_{j\in B_k(0)} \mathbb{E}\left[\frac{1}{N^k}\left(N^k\tau - \vartheta^{j,N,t_\alpha}\right) \mathbb{1}_{\left\{\xi^{j,N,t_\alpha}_{N^k\tau}=i\right\}}\right]\right] \\
=\ & 2N\theta_y \cdot \sum_{\alpha\in\mathsf{A}} \psi_\alpha \sum_{j\in B_k(0)} \mathbb{E}\left[\frac{1}{N^k}\left(N^k\tau - \vartheta^{j,N,t_\alpha}\right)\right. \\
& \qquad \left.\cdot \mathbb{1}_{\left\{\xi^{j,N,t_\alpha}_{N^k\tau}=i\right\}} \mathbb{1}_{\left\{\vartheta^{j,N,t_\alpha}=N^k(T-t_\alpha)\right\}}\right] && (9.320) \\
& +2N\theta_y \cdot \sum_{\alpha\in\mathsf{A}} \psi_\alpha \sum_{j\in B_k(0)} \mathbb{E}\left[\frac{1}{N^k}\left(N^k\tau - \vartheta^{j,N,t_\alpha}\right)\right. \\
& \qquad \left.\cdot \mathbb{1}_{\left\{\xi^{j,N,t_\alpha}_{N^k\tau}=i\right\}} \mathbb{1}_{\left\{\vartheta^{j,N,t_\alpha}\neq N^k(T-t_\alpha)\right\}}\right] && (9.321)
\end{aligned}$$

In case (9.320), the random walk has not jumped over a distance of k or larger. It can be found at i with a probability of at most $\exp\left(-c_0 N^k \tau\right) + \frac{1}{N^{k-1}}$ given that there was no jump. In case (9.321), there has been a jump. Therefore, the random walk is at i with probability of less than $\frac{1}{N^k}$ in that case. We therefore get:

$$\begin{aligned}
& \limsup_{N\to\infty} 2N\theta_y \cdot \sum_{\alpha\in\mathsf{A}} \psi_\alpha \\
& \qquad \cdot \sum_{j\in B_k(0)} \mathbb{E}\left[\frac{1}{N^k}\left(N^k\tau - \vartheta^{j,N,t_\alpha}\right) \mathbb{1}_{\left\{\xi^{j,N,t_\alpha}_{N^k\tau}=i\right\}} \mathbb{1}_{\left\{\vartheta^{j,N,t_\alpha}=N^k(T-t_\alpha)\right\}}\right] \\
& +\limsup_{N\to\infty} 2N\theta_y \cdot \sum_{\alpha\in\mathsf{A}} \psi_\alpha \\
& \qquad \cdot \sum_{j\in B_k(0)} \mathbb{E}\left[\frac{1}{N^k}\left(N^k\tau - \vartheta^{j,N,t_\alpha}\right) \mathbb{1}_{\left\{\xi^{j,N,t_\alpha}_{N^k\tau}=i\right\}} \mathbb{1}_{\left\{\vartheta^{j,N,t_\alpha}\neq N^k(T-t_\alpha)\right\}}\right] \\
\leq\ & \limsup_{N\to\infty} 2N\theta_y \sum_{\alpha\in\mathsf{A}} \psi_\alpha \\
& \qquad \cdot \sum_{j\in B_k(0)} \tau\left(\exp\left(-c_0 N^k\tau\right) + \frac{1}{N^{k-1}}\right) \mathbb{P}\left[\vartheta^{j,N,t_\alpha} = N^k(T-t_\alpha)\right] && (9.322) \\
& +\limsup_{N\to\infty} 2N\theta_y \sum_{\alpha\in\mathsf{A}} \psi_\alpha \sum_{j\in B_k(0)} \frac{1}{N^{2k}} \mathbb{E}\left[N^k\tau - \vartheta^{j,N,t_\alpha}\right] \quad . && (9.323)
\end{aligned}$$

We know that jumps over distance k happen with rate $\frac{c_{k-1}}{N^{k-1}}$. Hence

$$\mathbb{E}\left[N^k{}_T - \vartheta^{j,N,t_\alpha}\right] \leq \frac{N^{k-1}}{c^{k-1}} \quad . \tag{9.324}$$

We can conclude that $\mathbb{P}\left[\vartheta^{j,N,t_\alpha} = N^k\left(T - t_\alpha\right)\right]$ gets exponentially small as $N \to \infty$, hence we can neglegt the term (9.322) as $N \to \infty$. Calculating the limit of (9.323) we get:

$$\begin{aligned}
& \limsup_{N\to\infty} 2N\theta_y \sum_{\alpha\in\mathsf{A}} \psi_\alpha \sum_{j\in B_k(0)} \frac{1}{N^{2k}} \mathbb{E}\left[N^k{}_T - \vartheta^{j,N,t_\alpha}\right] \\
\leq\ & \limsup_{N\to\infty} 2N\theta_y \sum_{\alpha\in\mathsf{A}} \psi_\alpha \sum_{j\in B_k(0)} \frac{1}{N^{2k}} \left(\frac{N^{k-1}}{c^{k-1}}\right) \\
=\ & \limsup_{N\to\infty} 2N\theta_y \sum_{\alpha\in\mathsf{A}} \psi_\alpha \frac{1}{N^k} \left(\frac{N^{k-1}}{c^{k-1}}\right) \\
\leq\ & \limsup_{N\to\infty} 2N\theta_y \frac{1}{N^k} \left(\frac{N^{k-1}}{c^{k-1}}\right) = \frac{2\theta_y}{c^{k-1}} < \infty \quad .
\end{aligned} \tag{9.325}$$

This completes the proof of (9.162).

9.2.9 Proof of (9.163)

In this subsection, we proof (9.163), namely

$$\lim_{N\to\infty} \varrho^{\Upsilon}\left(\mathcal{L}\left\{\left(\bar{\mathfrak{X}}^{N,\infty,k}_{N^k(t+\bar{\ell}_N)}, \bar{\mathfrak{Y}}^{N,\infty,k}_{N^k(t+\bar{\ell}_N)}\right)_{t\geq 0}\right\} \quad , \quad \int \hat{\Psi}^{\theta^*}_{c_k} \quad \mathcal{L}\left\{\vartheta^N_{k+1}\right\}(d\theta^*)\right) = 0 \quad . \tag{9.326}$$

Recall the definition of the laws of non-stationary processes

$$\mathcal{L}\left\{\left(\bar{\mathfrak{X}}^{N,\infty,k}_{N^k(t+\bar{\ell}_N)}, \bar{\mathfrak{Y}}^{N,\infty,k}_{N^k(t+\bar{\ell}_N)}\right)_{t\geq 0}\right\}$$

provided by (9.159), namely

$$\mathcal{L}\left\{\left(\bar{\mathfrak{X}}^{N,\infty,k}_{N^k t}, \bar{\mathfrak{Y}}^{N,\infty,k}_{N^k t}\right)_{t\geq 0}\right\} := \int \Psi^{\theta^*}_{c_k} \quad \mathcal{L}\left\{\vartheta^N_{k+1}\right\}(d\theta^*) \quad . \tag{9.327}$$

Next, consider the following family of processes. Let

$$\left(\mathcal{L}\left\{\left(\mathfrak{X}^{(\theta)}_t, \mathfrak{Y}^{(\theta)}_t\right)_{t\geq 0}\right\}\right)_{\theta\in\left[0, \sup\limits_{N\in\mathbb{N}} \vartheta^N_{k+1}\right]} \tag{9.328}$$

be defined by

$$\mathcal{L}\left\{\left(\mathfrak{X}^{(\theta)}_t, \mathfrak{Y}^{(\theta)}_t\right)_{t\geq 0}\right\} := \Psi^{\theta}_{c_k} \quad . \tag{9.329}$$

By considerations carried out in the proof of Lemma 2, p. 44 it is sufficient to prove:

Lemma 13

1. *For all* $\theta\in\left[0, \sup\limits_{N\in\mathbb{N}} \vartheta^N_{k+1}\right]$, *we have*

$$\lim_{s\to\infty} \varrho\left(\mathcal{L}\left\{\left(\mathfrak{X}^{(\theta)}_{t+s}, \mathfrak{Y}^{(\theta)}_{t+s}\right)_{t\geq 0}\right\}, \hat{\Psi}^{\theta}_{c_k}\right) = 0 \quad . \tag{9.330}$$

2. *The limiting property (9.330) holds uniformly on* $\left[0, \sup\limits_{N\in\mathbb{N}} \vartheta^N_{k+1}\right]$.

Proof. For 2. there is nothing to prove, since $\left[0, \sup\limits_{N\in\mathbb{N}} \vartheta^N_{k+1}\right]$ is a compact set.

As to the rest of the proof, we refer to Appendix E. The considerations carried out there yield that the limit of $\mathcal{L}\left\{\left(\mathfrak{X}^{(\theta)}_{t+s}, \mathfrak{Y}^{(\theta)}_{t+s}\right)_{t\geq 0}\right\}$ exists uniquely and that it is a stationary process. ■

9.2.10 Proof of (9.164)

We have to prove

$$\lim_{N\to\infty} \varrho^{\Upsilon}\left(\int \hat{\Psi}^{\theta^*}_{c_k} \quad \mathcal{L}\left\{\vartheta^N_{k+1}\right\}(d\theta^*) \quad , \quad \int \hat{\Psi}^{\theta^*}_{c_k} \quad \mathcal{L}\left\{\vartheta_{k+1}\right\}(d\theta^*)\right) = 0 \quad . \tag{9.331}$$

It suffices to prove (by Lemma 2, p. 44) that the mapping

$$\begin{array}{ccc} \mathbb{R}_+ & \longrightarrow & \mathcal{M}^1\left(\mathcal{C}\left(\mathbb{R}_+, \mathbb{R}_+ \times \mathbb{R}_+\right)\right) \\ \vartheta & \longmapsto & \hat{\Psi}^{\vartheta}_{c_k} \end{array} \tag{9.332}$$

is continuous with respect to the topology given by weak convergence on the finite set Υ. Remember that this topology has the metrization ϱ^{Υ}. We only give a sketch of the proof, since similar problems have occurred earlier.

First of all, the claim that (9.332) is continuous corresponds to Lemma 9 on page 145. There, the corresponding continuity property for processes on the lattice is proven. Except for one step, Lemma 9 and the continuity of (9.332) can be proven similarly. Namely, the transience argument used to show that (7.250) vanishes P_{cat}-a.s. in the limit has to be replaced by a calculation regarding the Green's function on the hierarchical group.

9.2.11 Tightness

We have to prove that the sequence

$$\left(\mathcal{L}\left\{\left(\bar{X}^{N,k}_{N^j s(N)+N^k t}(0), \bar{Y}^{N,k}_{N^j s(N)+N^k t}(0)\right)_{t\geq 0}\right\}\right)_{N\in\mathbb{N}} \tag{9.333}$$

is a tight sequence. We have already proven (compare 9.1.7, p. 200) that the sequence

$$\left(\mathcal{L}\left\{\left(\bar{X}^{N,k}_{N^k t}(0), \bar{Y}^{N,k}_{N^k t}(0)\right)_{t\geq 0}\right\}\right)_{N\in\mathbb{N}} \tag{9.334}$$

is tight.

Similar to the proof given there, it suffices to prove that there exist $C < \infty, \tilde{p} > 2$ such that for $u, t \in \mathbb{R}_+$ with $u + 1 > t > u > 0$ we have

$$\limsup_{N\to\infty} \mathbf{E}\left[\left|\bar{X}^{N,k}_{N^j s(N)+N^k t}(0) - \bar{X}^{N,k}_{N^j s(N)+N^k u}(0)\right|^{\tilde{p}}\right] \leq C\left|t-u\right|^{\frac{\tilde{p}}{2}} \quad , \tag{9.335}$$

$$\limsup_{N\to\infty} \mathbf{E}\left[\left|\bar{Y}^{N,k}_{N^j s(N)+N^k t}(0) - \bar{Y}^{N,k}_{N^j s(N)+N^k u}(0)\right|^{\tilde{p}}\right] \leq C\left|t-u\right|^{\frac{\tilde{p}}{2}} \quad . \tag{9.336}$$

Following the line of the argumentation carried out in 9.1.7, if suffices to verify

$$\limsup_{N\to\infty} \sup_{\varrho\in[u,t]} \mathbf{E}\left[\left(X^N_{N^j s(N)+N^k \varrho}(0)\right)^{\tilde{p}} + \left(Y^N_{N^j s(N)+N^k \varrho}(0)\right)^{\tilde{p}}\right] < \infty \quad . \tag{9.337}$$

This follows immediately from equation (B.3), which is contained in Lemma 14, p. 247.

Appendix A

A random walk property

We want to prove the following proposition:

Proposition 3 *Assume*

$$a_t\,(i,j) > 0 \tag{A.1}$$

for all $t > 0$, $i, j \in \mathbb{Z}^d$ *and*

$$\sum_{i\in\mathbb{Z}^d} |i|^2\, a_t(0,i) < \infty \quad . \tag{A.2}$$

Then the limit

$$\lim_{T\to\infty} \int_0^T \left(\hat{a}_s\,(0,0) - \hat{a}_s\,(x,y)\right) ds \tag{A.3}$$

exists in $\mathbb{R}$ *for arbitrary* $x, y \in \mathbb{Z}^d$.

Proof. For $d \geq 3$ the Green function is finite (transient case). Therefore, there is almost nothing to prove in this case.

We henceforth focus on the case $d \in \{1,2\}$ .The claimed property is closely related to the existence of a potential kernel for random walk, as explained in [Spi76]. There (III.12.P1, p.121 and VII.28.P8, p.351) we can find the proof of the following (in our notation):

Let $(\alpha_n)_{n\in\mathbb{N}_0\cup\{-1\}}$ be defined by

$$\alpha_{-1} := 0 \quad , \tag{A.4}$$

$$\alpha_n := \sum_{k=0}^{n} \left(\hat{a}^{(k)}\,(0,0) - \hat{a}^{(k)}\,(x,y)\right) \quad , \tag{A.5}$$

then

$$\lim_{n\to\infty} \alpha_n \text{ exists for arbitrary } x, y \in \mathbb{Z}^d. \tag{A.6}$$

We have to consider

$$\hat{A}_T := \int_0^T \left(\hat{a}_t\,(0,0) - \hat{a}_t\,(x,y)\right) dt \tag{A.7}$$

for arbitrary $T \in \mathbb{R}_+$. As T is finite by assumption, this integral is well-defined. Expansion of the transition probabilities yields:

$$\hat{A}_T = \int_0^T \sum_{k=0}^{\infty} \exp(-t) \frac{t^k}{k!} \left(\hat{a}^{(k)}(0,0) - \hat{a}^{(k)}(x,y) \right) dt \quad . \tag{A.8}$$

With

$$\begin{aligned} &\sum_{k=0}^{\infty} \exp(-t) \frac{t^k}{k!} \left| \hat{a}^{(k)}(0,0) - \hat{a}^{(k)}(x,y) \right| \\ &\leq \sum_{k=0}^{\infty} \exp(-t) \frac{t^k}{k!} \cdot 1 = 1 \quad , \end{aligned} \tag{A.9}$$

the integrated sum converges absolutely. We therefore can interchange summation and integration:

$$\hat{A}_T = \sum_{k=0}^{\infty} \left(\int_0^T \exp(-t) \frac{t^k}{k!} dt \right) \left(\hat{a}^{(k)}(0,0) - \hat{a}^{(k)}(x,y) \right) \quad . \tag{A.10}$$

Let the sequence $\left(r_k^T\right)_{k \in \mathbb{N}_0}$ be defined by

$$r_k^T := \int_0^T \exp(-t) \frac{t^k}{k!} dt \quad . \tag{A.11}$$

We need the following facts:

- For all $k \in \mathbb{N}_0$, we have
$$\lim_{T \to \infty} r_k^T = 1 \quad . \tag{A.12}$$
This is clear, if we recall that the moments of the exponential distribution are given by the factorials.

- For each $T > 0$, the sequence $\left(r_k^T\right)_{k \in \mathbb{N}_0}$ is decreasing in k.
This can be verified as follows: evidently, we have
$$\exp(-t) \frac{t^k}{k!} - \exp(-t) \frac{t^{k+1}}{(k+1)!} \begin{array}{lll} > 0 & for & t < k+1 \\ = 0 & for & t = k+1 \\ < 0 & for & t > k+1 \end{array} \quad . \tag{A.13}$$
This yields:

Case 1 $T \leq k+1$

$$r_k^T - r_{k+1}^T = \int_0^T \left(\exp(-t) \frac{t^k}{k!} - \exp(-t) \frac{t^{k+1}}{(k+1)!} \right) dt > 0 \quad . \tag{A.14}$$

Case 2 $T > k + 1$

$$\begin{aligned} r_k^T - r_{k+1}^T &= \int_0^T \exp(-t)\frac{t^k}{k!}dt - \int_0^T \exp(-t)\frac{t^{k+1}}{(k+1)!}dt \\ &= \left(1 - \int_T^\infty \exp(-t)\frac{t^k}{k!}dt\right) - \left(1 - \int_T^\infty \exp(-t)\frac{t^{k+1}}{(k+1)!}dt\right) \\ &= -\left(\int_T^\infty \left(\exp(-t)\frac{t^k}{k!} - \exp(-t)\frac{t^{k+1}}{(k+1)!}\right)dt\right) \\ &> 0 \quad . \end{aligned} \tag{A.15}$$

In both cases, we have applied (A.13).

The two properties of $\left(r_k^T\right)$ mentioned above allow us to define a sequence of random variables as follows:

On a probability space $(\Omega, \mathfrak{A}, P)$ let $(\xi_T)_{T\geq 0}$ denote a family of $\mathbb{N}_0 \cup \{-1\}$-valued random variables with

$$P[\xi_T = -1] := e^{-T} \quad , \tag{A.16}$$

$$P[\xi_T = n] := \int_0^T \exp(-t)\frac{t^n}{n!}dt - \int_0^T \exp(-t)\frac{t^{n+1}}{(n+1)!}dt \quad (n \in \mathbb{N}_0) . \tag{A.17}$$

Note that by (A.12) the sum of probabilities indeed equals 1. We can rewrite (A.12) as weak convergence property, namely

$$\forall C > 0 : \lim_{T\to\infty} P[\xi_T > C] = 1 \quad . \tag{A.18}$$

With aid of the random variables just introduced, we get the following representation of (A.10):

$$\hat{A}_T = \mathbf{E}\left[\alpha_{\xi_T}\right] \quad , \tag{A.19}$$

using the convention

$$\sum_{k=0}^{-1}\left(\int_0^T \exp(-t)\frac{t^k}{k!}dt\right)\left(\hat{a}^{(k)}(0,0) - \hat{a}^{(k)}(x,y)\right) = 0 \quad . \tag{A.20}$$

Since $(\alpha_n)_{n\in\mathbb{N}_0\cup\{-1\}}$ is bounded, we can conclude from (A.19), the weak convergence result (A.18) and the discrete time result (A.6) that the limit (A.3) exists in $\mathbb{R}$. ■

Appendix B

About uniform boundedness of higher moments on the torus and on the hierarchical group

In the following, we prove two lemmas at once, namely the uniform boundedness of moments on the torus and on the hierarchical group. Since the proofs contain similar steps and share quite a lot of ideas and calculations, we have chosen a synoptic layout at the places where different steps have to be carried out for the two situations considered. Whenever we speak about the torus case $\Lambda = \mathbb{T}_N^d$ alone, we will signify this by creating a box. If we consider the hierarchical group case $\Lambda = \Omega_{L,N}$, we use a grey background.

In the following, we assume $d \geq 3$.	There is no corresponding assumption in the hierarchical group case.[1]

[1] Note that for $L = \infty$, the condition for transience is $\sum \frac{1}{c_k} < \infty$.

For the tightness proofs contained in the proof of the Finite System Scheme, we need uniform boundedness of a moment greater than two in rescaled time. A similar statement can be found in a paper by Cox, Greven and Shiga [CGS95] who consider the Finite System Scheme for a class of linearly interacting diffusions with $\mathbb{R}$-valued components. They prove that for the model they consider moments remain uniformly bounded in rescaled time if they exist at time zero.

In Section 1.2.4 we assumed

$$\sup_{N\in\mathbb{N}} \mathbf{E}\left[\left(X_0^N(0)\right)^p + \left(Y_0^N(0)\right)^p\right] < \infty \quad . \tag{B.1}$$

The following lemma shows that this property is preserved:

Lemma 14 *For arbitrary $t \in \mathbb{R}_+$ and $k \in \{1, ..., L\}$ the following is true:*

$$\boxed{\limsup_{N\to\infty} \sup_{u\in[0,t]} \mathbf{E}\left[\left(X_{\beta_N u}^N(0)\right)^p\right] < \infty \quad ,} \tag{B.2}$$

$$\limsup_{N\to\infty} \sup_{u\in[0,t]} \mathbf{E}\left[\left(X^N_{N^k u}(0)\right)^p\right] < \infty \quad . \tag{B.3}$$

Remark 22 *Recall that we have translation invariance as assumed by (1.11). Hence*

$$\mathbf{E}\left[\left(X^N_u(0)\right)^p\right] = \mathbf{E}\left[\left(X^N_u(i)\right)^p\right] \quad \boxed{\left(i\in\mathbb{T}^d_N, u\geq 0\right)} \quad \left(i\in\Omega_{L,N}, u\geq 0\right) \tag{B.4}$$

and for all $N\in\mathbb{N}$

$$\boxed{\sup_{u\in[0,t]} \mathbf{E}\left[\left(X^N_{\beta_N u}(0)\right)^p\right] = \sup_{u\in[0,t]} \mathbf{E}\left[\left(X^N_{\beta_N u}(i)\right)^p\right] \quad \left(i\in\mathbb{T}^d_N\right).} \tag{B.5}$$

$$\sup_{u\in[0,t]} \mathbf{E}\left[\left(X^N_{N^k u}(0)\right)^p\right] = \sup_{u\in[0,t]} \mathbf{E}\left[\left(X^N_{N^k u}(i)\right)^p\right] \quad \left(i\in\Omega_{L,N}\right). \tag{B.6}$$

Proof. Since $[0,t]$ is a compact set, it suffices to prove

$$\boxed{\limsup_{N\to\infty} \mathbf{E}\left[\left(X^N_{\beta_N t}(0)\right)^p\right] < \infty} \tag{B.7}$$

and

$$\limsup_{N\to\infty} \mathbf{E}\left[\left(X^N_{N^k t}(0)\right)^p\right] < \infty \quad . \tag{B.8}$$

We take the following decomposition (compare Lemma 4, p. 69) as starting point of our calculations:

$$\begin{aligned} X^N_{\beta_N t}(0) &= \sum_{j\in\Lambda} a^N_{c_x\beta_N t}(j,0)\, X^N_0(j) \\ &\quad + \sum_{j\in\Lambda} \int_0^{\beta_N t} a^N_{\beta_N t-\varrho}(j,0) \sqrt{2\gamma_x X^N_\varrho(j)\, Y^N_\varrho(j)} dB_\varrho(j) \end{aligned} \tag{B.9}$$

We consider the p-th moment, namely

$$\begin{aligned} \mathbf{E}\Bigg[\Bigg(&\sum_{j\in\Lambda} a^N_{c_x\beta_N t}(j,0)\, X^N_0(j) \\ &+ \sum_{j\in\Lambda} \int_0^{\beta_N t} a^N_{c_x(\beta_N t-\varrho)}(j,0) \sqrt{2\gamma_x X^N_\varrho(j)\, Y^N_\varrho(j)} dB_\varrho(j)\Bigg)^p\Bigg] \quad . \end{aligned} \tag{B.10}$$

To decompose this expectation, we need the following fact:

Lemma 15 *For $p > 2$ there exists $C_p \in \mathbb{R}_+$ with*

$$(\alpha+\beta)^p \leq (p-1)\,\alpha^p + C_p\beta^p \quad \forall \alpha,\beta \geq 0 \quad . \tag{B.11}$$

Proof. For $\alpha = 0$ or $\beta = 0$ there is nothing to prove. We therefore assume $\alpha, \beta > 0$. The claim is equivalent to

$$\exists C_p \in \mathbb{R}_+ : \left(\frac{\alpha}{\beta}+1\right)^p \leq (p-1)\left(\frac{\alpha}{\beta}\right)^p + C_p \quad \forall \alpha,\beta > 0 \quad . \tag{B.12}$$

It is sufficient to verify

$$\sup_{x>0}\left(x^p\left(\left(\frac{x+1}{x}\right)^p - (p-1)\right)\right) < \infty \quad . \tag{B.13}$$

By

$$\lim_{x\to\infty}\left(\left(\frac{x+1}{x}\right)^p - (p-1)\right) = 2-p < 0 \tag{B.14}$$

and $x^p \geq 0$ for $x \in \mathbb{R}_+$ we obtain

$$\limsup_{x\to\infty}\left(x^p\left(\left(\frac{x+1}{x}\right)^p - (p-1)\right)\right) < 0 \quad . \tag{B.15}$$

Since $x \longmapsto x^p \cdot \left(\left(\frac{x+1}{x}\right)^p - (p-1)\right)$ is continous on $(0,\infty)$, this is sufficient for (B.13). ■

We get: (with $\beta_N := N^k$ in the hierarchical group case):

$$\begin{aligned}
&\mathbf{E}\left[\left(\sum_{j\in\Lambda} a^{N\cdot}_{c_x\beta_N t}(j,0)\,X_0^N(j)\right.\right.\\
&\qquad\left.\left.+\sum_{j\in\Lambda}\int_0^{\beta_N t} a^N_{c_x(\beta_N t-\varrho)}(j,0)\sqrt{2\gamma_x X_\varrho^N(j)\,Y_\varrho^N(j)}dB_\varrho(j)\right)^p\right]\\
&\leq C_p\mathbf{E}\left[\left(\sum_{j\in\Lambda} a_{c_x\beta_N t}(j,0)\,X_0^N(j)\right)^p\right]\\
&+(p-1)\,\mathbf{E}\left[\left|\sum_{j\in\Lambda}\int_0^{\beta_N t} a^N_{c_x(\beta_N t-\varrho)}(j,0)\sqrt{2\gamma_x X_\varrho^N(j)\,Y_\varrho^N(j)}dB_\varrho(j)\right|^p\right]
\end{aligned} \tag{B.16}$$

By Jensen's inequality ($p > 1$), the first part of the r.h.s. of (B.16) can be bounded above as follows

$$\begin{aligned}
C_p\mathbf{E}\left[\left(\sum_{j\in\Lambda} a^N_{c_x\beta_N t}(j,0)\,X_0^N(j)\right)^p\right] &\leq C_p\mathbf{E}\left[\sum_{j\in\Lambda} a_{c_x\beta_N t}(j,0)\left(X_0^N(j)\right)^p\right]\\
&= C_p\mathbf{E}\left[\left(X_0^N(0)\right)^p\right] \quad .
\end{aligned} \tag{B.17}$$

By assumption (B.1) there exists a uniform upper bound K_p:

$$\sup_{N\in\mathbb{N}} C_p\mathbf{E}\left[\left(X_0^N(0)\right)^p\right] < K_p < \infty \quad . \tag{B.18}$$

Next, we consider the second term of the r. h. s. of (B.16), namely

$$(p-1)\,\mathbf{E}\left[\left|\sum_{j\in\Lambda}\int_0^{\beta_N t} a^N_{c_x(\beta_N t-\varrho)}(j,0)\sqrt{2\gamma_x X^N_\varrho(j)\,Y^N_\varrho(j)}dB_\varrho(j)\right|^p\right] \quad . \tag{B.19}$$

We can regard the term

$$\sum_{j\in\Lambda}\int_0^{\beta_N t} a^N_{c_x(\beta_N t-\varrho)}(j,0)\sqrt{2\gamma_x X^N_\varrho(j)\,Y^N_\varrho(j)}dB_\varrho(j) \tag{B.20}$$

as evaluation of the martingale $(M_t)_{t\geq 0}$ given by

$$M_t := \sum_{j\in\Lambda}\int_0^{t} a^N_{c_x(\beta_N t-\varrho)}(j,0)\sqrt{2\gamma_x X^N_\varrho(j)\,Y^N_\varrho(j)}dB_\varrho(j) \tag{B.21}$$

at $\beta_N t$. The fact that M is a local martingale enables us to apply the Burkholder-Davis-Gundy inequality (compare [KS], III.3.28). We get:

$$\sup_{r\in[0,R]} \mathbf{E}\left[|M_r|^p\right] \leq (p-1)^p\,\mathbf{E}\left[\left(\langle M\rangle_R\right)^{\frac{p}{2}}\right] \tag{B.22}$$

($\langle M\rangle$ denotes the increasing process). This yields

$$\begin{aligned}
&\mathbf{E}\left[\left(X^N_{\beta_N t}(0)\right)^p\right]\\
&\leq K_p + (p-1)^p\,(2\gamma_x)^{\frac{p}{2}}\\
&\qquad\cdot\mathbf{E}\left[\left(\sum_{j\in\Lambda}\int_0^{\beta_N t}\left(a^N_{c_x(\beta_N t-\varrho)}(j,0)\right)^2 X^N_\varrho(j)\,Y^N_\varrho(j)\,d\varrho\right)^{\frac{p}{2}}\right] \quad .
\end{aligned} \tag{B.23}$$

Define:

$$\alpha = 2\cdot\left(1-\frac{2}{p}\right) \quad , \tag{B.24}$$

$$q = \frac{1}{1-\frac{2}{p}} \quad . \tag{B.25}$$

This yields:

$$\alpha q = 2, \quad \frac{p}{2q} = \frac{p}{2} - 1 \quad . \tag{B.26}$$

In order to apply Hölder's inequality to get an upper bound of

$$\sum_{j\in\Lambda}\int_0^{\beta_N t}\left(a^N_{c_x(\beta_N t-\varrho)}(j,0)\right)^2 X^N_\varrho(j)\,Y^N_\varrho(j)\,d\varrho \tag{B.27}$$

we write (in continuation of (B.23)):

$$
\begin{aligned}
&\mathbf{E}\left[\left(X^N_{\beta_N t}(0)\right)^p\right] \\
&\leq K_p \\
&+ (p-1)^p (2\gamma_x)^{\frac{p}{2}} \\
&\quad \cdot \mathbf{E}\left[\left(\sum_{j\in\not\geq}\int_0^{\beta_N t}\left(a^N_{c_x(\beta_N t-\varrho)}(j,0)\right)^{\alpha}\left(a^N_{\beta_N t-\varrho}(j,0)\right)^{2-\alpha} X^N_\varrho(j)\, Y^N_\varrho(j)\, d\varrho\right)^{\frac{p}{2}}\right] \quad .
\end{aligned} \tag{B.28}
$$

Since (compare (B.25))

$$
\frac{1}{\frac{p}{2}} + \frac{1}{q} = 1 \quad , \tag{B.29}
$$

we obtain by Hölder's inequality

$$
\begin{aligned}
&\mathbf{E}\left[\left(\sum_{j\in\Lambda}\int_0^{\beta_N t}\left(a^N_{c_x(\beta_N t-\varrho)}(j,0)\right)^{\alpha}\left(a^N_{c_x(\beta_N t-\varrho)}(j,0)\right)^{2-\alpha} X^N_\varrho(j)\, Y^N_\varrho(j)\, d\varrho\right)^{\frac{p}{2}}\right] \\
&\leq \mathbf{E}\left[\left(\left(\sum_{j\in\Lambda}\int_0^{\beta_N t}\left(a^N_{c_x(\beta_N t-\varrho)}(j,0)\right)^{\alpha q} d\varrho\right)^{\frac{1}{q}}\right.\right. \\
&\qquad \left.\left.\cdot \left(\sum_{j\in\Lambda}\int_0^{\beta_N t}\left(\left(a^N_{c_x(\beta_N t-\varrho)}(j,0)\right)^{2-\alpha} X^N_\varrho(j)\, Y^N_\varrho(j)\right)^{\frac{p}{2}} d\varrho\right)^{\frac{1}{\frac{p}{2}}}\right)^{\frac{p}{2}}\right]
\end{aligned} \tag{B.30}
$$

We evaluate the product[2] and get

$$= \mathbf{E}\left[\left(\sum_{j\in\Lambda}\int_0^{\beta_N t}\left(a^N_{c_x(\beta_N t-\varrho)}(j,0)\right)^{\alpha q} d\varrho\right)^{\frac{p}{2q}}\right.$$

$$\left.\cdot\left(\sum_{j\in\Lambda}\int_0^{\beta_N t}\left(\left(a^N_{c_x(\beta_N t-\varrho)}(j,0)\right)^{2-\alpha} X^N_\varrho(j)\, Y^N_\varrho(j)\right)^{\frac{p}{2}} d\varrho\right)\right]$$

$$\overset{(B.24)}{=}\left(\sum_{j\in\Lambda}\int_0^{\beta_N t}\left(a^N_{c_x(\beta_N t-\varrho)}(j,0)\right)^2 d\varrho\right)^{\frac{p}{2}-1}$$

$$\cdot\,\mathbf{E}\left[\left(\sum_{j\in\Lambda}\int_0^{\beta_N t}\left(a^N_{c_x(\beta_N t-\varrho)}(j,0)\right)^{\left(2-2\cdot\left(1-\frac{2}{p}\right)\right)\frac{p}{2}}\left(X^N_\varrho(j)\, Y^N_\varrho(j)\right)^{\frac{p}{2}} d\varrho\right)\right]$$

$$=\left(\sum_{j\in\Lambda}\int_0^{\beta_N t}\left(a^N_{c_x(\beta_N t-\varrho)}(j,0)\right)^2 d\varrho\right)^{\frac{p}{2}-1}$$

$$\cdot\int_0^{\beta_N t}\sum_{j\in\Lambda}\left(a^N_{c_x(\beta_N t-\varrho)}(j,0)\right)^2 \mathbf{E}\left[\left(\left(X^N_\varrho(j)\, Y^N_\varrho(j)\right)^{\frac{p}{2}}\right)\right] d\varrho \qquad \text{(B.31)}$$

Since we have translation invariance, we can write:

$$=\left(\sum_{j\in\Lambda}\int_0^{\beta_N t}\left(a^N_{c_x(\beta_N t-\varrho)}(j,0)\right)^2 d\varrho\right)^{\frac{p}{2}-1}$$

$$\cdot\int_0^{\beta_N t}\sum_{j\in\Lambda}\left(a^N_{c_x(\beta_N t-\varrho)}(j,0)\right)^2 \mathbf{E}\left[\left(\left(X^N_\varrho(0)\, Y^N_\varrho(0)\right)^{\frac{p}{2}}\right)\right] d\varrho \qquad \text{(B.32)}$$

Applying the considerations carried out in the box on page 78 ("*How to calculate* $\sum_{z\in\mathbb{Z}^d}\int_0^t a_{c_x s}(\beta,z)\, a_{c_x}$
we get:

$$=\left(\int_0^{\beta_N t}\hat{a}^N_{2c_x\varrho}(0,0)\, d\varrho\right)^{\frac{p}{2}-1}\int_0^{\beta_N t}\hat{a}^N_{2c_x(\beta_N t-\varrho)}(0,0)\,\mathbf{E}\left[\left(\left(X^N_\varrho(0)\, Y^N_\varrho(0)\right)^{\frac{p}{2}}\right)\right] d\varrho\ . \qquad \text{(B.33)}$$

Next, we discuss the Green's functions occuring in (B.33).

---

[2] "$\left(\left(\ \right)^{\frac{1}{q}}\left(\ \right)^{\frac{1}{\frac{\tilde{p}}{2}}}\right)^{\frac{\tilde{p}}{2}} = \left(\ \right)^{\frac{\tilde{p}}{2q}}\left(\ \right)^1$ "

For the torus case, we know from [CGS95], Lemma 2.1c that

$$\lim_{N\to\infty} \int_0^{\beta_N \tilde{s}} \hat{a}_u^N(0,0)\,du = \lim_{N\to\infty} \int_0^{\beta_N \tilde{s}} \hat{a}_u(0,0)\,du + \tilde{s} \quad (\text{for } \tilde{s} > 0). \tag{B.34}$$

holds.

The corresponding inequality of the hierarchical group case can be found in chapter 9.
There (equation (9.299), page 235) we prove:

$$\limsup_{N\to\infty} \int_0^{\beta_N \tilde{s}} \hat{a}_u^N(0,0)\,du \leq \sum_{\ell=0}^{k} \frac{1}{c_\ell} \quad . \tag{B.35}$$

We apply this facts to (B.33). Both for the lattice case as for the hierarchical group case we get:[3]

$$\begin{aligned} &\exists \hat{K} < 0 : \\ &\mathbf{E}\left[\left(X_{\beta_N t}^N(0)\right)^p\right] \\ &\leq \hat{K}(1+t) \int_0^{\beta_N t} \hat{a}_{2(\beta_N t - s)}^N(0,0) \left(\mathbf{E}\left[\left(X_{\beta_N s}^N(0)\right)^p\right] + \mathbf{E}\left[\left(Y_{\beta_N s}^N(0)\right)^p\right]\right) ds \end{aligned} \tag{B.36}$$

We need the following:

$$\limsup_{N\to\infty} \sup_{u\in[0,t]} \mathbf{E}\left[\left(Y_{\beta_N u}^N(0)\right)^p\right] < \infty \quad . \tag{B.37}$$

This was proved by Cox, Greven and Shiga ([CGS95], Lemma 2.2d)

The uniform boundedness property (B.37) of the catalyst can be obtained by carrying out the calculations of this proof for a reactant with constant catalyst.
In particular, the following holds:
$\exists K < 0 : \mathbf{E}\left[\left(Y_{\beta_N t}^N(0)\right)^p\right] \leq K(1+t) \int_0^{\beta_N t} \hat{a}_{2(\beta_N t - s)}^N(0,0) \left(\mathbf{E}\left[\left(Y_{\beta_N s}^N(0)\right)^p\right]\right) ds.$

We get:

$$\begin{aligned} &\exists K < 0 : \\ &\mathbf{E}\left[\left(X_{\beta_N t}^N(0)\right)^p\right] \\ &\leq K(1+t) \int_0^{\beta_N t} \hat{a}_{2(\beta_N t - s)}^N(0,0) \left(\mathbf{E}\left[\left(X_{\beta_N s}^N(0)\right)^p\right] + 1\right) ds \quad . \end{aligned} \tag{B.38}$$

[3] In the sequel we set $c_x = 1$. This does not change the results, but allows us to discuss the torus case and the hierarchical group case jointly.

Let

$$M_N := \int_0^{\beta_N t} \hat{a}^N_{2s}(0,0)\, ds \quad . \tag{B.39}$$

By (B.34) resp (B.35), we have

$$\limsup_{N\to\infty} M_N < \infty \quad . \tag{B.40}$$

Therefore we can write:

$$\begin{aligned} &\exists K < 0 : \\ &\mathbf{E}\left[\left(X^N_{\beta_N t}(0)\right)^p\right] \\ &\leq K(1+t) \int_0^{\beta_N t} \hat{a}^N_{2(\beta_N t - s)}(0,0) \left(\mathbf{E}\left[\left(X^N_{\beta_N s}(0)\right)^p\right]\right) ds \quad . \end{aligned} \tag{B.41}$$

Consider the transformation of time

$$s \longmapsto f_N(s) := \int_0^s \hat{a}^N_{2(\beta_N t - u)}(0,0)\, du \tag{B.42}$$

By assumption (compare (1.3)), $a(\cdot,\cdot)$ is irreducible. Therefore, the term to be integrated is strictly positive. Hence the considered transformation of time is a bijective map from $[0, \beta_N t]$ to $[0, M_N]$.

Define

$$\Xi^N_u := \mathbf{E}\left[\left(X^N_{f_N^{-1}(u)}(0)\right)^p\right] \quad , \quad (u \in [0, M_N]) . \tag{B.43}$$

Then, applying the transformation formula to (B.41), we get:

$$\begin{aligned} \Xi^N_u &\leq K\left(1 + \frac{f_N^{-1}(u)}{\beta_N}\right) \int_0^u \Xi^N_s\, ds \\ &\leq K(1+t) \int_0^u \Xi^N_s\, ds \quad . \end{aligned} \tag{B.44}$$

We are now in the situation that we can apply Gronwall's inequality[4] (compare [KS], 5.2.7,

[4] Citation from [KS]:
Suppose that the continuous function $g(t)$ satisfies

$$0 \leq g(t) \leq \alpha(t) + \beta \int_0^t g(s) ds; \quad 0 \leq t \leq T$$

with $\beta \geq 0$ and $\alpha : [0,T] \to \mathbb{R}$ integrable. Then

$$g(t) \leq \alpha(t) + \beta \int_0^t \alpha(s) e^{\beta(t-s)} ds; \quad 0 \leq t \leq T.$$

p. 287-288). We get:

$$\begin{aligned}\Xi_u^N &\leq K + (1+t) \int_0^u K e^{K(1+t)(u-s)}\, ds \\ &\leq K \left(e^{(1+t)u} \right) \quad .\end{aligned} \tag{B.45}$$

In particular, this yields

$$\begin{aligned}\limsup_{N\to\infty} \mathbf{E}\left[\left(X_{\beta_N t}^N (0) \right)^p \right] &= \limsup_{N\to\infty} \Xi_{M_N}^N \\ &\leq \limsup_{N\to\infty} \left(K \left(e^{(1+t)M_N} \right) \right) \quad .\end{aligned} \tag{B.46}$$

By (B.40) this is finite. This completes the proof. ■

Appendix C

Boundedness of higher moments (infinite lattice)

Consider the process (X, Y) as defined by the equations (1.6), (1.7) with the initial conditions described in 1.2.4. Assuming

$$d \geq 3 \quad ,$$

we claim that the following is true:

Lemma 16 *Assume that for* $p > 2$ *the initial condition fulfills*

$$\mathbf{E}\left[(X_0(0))^p + (Y_0(0))^p\right] < \infty \quad . \tag{C.1}$$

Then the following holds

$$\sup_{u \in \mathbb{R}_+} \mathbf{E}\left[(X_u(0))^p\right] < \infty \quad , \tag{C.2}$$

$$\sup_{u \in \mathbb{R}_+} \mathbf{E}\left[(Y_u(0))^p\right] < \infty \quad . \tag{C.3}$$

Proof. Similarly to (B.33), there exists $K_p < \infty$ s. t. for $u \in \mathbb{R}_+$ we have

$$\begin{aligned}
&\mathbf{E}\left[(X_u(0))^p\right] \\
&\leq K_p \\
&\quad + (p-1)^p (2\gamma_x)^{\frac{p}{2}} \\
&\quad \cdot \left(\int_0^u \hat{a}_{2c_x\varrho}(0,0)\, d\varrho\right)^{\frac{p}{2}-1} \int_0^u \hat{a}_{2c_x(u-\varrho)}(0,0)\, \mathbf{E}\left[\left((X_\varrho(0)\, Y_\varrho(0))^{\frac{p}{2}}\right)\right] d\varrho \quad . \\
&\leq K_p \\
&\quad + (p-1)^p (2\gamma_x)^{\frac{p}{2}} \\
&\quad \cdot \left(\int_0^u \hat{a}_{2c_x\varrho}(0,0)\, d\varrho\right)^{\frac{p}{2}-1} \int_0^u \hat{a}_{2c_x(u-\varrho)}(0,0)\, \frac{\mathbf{E}\left[(X_\varrho(0))^p\right] + \mathbf{E}\left[((Y_\varrho(0))^p)\right]}{2} d\varrho \quad .
\end{aligned} \tag{C.4}$$

and

$$\begin{aligned}
&\mathbf{E}\left[(Y_u(0))^p\right] \\
&\leq K_p \\
&\quad + (p-1)^p \left(2\gamma_y\right)^p \\
&\quad \cdot \left(\int_0^u \hat{a}_{2c_y\varrho}(0,0)\, d\varrho\right)^{\frac{p}{2}-1} \int_0^u \hat{a}_{2c_x(u-\varrho)}(0,0)\, \mathbf{E}\left[\left((Y_\varrho(0))^{\frac{p}{2}}\right)\right] d\varrho \quad . \qquad \text{(C.5)}
\end{aligned}$$

We consider the case $d \geq 3$. Therefore we have

$$\int_0^\infty \hat{a}_{2\varrho}(0,0)\, d\varrho =: A < \infty \quad . \qquad \text{(C.6)}$$

This allows us to proceed similarly to the proof of Lemma 14. We give a sketch what has to be done:

- Carry out a transformation of time similar to (B.42).
- Apply Gronwall's inequality

Copying the proof of Lemma 14, the claim follows. ■

Appendix D

Remarks on the zero-dimensional model

This appendix is a supplement to Section 2.1.1, where we discussed the process given by the system of stochastic differential equations

$$dX_t = \sqrt{2\gamma_x X_t Y_t}\, dB_t^x \quad , \tag{D.1}$$

$$dY_t = \sqrt{2\gamma_y Y_t}\, dB_t^y \quad , \tag{D.2}$$

and initial conditions X_0, Y_0.

In particular, we provide a proof of (2.8). This proof requires some of the methods introduced in Chapter 5.

In addition to that, we provide the source code of the program used for the simulations of the particle model.

D.1 Probability of survival of the reactant

We prove the following Proposition:

Proposition 4 *The probability that the reactant dies out in the long run is given by the following expression:*

$$\lim_{t\to\infty} \mathbf{P}\left[X_t = 0\right] = \frac{1}{\sqrt{\frac{4\gamma_y}{\gamma_x}\frac{X_0}{Y_0^2} + 1}} \quad . \tag{D.3}$$

Proof. First, we focus on the distribution of the life time of the catalyst. The log-Laplacian u_t^ℓ ($\ell > 0$) - for which

$$\mathbf{E}\left[\exp\left(-\ell Y_t\right)\right] = \exp\left(-u_t^\ell Y_0\right) \tag{D.4}$$

holds - solves the differential equation:

$$u_0^\ell = \ell \quad , \tag{D.5}$$

$$\frac{\partial}{\partial t} u_t^\ell = -\gamma_y \left(u_t^\ell\right)^2 \quad . \tag{D.6}$$

This ordinary differential equation can be solved explicitly:

$$u_t^\ell = \frac{\ell}{1+\ell\gamma_y t} \quad .$$

Hence we get

$$\mathbf{E}\left[\exp\left(-\ell Y_t\right)\right] = \exp\left(-\frac{\ell Y_0}{1+\ell\gamma_y t}\right) \quad . \tag{D.7}$$

Letting $\ell \to \infty$, we can calculate probability that the catalyst dies out:

$$\mathbf{P}\left[Y_t = 0\right] = \exp\left(-\frac{Y_0}{\gamma_y t}\right) \quad . \tag{D.8}$$

This goes well together with the survival probability of the critical Galton-Watson process (finite variance branching), which is as well $O\left(\frac{1}{t}\right)$ for $t \to \infty$.

We calculate the survival probability of the reactant as follows: In distribution, (2.6) is equal to

$$X_t = X_0 + \int\limits_0^{\int_0^t 2\gamma_x Y_\tau d\tau} \sqrt{X_s}\, dB_s^x \quad . \tag{D.9}$$

This yields

$$var_{reac}^Y\left(X_t\right) = \theta_x \int\limits_0^t 2\gamma_x Y_\tau d\tau \quad . \tag{D.10}$$

We next calculate the distribution of a multiple of the right hand side of (D.10), namely $\int_0^t Y_\tau d\tau$. Let w_t^ℓ $(\ell > 0)$ be given by the ordinary differential equation

$$w_0^\ell = 0 \tag{D.11}$$

$$\frac{\partial}{\partial t} w_t^\ell = -\gamma_y \left(w_t^\ell\right)^2 + \ell \quad . \tag{D.12}$$

Then (for arguments similar to those used by Iscoe when calculating weighted occupation times [Isc86], compare also Proposition 1, p. 49) we have

$$-w_t^\ell Y_0 = \log \mathbf{E}\left[\exp\left(-\ell \int\limits_0^t Y_\tau d\tau\right)\right] \quad . \tag{D.13}$$

It is possible to calculate w_t^ℓ explicitly. Equation (D.11) can be explicitely solved:

$$w_t^\ell = \tanh\left(\sqrt{\ell\gamma_y} t\right)\sqrt{\frac{\ell}{\gamma_y}} \quad . \tag{D.14}$$

Being interested in the limit $t \to \infty$, we get:

$$w_\infty^\ell := \lim_{t\to\infty} w_t^\ell = \sqrt{\frac{\ell}{\gamma_y}} \quad . \tag{D.15}$$

Hence $\int_0^\infty Y_\tau d\tau$ is $\frac{1}{2}$-stable distributed. The density is given by

$$\frac{\mathbf{P}\left[\int_0^\infty Y_\tau d\tau \in dt\right]}{dt} = \frac{Y_0}{2\sqrt{\gamma_y}\sqrt{\pi}t^{\frac{3}{2}}} e^{-\frac{Y_0^2}{4\gamma_y t}} \quad . \tag{D.16}$$

As an immediate consequence we get that the quenched variance - that means the variance of the reactant given the catalyst - can be identified with a random variable which is a multiple of a random variable with $\frac{1}{2}$-stable distribution. The log-Laplace transform of the distribution of the reactant can be obtained by integration:

$$\lim_{t\to\infty} \mathbf{E}\left[\exp\left(-\ell X_t\right)\right] = \int_0^\infty \exp\left(-\frac{\ell X_0}{1+\ell\gamma_x t}\right) \frac{Y_0}{2\sqrt{\gamma_y}\sqrt{\pi}t^{\frac{3}{2}}} e^{-\frac{Y_0^2}{4\gamma_y t}} dt \quad . \tag{D.17}$$

Letting ℓ to infinity, we get the probability that the reactant dies out:

$$\begin{aligned} P\left[X_t = 0\right] &= \lim_{\ell\to\infty} \int_0^\infty \exp\left(-\frac{\ell X_0}{1+\ell\gamma_x t}\right) \frac{Y_0}{2\sqrt{\gamma_y}\sqrt{\pi}t^{\frac{3}{2}}} e^{-\frac{Y_0^2}{4\gamma_y t}} dt \\ &= \frac{Y_0}{2\sqrt{\gamma_y}\sqrt{\pi}} \int_0^\infty \exp\left(-\left(\frac{X_0}{\gamma_x} + \frac{Y_0^2}{4\gamma_y}\right)\frac{1}{t}\right) t^{-\frac{3}{2}} dt \\ &= \frac{Y_0}{2\sqrt{\gamma_y}\sqrt{\pi}} \frac{\sqrt{\pi}}{\sqrt{\frac{X_0}{\gamma_x} + \frac{Y_0^2}{4\gamma_y}}} \\ &= \frac{1}{\sqrt{\frac{4\gamma_y}{\gamma_x}\frac{X_0}{Y_0^2} + 1}} \quad . \end{aligned} \tag{D.18}$$

■

D.2 Source code for simulating the Catalytic Branching Process

We do not provide a simulation of the diffusion process with evolution equation

$$dX_t = \sqrt{2\gamma_x X_t Y_t}\, dB_t^x \quad , \tag{D.19}$$

$$dY_t = \sqrt{2\gamma_y Y_t}\, dB_t^y \quad . \tag{D.20}$$

Instead, we consider the particle analogue, namely the Catalytic Branching Process. This is the catalyst-reactant particle process where the catalyst simply branches with rate one, where the branching rate of the reactant is given by the number of catalysts present.

The program we have written calculates samples of paths, which are saved as a file which can be read and visualized by a plotting software (gnuplot).

Remark 23 *By "`$totalnr=1+$reac;`" (line 5) we enforce that at every step either at birth event or a death event takes place.*

```
srand();                                  #initialization of random number generator (takes time as seed)
$cat=15; $reac=100;                       #initial state of catalyst and reactant
open(DAT, ">daten.dat");                  #points of the cat/reac path are to be written in 'daten.dat'
while($cat!=0){                           #stop when the catalyst reaches zero
        $randomnr=rand();                 #next random value in [0,1]
        $totalnr=1+$reac;                 #$totalnr is the rate that something happens

                                          #with rate 1/(2*$totalnr) there occurs a birth event of the catalyst
    if($randomnr <= 1/(2*$totalnr)) {$cat=$cat+1;}

                                          #with rate 1/(2*$totalnr) there occurs a death event of the catalyst
    elsif($randomnr<=1/(1.*$totalnr)) {$ cat--;}

                                          #in half of the remaining cases there occurs a birth event of the reactant
    elsif($randomnr<=1/(1.*$totalnr)+$ reac/(2.*$totalnr)) {$reac++;}

                                          #in the remaining cases there occurs a death event of the reactant
    else    {$reac--;}

                                          #add the position of catalyst and reactant to daten.dat
    print (DAT "$cat $reac \n");
}
close(DAT);                               #close data stream
                                          #as a control the user obtains the final state of the process
print ("Catalyst at the end: $cat, Reactant at the end: $reac \n");
```

Figure D.1: Perl-Skript used for simulation

It is worth mentioning that the expectation of the time needed to finish the calculation of the path is infinite.

Appendix E

A property of the Mean-Field limit – Existence of a stationary process

In the sequel, we consider the $\mathbb{R}_+ \times \mathbb{R}_+$-valued process $(Z_t)_{t\geq 0} = (X_t, Y_t)_{t\geq 0}$ defined as strong solution of

$$
\begin{aligned}
X_0 &= \theta_x, && \text{(E.1)}\\
Y_0 &= \theta_y, && \text{(E.2)}\\
dX_t &= c\,(\theta_x - X_t)\,dt + \sqrt{2\gamma X_t Y_t}\,dB_t^x, && \text{(E.3)}\\
dY_t &= c\,(\theta_y - Y_t)\,dt + \sqrt{2\gamma Y_t\,(i)}\,dB_t^y && \text{(E.4)}
\end{aligned}
$$

with $c > 0, \gamma, \theta_x, \theta_y \geq 0$; $(B_t^x, B_t^y)_{t\geq 0}$ denotes planar Brownian motion. We claim that the following holds

Proposition 5 *There exists an unique stationary process with law $\hat{\Psi}_c^\theta$ for which we have*

$$\mathcal{L}\left\{(Z_{s+t})_{t\geq 0}\right\} \overset{s\to\infty}{\Longrightarrow} \hat{\Psi}_c^\theta \quad . \qquad \text{(E.5)}$$

E.1 Proof of the proposition

E.1.1 Tools

The proof uses the same methods as the proof of the second part of Theorem 0 ("High dimensional case"). In addition to that, the duality relations introduced in Chapter 8 will play a main role.

E.1.2 Moments

We immediately get from the SDE governing the system:

$$
\begin{aligned}
\mathbf{E}\,[X_t] &= \theta_x \quad (t \geq 0)\,, && \text{(E.6)}\\
\mathbf{E}\,[Y_t] &= \theta_y \quad (t \geq 0)\,. && \text{(E.7)}
\end{aligned}
$$

Applying Ito's formula, one obtains the following expressions for the second moments:

$$\mathbf{E}\left[(X_t)^2\right] = (\theta_x)^2 + \theta_x\theta_y\frac{\gamma}{c}\cdot(1-\exp(-2ct)) \quad (t \geq 0)\,, \tag{E.8}$$

$$\mathbf{E}\left[(Y_t)^2\right] = (\theta_y)^2 + \theta_y\frac{\gamma}{c}\cdot(1-\exp(-2ct)) \quad (t \geq 0)\,. \tag{E.9}$$

Considering higher moments, there is the following lemma:

Lemma 17 *Let $p > 2$. Assume*

$$\mathbf{E}\left[(X_0)^p\right] + \mathbf{E}\left[(Y_0)^p\right] < \infty \quad . \tag{E.10}$$

Then we have

$$\sup_{t\in\mathbb{R}_+}\left(\mathbf{E}\left[(X_t)^p\right] + \mathbf{E}\left[(Y_t)^p\right]\right) < \infty \quad . \tag{E.11}$$

Proof. Since – by (E.8) and (E.9) – we have uniform boundedness of second moments we may suppose that we already know (compare (E.6),(E.7)):

$$\sup_{t\in\mathbb{R}_+}\left(\mathbf{E}\left[(X_t)^{p-1}\right] + \mathbf{E}\left[(Y_t)^{p-1}\right]\right) < \infty \quad . \tag{E.12}$$

Applying Ito's formula to (E.4), we get:

$$\begin{aligned}\frac{\partial}{\partial t}\mathbf{E}\left[(Y_t)^p\right] &= pc\theta_y\mathbf{E}\left[(Y_t)^{p-1}\right] \\ &\quad -pc\mathbf{E}\left[(Y_t)^p\right] \\ &\quad +\gamma\frac{p\,(p-1)}{2}\mathbf{E}\left[(Y_t)^{p-1}\right] \quad .\end{aligned} \tag{E.13}$$

With (E.12), this yields

$$\sup_{t\in\mathbb{R}_+}\left(\mathbf{E}\left[(Y_t)^{p-1}\right]\right) < \infty \quad . \tag{E.14}$$

Applying Ito's formula to (E.3), we get:

$$\begin{aligned}\frac{\partial}{\partial t}\mathbf{E}\left[(X_t)^p\right] &= pc\theta_x\mathbf{E}\left[(X_t)^{p-1}\right] \\ &\quad -pc\mathbf{E}\left[(X_t)^p\right] \\ &\quad +\gamma\frac{p\,(p-1)}{2}\mathbf{E}\left[(X_t)^{p-1}\,Y_t\right] \quad .\end{aligned} \tag{E.15}$$

Applying Hölder's inequality, we get

$$\begin{aligned}\frac{\partial}{\partial t}\mathbf{E}\left[(X_t)^p\right] &\leq pc\theta_x\mathbf{E}\left[(X_t)^{p-1}\right] \\ &\quad -pc\mathbf{E}\left[(X_t)^p\right] \\ &\quad +\gamma\frac{p\,(p-1)}{2}\left(\mathbf{E}\left[(X_t)^p\right]\right)^{\frac{p-1}{p}}\left(\mathbf{E}\left[Y_t^p\right]\right)^{\frac{1}{p}} \quad .\end{aligned} \tag{E.16}$$

Next, we apply that $\forall\delta > 0\exists K_\delta < \infty$ s.t for all $x \geq 0$:

$$x^{\frac{p-1}{p}} \leq K_\delta + \delta x \quad . \tag{E.17}$$

We get:

$$\begin{aligned}\frac{\partial}{\partial t}\mathbf{E}\left[(X_t)^p\right] &\leq pc\theta_x\mathbf{E}\left[(X_t)^{p-1}\right] \\ &-pc\mathbf{E}\left[(X_t)^p\right] \\ &+\gamma\frac{p(p-1)}{2}\left(K_\delta+\delta\mathbf{E}\left[(X_t)^p\right]\right)\left(\mathbf{E}\left[Y_t^p\right]\right)^{\frac{1}{p}} \quad .\end{aligned} \tag{E.18}$$

In the sequel, assume

$$\delta < \frac{2c}{\gamma(p-1)\left(\mathbf{E}\left[Y_t^p\right]\right)^{\frac{1}{p}}} \quad .$$

By construction, it is possible to choose $C<\infty$ s.t.

$$C > \sup_{t>0}\left(pc\theta_x\mathbf{E}\left[(X_t)^{p-1}\right]+\left(\mathbf{E}\left[Y_t^p\right]\right)^{\frac{1}{p}}+\frac{1}{pc-\gamma\frac{p(p-1)}{2}\delta\left(\mathbf{E}\left[Y_t^p\right]\right)^{\frac{1}{p}}}\right) \quad . \tag{E.19}$$

This yields

$$\frac{\partial}{\partial t}\mathbf{E}\left[(X_t)^p\right] \leq C-\frac{1}{C}\mathbf{E}\left[(X_t)^p\right] \quad . \tag{E.20}$$

Solving the corresponding ordinary differential equation, we get

$$\mathbf{E}\left[(X_t)^p\right] \leq C^2+\exp\left(-\frac{t}{C}\right)\cdot\left(\mathbf{E}\left[(X_0)^p\right]-C^2\right) \quad . \tag{E.21}$$

Hence we have

$$\limsup_{t\to\infty}\mathbf{E}\left[(X_t)^p\right] \leq C^2 \quad . \tag{E.22}$$

This implies the claim, hence we are done. ■

E.1.3 Main part of the proof

Now let $(\varsigma_n)\in\mathbb{R}_+^{\mathbb{N}}$ denote an increasing sequence with $\lim_{n\to\infty}\varsigma_n=\infty$. By uniform integrability (note that we have an uniform upper bound for second moments) we get the following:

Any accumulation point (w. r. t. the weak topology) of $(\mathcal{L}\{(X_{\varsigma_n},Y_{\varsigma_n})\})$ is a probability measure on $\mathbb{R}_+\times\mathbb{R}_+$ with intensity (θ_x,θ_y).

We claim that the following holds:

Claim 9 *For given $\theta_x,\theta_y>0$ there exists a stationary process with law*

$$\vartheta_{(\theta_x,\theta_y)}\in\mathcal{M}^1\left(\mathcal{C}\left(\mathbb{R}_+,\mathbb{R}_+\times\mathbb{R}_+\right)\right) \tag{E.23}$$

such that for any initial condition $\mathcal{L}\{(X_0,Y_0)\}$ with

- $$\mathbf{E}\left[X_0\right]=\theta_x \quad , \tag{E.24}$$
- $$\mathbf{E}\left[Y_0\right]=\theta_y \quad , \tag{E.25}$$

- $$\mathbf{E}\left[X_0 Y_0\right] = \theta_x \theta_y \quad , \tag{E.26}$$

- $$\exists \alpha > 0 : \mathbf{E}\left[(X_0)^{2+\alpha} + (Y_0)^{2+\alpha}\right] < \infty \tag{E.27}$$

we have convergence of the laws of paths, namely

$$\mathcal{L}\left\{(X_{t+s}, Y_{t+s})_{s\geq 0}\right\} \quad \overset{t\to\infty}{\Longrightarrow} \quad \vartheta_{(\theta_x,\theta_y)} \quad . \tag{E.28}$$

Definition 11 *Let $\tilde{S}_{(\theta_x,\theta_y)} \subset \mathcal{M}^1(\mathbb{R}_+ \times \mathbb{R}_+)$ denote the set of initial conditions $\mathcal{L}\{(X_0, Y_0)\}$ for which (E.24), (E.25), (E.26) and (E.27) hold.*

Remark 24 *Note that $\tilde{S}_{(\theta_x,\theta_y)}$ is preserved under the evolution semigroup defined by (E.3) and (E.4).*

1. The proof consists of three parts. We first give a survey what will be proved in each of the three parts.

I. *Convergence of finite dimensional distributions*
Let $n \in \mathbb{N}$, $0 \leq t_1 < t_2 < ... < t_n < \infty$. Let

$$\mathcal{T} := (t_1, ..., t_n) \in \mathbb{R}_+^n \quad . \tag{E.29}$$

We have to show that for any choice of $n, \mathcal{T}$ there exists a unique measure

$$\mu^{\mathcal{T}}_{(\theta_x,\theta_y)} \in \mathcal{M}^1\left((\mathbb{R}_+ \times \mathbb{R}_+)^{\otimes n}\right) \tag{E.30}$$

with[1]

$$\mathcal{L}\left\{(X_{s+t}, Y_{s+t})_{t\in\mathcal{T}}\right\} \quad \overset{s\to\infty}{\Longrightarrow} \quad \mu^{\mathcal{T}}_{(\theta_x,\theta_y)} \quad . \tag{E.31}$$

In particular, we claim that there exists $\mu^{(0)}_{(\theta_x,\theta_y)} \in \mathcal{M}^1(\mathbb{R}_+ \times \mathbb{R}_+)$ with

$$\mathcal{L}\left\{(X_s, Y_s)\right\} \quad \overset{s\to\infty}{\Longrightarrow} \quad \mu^{(0)}_{(\theta_x,\theta_y)} \quad . \tag{E.32}$$

II. *Pathwise convergence*
We define $\vartheta_{(\theta_x,\theta_y)}$ as the law of the process given by (E.3), (E.4) and initial condition $\mu^{(0)}_{(\theta_x,\theta_y)}$.
We know from I. that $\mathcal{L}\left\{(X_{\varsigma+t}, Y_{\varsigma+t})_{t\geq 0}\right\}$ converges to $\vartheta_{(\theta_x,\theta_y)}$ as $\varsigma \to \infty$ in terms of finite dimensional distributions. Let $(\varsigma_i)_{i\in\mathbb{N}} \in \mathbb{R}_+^{\mathbb{N}}$ be a strictly increasing sequence with

$$\lim_{i\to\infty} \varsigma_i = \infty \quad . \tag{E.33}$$

[1] We assume $\mathcal{L}\{Z_0\} \in \tilde{S}_{(\theta_x,\theta_y)}$.

We have to show that $\mathcal{L}\left\{(X_{\varsigma_i+t}, Y_{\varsigma_i+t})_{t\geq 0}\right\}$ converges to $\vartheta_{(\theta_x,\theta_y)}$ in terms of weak convergence on $\mathcal{M}^1\left(\mathcal{C}\left(\mathbb{R}_+, \mathbb{R}_+ \times \mathbb{R}_+\right)\right)$. We know from (E.31) that the only candidate for being an accumulation point of

$$\left(\mathcal{L}\left\{(X_{\varsigma_i+t}, Y_{\varsigma_i+t})_{t\geq 0}\right\}\right)_{i\in\mathbb{N}} \tag{E.34}$$

is $\vartheta_{(\theta_x,\theta_y)}$. It remains to verify the existence of accumulation points of the set

$$\mathcal{L}^{(\varsigma_i)} := \left\{\mathcal{L}\left\{(X_{\varsigma+t}, Y_{\varsigma+t})_{t\geq 0}\right\}, \varsigma \in \{\varsigma_1, \varsigma_2, ...\}\right\} \quad .$$

As there exist accumulation points if $\mathcal{L}^{(\varsigma_i)}$ is relatively compact with respect to the underlying topology induced by weak convergence on the set of probability measures on the pathspace $\mathcal{M}^1\left(\mathcal{C}\left(\mathbb{R}_+, \mathbb{R}_+ \times \mathbb{R}_+\right)\right)$, and as relative compactness of sets containing probability laws on Polish spaces is equivalent to the tightness property of these laws, the following claim is sufficient for (E.28).

Claim 10 *We claim:*

$$\mathcal{L}^{(\varsigma_i)} \textit{ is tight.} \tag{E.35}$$

III. *Stationarity:*
Consider the measure $\mu^{(0)}_{(\theta_x,\theta_y)}$ introduced in (E.32). It is uniquely characterized as weak limit point of $\mathcal{L}\{(X_s, Y_s)\}$ as $s \to \infty$. We claim:

Claim 11 *The measure $\mu^{(0)}_{(\theta_x,\theta_y)}$ is stationary with respect to the evolution semigroup corresponding to the evolution equations (E.3) and (E.4).*

Proof of I. - Convergence of finite dimensional distributions

Thoughout the proof, we will change our point of view: Instead of using time $t = 0$ as time of initialization, we rather start the processes at times $T < 0$ and compare their distributions at times $-t_1, -t_2, ..., -t_n$ letting $T \to -\infty$.
Let $(\mathfrak{X}, \mathfrak{Y})$ be a random variable with values in $\mathbb{R}_+ \times \mathbb{R}_+$ and $\mathcal{L}\{(\mathfrak{X}, \mathfrak{Y})\} \in \tilde{S}_{(\theta_x,\theta_y)}$.

This random variable serves as initialization of processes $\left(X_t^T, Y_t^T\right)_{t\in[T,0]}$. Let these processes solve the evolution equations (E.3) and (E.4).

Namely, we can write

$$\left(X_T^T, Y_T^T\right) = (\mathfrak{X}, \mathfrak{Y}) \quad , \tag{E.36}$$

and let for $t \in (T, \infty)$ the evolution be given by the system of stochastic differential equations

$$dX_t^T = c\left(\theta_x - X_t^T\right) dt + \sqrt{2\gamma X_t^T Y_t^T}\, dB_t^x, \tag{E.37}$$

$$dY_t^T = c\left(\theta_y - Y_t^T\right) dt + \sqrt{2\gamma Y_t^T}\, dB_t^y \quad . \tag{E.38}$$

Let $\tilde{\mathcal{T}} := \{-t_n, ..., -t_1\}$. We have to prove that there exists

$$\mu^{\tilde{\mathcal{T}}}_{(\theta_x,\theta_y)} \in \mathcal{M}^1\left(\left(\mathbb{R}_+ \times \mathbb{R}_+\right)^{\otimes n}\right) \tag{E.39}$$

with

$$\mathcal{L}\left\{\left(X_\tau^T, Y_\tau^T\right)_{\tau\in\tilde{T}}\right\} \quad \overset{T\to-\infty}{\Longrightarrow} \quad \mu^{\tilde{T}}_{(\theta_x,\theta_y)} \quad . \tag{E.40}$$

Consider $(\alpha_k, \beta_k)_{k\in\{1,\dots,n\}} \in \mathbb{R}_+^n$. Consider the family

$$\left\{\mathcal{L}\left\{\sum_{k=1}^{n}\left(\alpha_k X_{-t_k}^T + \beta_k Y_{-t_k}^T\right)\right\}\right\}_{T<0} \quad . \tag{E.41}$$

We have to verify that we have weak convergence of (E.41) as $T \to -\infty$ for any choice of $(\alpha_k, \beta_k)_{k\in\{1,\dots,n\}} \in \mathbb{R}_+^n$ Since we consider a family of non-negative random variables, and since expectations are uniformly bounded, namely

$$\limsup_{T<0} \mathbf{E}\left[\sum_{k=1}^{n}\left(\alpha_k X_{-t_k}^T + \beta_k Y_{-t_k}^T\right)\right] = \sum_{k=1}^{n}\left(\alpha_k\theta_x + \beta_k\theta_y\right) < \infty \quad , \tag{E.42}$$

the family $\left\{\mathcal{L}\left\{\sum_{k=1}^{n}\left(\alpha_k X_{-t_k}^T + \beta_k Y_{-t_k}^T\right)\right\}\right\}_{T<0}$ is a tight family of probability measures. Then the limit (as $T \to -\infty$) of the Laplace transforms

$$\mathbf{E}\left[\exp\left(-\lambda\sum_{k=1}^{n}\left(\alpha_k X_{-t_k}^T + \beta_k Y_{-t_k}^T\right)\right)\right] \quad ,(\lambda > 0) \tag{E.43}$$

– if it exists – is again a Laplace transform of a probability measure on $\mathbb{R}_+$. (compare the footnote on page 87) Since this limit Laplace transform characterizes the limit probability law, it only remains to prove that the limit of (E.43) exists as $T \to -\infty$ for all choices of $(\alpha_k, \beta_k)_{k\in\{1,\dots,n\}} \in (\mathbb{R}_+ \times \mathbb{R}_+)^n$.

At this point, we can make use of the fact that the catalyst converges weakly on the path space: There exists $\vartheta_{\theta_y} \in \mathcal{M}^1\left(\mathcal{C}\left(\mathbb{R}_+, \mathbb{R}_+\right)\right)$ with

$$\mathcal{L}\left\{\left(Y_{t+s}\right)_{s\geq 0}\right\} \underset{t\to\infty}{\Longrightarrow} \vartheta_{\theta_y} \quad . \tag{E.44}$$

This is one of the basic properties of Feller's branching diffusion. It can be proven by coupling arguments – or alternatively by the proof we give below for the reactant, replacing the catalyst by a constant value. As law of a stationary process, ϑ_{θ_y} can be extended canonically to

$$\vartheta^{\pm}_{\theta_y} \in \mathcal{M}^1\left(\mathcal{C}\left(\mathbb{R}_+, \mathbb{R}_+\right)\right), \tag{E.45}$$

($\pm$ denotes that the paths are indexed by $\mathbb{R}$) describing the stationary process living on $(-\infty, +\infty)$. [This is an immediate consequence of the Kolmogoroff extension theorem. By time shift, it is possible to construct consistent families of marginal distributions even for negative times. Then the process can be constructed as a projective limit.] In terms of our setup to start processes at times $T < 0$, letting $T \to -\infty$, we get the following reformulation of (E.44):

$$\mathcal{L}\left\{\left(Y_{t\vee T}^T\right)_{t\in\mathbb{R}}\right\} \underset{T\to-\infty}{\Longrightarrow} \vartheta^{\pm}_{\theta_y} \quad . \tag{E.46}$$

Remark 25 *Simplifying notation, we will not write down technical cut-offs like "$t \vee T$" in future. Compare the remark on page 87 for further explanation.*

According to the Skorohod representation (compare page 40) there exists a probability space with probability measure denoted[2] by $\mathbb{P}$, on which we can define the following two objects:

- a family of processes with state space $\mathbb{R}_+ \times \mathbb{R}_+$

$$\left\{ \left(\hat{X}_t^T, \hat{Y}_t^T \right)_{t \in \mathbb{R}} \right\}_{T<0} \tag{E.47}$$

- a limit catalyst

$$\left(\hat{Y}_t^{-\infty} \right)_{t \in \mathbb{R}} \tag{E.48}$$

with

$$\mathbb{L} \left\{ \left(\hat{Y}_t^{-\infty} \right)_{t \in \mathbb{R}} \right\} = \vartheta_{\theta_y}^{\pm} \quad . \tag{E.49}$$

These objects shall satisfy

- *The laws of the processes coincide*

$$\forall T < 0 : \mathbb{L} \left\{ \left(\hat{X}_t^T, \hat{Y}_t^T \right)_{t \in \mathbb{R}} \right\} = \mathcal{L} \left\{ \left(X_t^T, Y_t^T \right)_{t \in \mathbb{R}} \right\} \tag{E.50}$$

- *Application of the Skorohod representation to the catalyst*

$$\begin{aligned} & & \forall \vartheta &> 0 : \\ \lim_{T \to -\infty} \sup_{r \in [-\vartheta, 0]} & \left| \hat{Y}_r^T (i) - \hat{Y}_r^{-\infty} (i) \right| &= 0 & \quad \mathbb{P}\text{-a.s.} \end{aligned} \tag{E.51}$$

We have claimed that (E.43) converges for all choices of λ, $(\alpha_k, \beta_k)_{k \in \{1,\dots,n\}}$. It is possible to absorbe λ in $(\alpha_k, \beta_k)_{k \in \{1,\dots,n\}}$. This way, we can reformulate the claimed convergence of (E.43) as follows:

Claim 12 *The limit*

$$\lim_{T \to -\infty} \mathbb{E} \left[\exp \left(- \sum_{k=1}^{n} \left(\alpha_k \hat{X}_{-t_k}^T + \beta_k \hat{Y}_{-t_k}^T \right) \right) \right] \tag{E.52}$$

exists.

2

a) We denote the law of a random variable X with respect to $\mathbb{P}$ by $\mathbb{L}$.

b) The catalyst-reactant decomposition is denoted as follows:

$$\mathbb{P}[\cdot] = \int \mathbb{P}_{reac}^{\hat{Y}} [\cdot] \quad \mathbb{P}_{cat} \left(d\hat{Y} \right)$$

c) Expectation is denoted by $\mathbb{E}$, quenched expectation by $\mathbb{E}_{reac}^{\hat{Y}}$.

Next, we make use of the catalyst-reactant setting (compare the discussion in the "Methods and Tools" Section carried out in 5.2). We consider the Laplace transforms describing the finite dimensional distributions of the reactant $\hat{X}^T$ constructed for given catalyst paths $\left\{\hat{Y}^T\right\}_{T<0}$, namely

$$\mathbb{E}_{reac}^{\hat{Y}}\left[\exp\left(-\sum_{k=1}^{n}\alpha_k\hat{X}_{-t_k}^T\right)\right] \quad . \tag{E.53}$$

Note that this suffices since by (E.51) the catalyst paths converge as $T \to -\infty$ (in the sense of compact convergence, as described by (E.51)) $\mathbb{P}$-a. s.

In Chapter 8 (compare especially 8.2) we lay out how $\mathbb{E}_{reac}^{\hat{Y}}\left[\exp\left(-\sum_{k=1}^{n}\alpha_k X_{-t_k}^T\right)\right]$ can be described using a dual picture. First of all, we have to introduce two auxiliary sites "Δ" and "Σ". We define

$$\begin{aligned}\hat{X}^T(\Delta) &\equiv \theta_x \quad , &(E.54)\\ \hat{X}_t^T(\Sigma) &= \sum_{k=1}^{n}\alpha_k\hat{X}_{-t_k}^T\cdot\mathbb{I}_{\{t\geq -t_k\}} \quad (t\in[T,\infty)) &(E.55)\end{aligned}$$

The following duality relation ($\mathbf{u}$ is defined below) is proven to hold for almost all $\hat{Y}$:

$$\begin{aligned}&\mathbb{E}_{reac}^{\hat{Y}}\left[\exp\left(-\sum_{k=1}^{n}\alpha_k\hat{X}_{-t_k}^T\right)\right]\\ &= \mathbb{E}_{reac}^{\hat{Y}}\left[\exp\left(-\mathbf{u}_{-T}^T\mathfrak{X}-\mathbf{u}_{-T}^T(\Delta)\hat{X}_T^T(\Delta)-\mathbf{u}_{-T}^T(\Sigma)\hat{X}_T^T(\Sigma)\right)\right] \quad . \qquad (E.56)\end{aligned}$$

The dual $\left(\mathbf{u}_t^T\right)_{t\in[T,0]}$ is given by its

- initial conditions

$$\begin{aligned}\mathbf{u}_0^T &= 0 \quad , &(E.57)\\ \mathbf{u}_0^T(\Delta) &= 0 \quad , &(E.58)\\ \mathbf{u}_0^T(\Sigma) &= 1 \quad . &(E.59)\end{aligned}$$

- jumps (left-continuous) for $t = t_k$, $(k \in \{1,...,n\})$

$$\begin{aligned}\lim_{s\to t+}\mathbf{u}_t^T &= \lim_{s\to t-}\mathbf{u}_t^T+\alpha_k \quad , &(E.60)\\ \lim_{s\to t+}\mathbf{u}_t^T(\Delta) &= \lim_{s\to t-}\mathbf{u}_t^T(\Delta) \quad , &(E.61)\\ \lim_{s\to t+}\mathbf{u}_t^T(\Sigma) &= \lim_{s\to t-}\mathbf{u}_t^T(\Sigma) \quad . &(E.62)\end{aligned}$$

- evolution equations for $t \in (0,-T)\setminus\{t_1,...,t_n\}$

$$\begin{aligned}\frac{\partial}{\partial t}\mathbf{u}_t^T &= -\gamma_x\hat{Y}_{-t}^T\left(\mathbf{u}_t^T\right)^2-c_x\mathbf{u}_t^T \quad , &(E.63)\\ \frac{\partial}{\partial t}\mathbf{u}_t^T(\Delta) &= c_x\mathbf{u}_t^T \quad , &(E.64)\\ \frac{\partial}{\partial t}\mathbf{u}_t^T(\Sigma) &= 0 \quad . &(E.65)\end{aligned}$$

A proof of (E.56) can be found on page 161.

We have to compare $\left(\mathbf{u}_t^T\right)$ with $\left(\mathbf{u}_t^{-\infty}\right)$, which we define next. Since $\left(\mathbf{u}_t^{-\infty}\right)$ corresponds to the limit case $T \to -\infty$, in particular with limit catalyst $\hat{Y}^{-\infty}$, the following characterization intuitively makes sense. Let $\left(\mathbf{u}_t^{-\infty}\right)$ be defined by its

- initial conditions

$$\begin{aligned} \mathbf{u}_0^{-\infty} &= 0 \quad , &(E.66)\\ \mathbf{u}_0^{-\infty}(\Delta) &= 0 \quad , &(E.67)\\ \mathbf{u}_0^{-\infty}(\Sigma) &= 1 \quad . &(E.68)\end{aligned}$$

- jumps (left-continuous) for $t = t_k$, $(k \in \{1, ..., n\})$

$$\begin{aligned} \lim_{s\to t+} \mathbf{u}_t^{-\infty} &= \lim_{s\to t-} \mathbf{u}_t^{-\infty} + \alpha_k \quad , &(E.69)\\ \lim_{s\to t+} \mathbf{u}_t^{-\infty}(\Delta) &= \lim_{s\to t-} \mathbf{u}_t^{-\infty}(\Delta) \quad , &(E.70)\\ \lim_{s\to t+} \mathbf{u}_t^{-\infty}(\Sigma) &= \lim_{s\to t-} \mathbf{u}_t^{-\infty}(\Sigma) \quad . &(E.71)\end{aligned}$$

- evolution equations for $t \in (0, \infty) \setminus \{t_1, ..., t_n\}$

$$\begin{aligned} \frac{\partial}{\partial t}\mathbf{u}_t^{-\infty} &= -\gamma_x \hat{Y}_{-t}^{-\infty}\left(\mathbf{u}_t^{-\infty}\right)^2 - c_x \mathbf{u}_t^{-\infty} \quad , &(E.72)\\ \frac{\partial}{\partial t}\mathbf{u}_t^{-\infty}(\Delta) &= c_x \mathbf{u}_t^{-\infty} \quad , &(E.73)\\ \frac{\partial}{\partial t}\mathbf{u}_t^{-\infty}(\Sigma) &= 0 \quad . &(E.74)\end{aligned}$$

We claim the following:

$$\begin{aligned} &\lim_{T\to-\infty} \mathbb{E}_{cat} \Bigg| \mathbb{E}_{reac}^{\hat{Y}} \left[\exp\left(-\mathbf{u}_{-T}^T \mathfrak{X} - \mathbf{u}_{-T}^T(\Delta) \underbrace{\hat{X}_T(\Delta)}_{=\theta_x} - \mathbf{u}_{-T}^T(\Sigma) \underbrace{\hat{X}_T(\Sigma)}_{=0} \right) \right] \\ &- \lim_{\tilde{T}\to-\infty} \mathbb{E}_{reac}^{\hat{Y}} \left[\exp\left(\left(-\mathbf{u}_{-\tilde{T}}^{-\infty}\theta_x - \mathbf{u}_{-\tilde{T}}^{-\infty}(\Delta)\,\theta_x - \mathbf{u}_{-\tilde{T}}^{-\infty}(\Sigma)\cdot 0 \right) \right) \right] \Bigg| = 0 \quad (E.75)\end{aligned}$$

Before proving (E.75) we still have to prove that limit that occurs in the second line is well-defined:

1. Equation (E.72) yields:
   - $\left(\mathbf{u}_t^{-\infty}\right)_{t\geq 0}$ is decreasing except at the finite number of points of discontinuity
   - $$\exists K \in \mathbb{R}_+ : \mathbf{u}_t^{-\infty} \leq K \exp(-c_x t) \quad \forall t \geq 0, \qquad (E.76)$$
   - $$\mathbf{u}_t^{-\infty} \geq 0 \quad \forall t \geq 0. \qquad (E.77)$$

Therefore, $\mathbf{u}_{-\tilde{T}}^{-\infty}\theta_x$ has a limit as $\tilde{T} \to -\infty$.

2. Equation (E.73) yields:

   - $\left(\mathbf{u}_t^{-\infty}(\Delta)\right)_{t\geq 0}$ is increasing
   -

$$\begin{aligned} \mathbf{u}_t^{-\infty}(\Delta) &\leq K \cdot \frac{1}{c_x}\left(1-\exp\left(-c_x t\right)\right) \\ &\leq K \cdot \frac{1}{c_x} < \infty \quad (t \geq 0)\,. \end{aligned} \tag{E.78}$$

Therefore, $\mathbf{u}_t^{-\infty}(\Delta)\,\theta_x$ is increasing with t and bounded, hence the considered limit exists.

The proof of (E.75) consists of two steps:

1. We have to verify

$$\lim_{T\to-\infty} \mathbb{E}_{cat}\left[\left|\exp\left(-\mathbf{u}_{-T}^{T}\mathfrak{X}\right) - \exp\left(-\mathbf{u}_{-T}^{T}\theta_x\right)\right|\right] = 0 \quad . \tag{E.79}$$

2. We have to prove

$$\lim_{T\to-\infty} \mathbb{E}_{cat}\left|\exp\left(-\left(\mathbf{u}_{-T}^{T} + \mathbf{u}_{-T}^{T}(\Delta)\right)\theta_x\right)\right. \tag{E.80}$$

$$\left. - \exp\left(-\left(\mathbf{u}_{-\tilde{T}}^{-\infty} + \mathbf{u}_{-\tilde{T}}^{-\infty}(\Delta)\right)\theta_x\right)\right| = 0 \quad . \tag{E.81}$$

Proof of 1.: With dominated convergence, it suffices to prove

$$\begin{aligned} &\lim_{T\to-\infty} \exp\left(-\left\langle \mathbf{u}_{-T}^{T}, \mathfrak{X}\right\rangle\right) \\ = &\lim_{T\to-\infty} \left[\exp\left(-\left\langle \mathbf{u}_{-T}^{T}, \theta_x \mathfrak{n}\right\rangle\right)\right] \quad \mathbb{P}_{cat} - a.s. \end{aligned} \tag{E.82}$$

This follows immediately from (E.76): Both l. h. s. and r. h. s. converge to 1 ($\mathbb{P}_{cat} - a.s.$).

Proof of 2.: Applying (E.76) it is enough to prove

$$\lim_{T\to-\infty} \mathbb{E}_{cat}\left|\mathbb{E}_{reac}^{\hat{Y}}\left[\exp\left(-\mathbf{u}_{-T}^{T}(\Delta)\right)\right] - \mathbb{E}_{reac}^{\hat{Y}}\left[\exp\left(-\mathbf{u}_{-\tilde{T}}^{-\infty}(\Delta)\right)\right]\right| = 0 \quad . \tag{E.83}$$

Let $\varepsilon > 0$. It suffices to verify the following:

$$\lim_{T\to-\infty} \sup_{r\in\mathbb{R}_+} \left(\mathbb{E}_{cat}\left[\left|\mathbf{u}_r^{T}(\Delta) - \mathbf{u}_r^{-\infty}(\Delta)\right| \wedge 1\right]\right) < 2\varepsilon \quad . \tag{E.84}$$

This is equivalent to the following statement

$$\lim_{T\to-\infty}\sup_{r\in\mathbb{R}_+}\left(\mathbb{E}_{cat}\left[\left|\int_0^r c_x\left(\mathbf{u}_t^T-\mathbf{u}_t^{-\infty}\right)dt\right|\wedge 1\right]\right)<2\varepsilon \quad . \tag{E.85}$$

Note that (E.76) can be generalized in the following way:

$$\exists K\in\mathbb{R}_+ : \sup_{\check{T}\in\mathbb{R}_-\cup\{-\infty\}} \mathbf{u}_t^{\check{T}}\le K\exp\left(-c_x t\right) \quad \forall t\ge 0, \tag{E.86}$$

This property of exponential decay yields that we can be sure that there exists $R_\varepsilon<\infty$ with

$$\lim_{T\to-\infty}\mathbb{E}_{cat}\left[\left|\int_{R_\varepsilon}^{\infty} c_x\left(\mathbf{u}_t^T-\mathbf{u}_t^{-\infty}\right)dt\right|\wedge 1\right]<\varepsilon \tag{E.87}$$

Hence, it is sufficient to show

$$\lim_{T\to-\infty}\mathbb{E}_{cat}\left[\int_0^{R_\varepsilon} c_x\left|\mathbf{u}_t^T-\mathbf{u}_t^{-\infty}\right|dt\wedge 1\right]<\varepsilon \tag{E.88}$$

Since the catalyst processes converge with probability 1 (compare (E.51)), we are free not not take into account the zero zet containing all cases for which the pathwise convergence of the catalyst "$\hat{Y}^T\to\hat{Y}^{-\infty}$" does not take occur.

$$\lim_{T\to-\infty}\int_0^{R_\varepsilon}\mathbb{E}_{cat}\left[c_x\left|\mathbf{u}_t^T-\mathbf{u}_t^{-\infty}\right|\cdot\mathbb{I}_{\{\hat{Y}^T\to\hat{Y}^{-\infty}\}}\right]dt<\varepsilon$$

It suffices to prove

$$\lim_{T\to-\infty}\sup_{t\in[0,R_\varepsilon]}\mathbb{E}_{cat}\left[c_x\left|\mathbf{u}_t^T-\mathbf{u}_t^{-\infty}\right|\cdot\mathbb{I}_{\{\hat{Y}^T\to\hat{Y}^{-\infty}\}}\right]<\frac{\varepsilon}{R_\varepsilon} \tag{E.89}$$

With compactness of $[0,R_\varepsilon]$, it is enough to carry out the proof pointwise.

Consider the difference $\left(\mathbf{u}_t^T-\mathbf{u}_t^{-\infty}\right)_{t\ge 0}$. Since by construction the jumps of $\left(\mathbf{u}_t^T\right)$ and $\left(\mathbf{u}_t^{-\infty}\right)$ coincide, this difference process is continuous and piecewise differentiable. It can be (a.s.) characterized by the following differential equation[3]:

$$\begin{aligned}\mathbf{u}_0^T-\mathbf{u}_0^{-\infty} &= 0 \quad , &(\text{E.90})\\ \frac{\partial}{\partial t}\left(\mathbf{u}_t^T-\mathbf{u}_t^{-\infty}\right) &= -c_x\left(\mathbf{u}_t^T-\mathbf{u}_t^{-\infty}\right)\\ &\quad+\gamma_x\left(\hat{Y}_{-t}^{-\infty}\left(\mathbf{u}_t^{-\infty}\right)^2-\hat{Y}_{-t}^T\left(\mathbf{u}_t^T\right)^2\right) &(\text{E.91})\end{aligned}$$

[3]As usual, $\frac{\partial}{\partial t}$ denotes only one-sided differentials at the points of non-differentiability.

We introduce a notation for the difference:

$$\mathbf{D}_t^T := \mathbf{u}_t^T - \mathbf{u}_t^{-\infty} \quad (t \geq 0). \tag{E.92}$$

Using the identity

$$ab - a'b' = \frac{1}{2}\left[\left(a + a'\right)\left(b - b'\right) + \left(a - a'\right)\left(b + b'\right)\right] \quad , \tag{E.93}$$

we get

$$\begin{aligned}\frac{\partial}{\partial t}\mathbf{D}_t^T &= -c_x \mathbf{D}_t^T \\ &\quad -\frac{1}{2}\gamma_x\left(\hat{Y}_{-t}^T + \hat{Y}_{-t}^{-\infty}\right)\mathbf{D}_t^T\left(\mathbf{u}_t^T + \mathbf{u}_t^{-\infty}\right) \\ &\quad +\frac{1}{2}\gamma_x\left(\hat{Y}_{-t}^{-\infty} - \hat{Y}_{-t}^T\right)\left(\left(\mathbf{u}_t^T\right)^2 + \left(\mathbf{u}_t^{-\infty}\right)^2\right) \quad \mathbb{P}_{cat}\text{-a.s.}\end{aligned} \tag{E.94}$$

We want to prove

$$\lim_{T \to -\infty} \sup_{t \in [0, R_\varepsilon]} \mathbb{E}_{cat}\left[c_x \left|\mathbf{u}_t^T - \mathbf{u}_t^{-\infty}\right| \cdot \mathbb{I}_{\{\hat{Y}^T \to \hat{Y}^{-\infty}\}}\right] < \frac{\varepsilon}{R_\varepsilon} \tag{E.95}$$

To approximate $x \longmapsto |x|$, which is not differentiable at zero, we again (compare (6.157)) use the following sequence of approximation functions.

$$g_n(x) := \frac{1}{n}\ln\cosh(nx) \quad (n \in \mathbb{N}). \tag{E.96}$$

Recall the following properties of $(g_n)_{n \in \mathbb{N}}$:

1. $g_n'(x) = \tanh(nx)$, $n \in \mathbb{N}$
2. $g_n(x) \leq x g_n'(x) \leq |x| \leq g_n(x) + \frac{\ln 2}{n}$, $n \in \mathbb{N}$.

Applying this to (E.94), we get:

$$\begin{aligned}\frac{\partial}{\partial t} g_n\left(\mathbf{D}_t^T\right) &= -c_x g_n'\left(\mathbf{D}_t^T\right) \cdot \mathbf{D}_t^T \\ &\quad -\frac{1}{2}\gamma_x\left(\hat{Y}_{-t}^T + \hat{Y}_{-t}^{-\infty}\right) g_n'\left(\mathbf{D}_t^T\right) \cdot \mathbf{D}_t^T\left(\mathbf{u}_t^T + \mathbf{u}_t^{-\infty}\right) \\ &\quad + g_n'\left(\mathbf{D}_t^T\right) \cdot \frac{1}{2}\gamma_x\left(\hat{Y}_{-t}^{-\infty} - \hat{Y}_{-t}^T\right)\left(\left(\mathbf{u}_t^T\right)^2 + \left(\mathbf{u}_t^{-\infty}\right)^2\right) \\ &\leq \frac{1}{2}\gamma_x\left|\hat{Y}_{-t}^{-\infty} - \hat{Y}_{-t}^T\right|\left(\left(\mathbf{u}_t^T\right)^2 + \left(\mathbf{u}_t^{-\infty}\right)^2\right) \quad .\end{aligned} \tag{E.97}$$

Note that we can conclude from $\sum_{k=1}^n a_k < \infty$ and (E.86) that there exists $A < \infty$ with

$$g_n\left(\mathbf{D}_t^T\right) \leq \int_0^t \left|\hat{Y}_{-s}^{-\infty} - \hat{Y}_{-s}^T\right| ds \quad (t \in [0, R_\varepsilon]). \tag{E.98}$$

With the pathwise convergence property of the catalyst (E.51) this yields the claim.

Proof of II. - Pathwise convergence

We have proved step I of our original three step program. For step II it remains to show that there is convergence not only in terms of finite dimensional distributions, but weakly on the pathspace. One has to show that for any $\vartheta < 0$ the family

$$\left\{\mathcal{L}\left\{\left(X^T_{t\vee T}, Y^T_{t\vee T}\right)_{t\in[\vartheta,0]}\right\}\right\}_{T<0} \tag{E.99}$$

is a tight family. We give a short sketch of the proof – is is quite similar to the tightness proof of page 6.2. The two crucial steps are the following

- Tightness of initial conditions (at $t = \vartheta$)
- The Kolmogorov criterion (see [EK86], Proposition 3.6.3) holds.

The first point follows from the finiteness of moments assumed for the initial condition. The verifiction of the Kolmogorov criterion relies on the uniform boundedness of $(2+\alpha)$th moments ($\alpha > 0$) as we have proven in Lemma 17, p. 264.

Proof of III. - Stationarity

We have to prove that $\mu^{(0)}_{(\theta_x,\theta_y)}$ is stationary. Recall that $\mu^{(0)}_{(\theta_x,\theta_y)}$ is uniquely given by the property (proven in I.) that for all processes with initial condition $\mathcal{L}\left\{\left(X^T_T, Y^T_T\right)\right\} \in \tilde{S}_{(\theta_x,\theta_y)}$ we have

$$\mathcal{L}\left\{\left(X^T_0, Y^T_0\right)\right\} \quad \overset{T\to-\infty}{\Longrightarrow} \quad \mu^{(0)}_{(\theta_x,\theta_y)} \quad . \tag{E.100}$$

Clearly, $\mu^{(0)}_{(\theta_x,\theta_y)}$ is stationary iff the process $\vartheta_{(\theta_x,\theta_y)}$ defined by its initial law $\mu^{(0)}_{(\theta_x,\theta_y)}$ and the evolution equations (E.3), (E.4) is stationary[4]. We can describe $\vartheta_{(\theta_x,\theta_y)}$ as weak limit point:

$$\mathcal{L}\left\{\left(X^T_t, Y^T_t\right)_{t\in\mathbb{R}}\right\} \quad \overset{T\to-\infty}{\Longrightarrow} \quad \vartheta_{(\theta_x,\theta_y)} \quad . \tag{E.102}$$

By construction, $\vartheta_{(\theta_x,\theta_y)}$ follows the evolution mechanism given by (E.3) and (E.4) and has marginal distribution $\mu^{(0)}_{(\theta_x,\theta_y)}$ at time zero (by (E.100)). To prove that $\vartheta_{(\theta_x,\theta_y)}$ is the law of a stationary process, it suffices to ensure that it is invariant under any finite shift of time. In other words, the distribution of paths is stationary with respect to any semigroup describing time-shift. It suffices to consider left-shifts, which are defined as follows:

Definition 12 *The family of* left-shifts $(\mu L_s)_{s\geq 0}$ of a probability measure μ *on* $\mathcal{C}(\mathbb{R}, \mathbb{R}_+)$ *is defined as follows:*

Consider a continuous process $(Z_t)_{t\in\mathbb{R}}$ *with* $\mu = \mathcal{L}\{Z\}$. *Then* μL_s $(s \geq 0)$ *is given by*

$$\mu L_s = \mathcal{L}\left\{(Z_{t+s})_{t\geq 0}\right\} \quad . \tag{E.103}$$

[4] A process $(\Xi_t)_{t\geq 0}$ is called stationary if

$$\mathcal{L}\left\{(\Xi_t)_{t\geq 0}\right\} = \mathcal{L}\left\{(\Xi_{t+s})_{t\geq 0}\right\} \tag{E.101}$$

for any $s > 0$.

We claim:

$$\forall s \geq 0 : \vartheta_{(\theta_x,\theta_y)} = \vartheta_{(\theta_x,\theta_y)} L_s \quad . \tag{E.104}$$

Clearly, there is a Markov semigroup (L_s) acting on

$$\boldsymbol{C} := \mathcal{C}_b\left(\mathcal{C}\left(\mathbb{R}, \mathbb{R}_+ \times \mathbb{R}_+\right), \mathbb{R}_+\right),$$

which describes left-shift. The domain $\boldsymbol{C}$ of this semigroup has the property that it is invariant under the semigroup:

$$\forall s \geq 0, f \in \boldsymbol{C} : L_s f \in \boldsymbol{C} \quad . \tag{E.105}$$

Consider $\mathfrak{Z} \in \tilde{S}_{(\theta_x,\theta_y)}$. Denote by Υ the law of the process that can be constructed taking $\mathfrak{Z}$ as initial law and (E.3),(E.4) as evolution equations (we construct the process as strong solution with respect to some driving Brownian motions). By (6.200), we know $f \in \mathcal{C}_b\left(\mathcal{C}\left(\mathbb{R}, \mathbb{R}_+ \times \mathbb{R}_+\right), \mathbb{R}_+\right)$ we have

$$\left\langle \vartheta_{(\theta_x,\theta_y)}, f \right\rangle = \lim_{t \to \infty} \left\langle \Upsilon, L_t f \right\rangle \quad . \tag{E.106}$$

We have to prove that it is allowed to choose $\vartheta_{(\theta_x,\theta_y)}$ as initial law of the historical process. Namely we have to prove

$$\mu^{(0)}_{(\theta_x,\theta_y)} \in \tilde{S}_{(\theta_x,\theta_y)} \quad . \tag{E.107}$$

This follows immediately by Ito's formula, calculating moments (compare E.1.2).

Finally, we get from (E.107) and (E.106):

$$\begin{aligned} \left\langle \vartheta_{(\theta_x,\theta_y)}, L_s f \right\rangle &= \lim_{t \to \infty} \left\langle \Upsilon, L_{s+t} f \right\rangle \\ &= \lim_{t \to \infty} \left\langle \Upsilon, L_t f \right\rangle \\ &= \left\langle \vartheta_{(\theta_x,\theta_y)}, f \right\rangle \quad . \end{aligned} \tag{E.108}$$

This completes the proof of stationarity of $\vartheta_{(\theta_x,\theta_y)}$ under the semigroup of left-shifts L.

□

Bibliography

[Ald89] D. Aldous: *Stopping times and tightness,* Annals of Probability, **17**, 586-595, 1989

[BCGH1] Baillon, Cox, Greven, den Hollander: *On the attracting orbit of a nonlinear transformation arising from renormalisation of hierarchically interacting diffusions - part I: the compact case,* Can. J. Math., **47**, 3-27, 1995

[BCGH2] Baillon, Cox, Greven, den Hollander: *On the attracting orbit of a nonlinear transformation arising from renormalisation of hierarchically interacting diffusions - part II: the non-compact case*, J. of funcional analysis, **146**, 236-298, 1997

[Bu] D. Burkholder: *Sharp inequalities for martingales and stochastic integrals,* Astérique **157-158**, 75-94, 1988

[CDP00] J. T. Cox, R. Durrett, E. Perkins: *Rescaled voter models converge to super-Brownian motion,* Ann. Prob. **28**(1): 185-234, 2000

[CFG96] J. T. Cox, K. Fleischmann, A. Greven: *Comparison of interacting diffusions and an application to their ergodic theory,* Prob. Theory and Rel. Fields **105**, 513-528, 1996

[CG90] J. T. Cox, A. Greven: *The longterm Behaviour of Some Finite Particle Systems,* Prob. Theory and Rel. Fields **85**, 195-237, 1990

[CG94] J. T. Cox, A. Greven: *Ergodic theorems for infinite systems of interacting diffusions,* Ann.Prob. **22**, 833-853, 1994

[CGS95] J. T. Cox, A. Greven, T. Shiga: *Finite and infinite systems of interacting diffusions,* Prob. Theory and Rel. Fields **103**, 165-197, 1995

[CGr85] J. T. Cox, D. Griffeath: *Occupation times for critical branching Brownian motions,* Ann Prob. **13**, 1108-1132, 1985

[CGr86] J. T. Cox, D. Griffeath: *Diffusive clustering in the two-dimensional voter model,* Ann. Prob. **14**(2),347-370, 1986

[CK02] J. T. Cox, A. Klenke: *Rescaled interacting diffusions converge to Super Brownian motion,* Annals of Applied probability, to appear, 2002

[CDG03] J.T. Cox, D. Dawson, A. Greven: *Mutually catalytic super branching random walks: Large finite systems and renormalization analysis*, Mem. AMS, to appear 2003

[D01] D. Dawson: *Summer 2001 Class Notes: Measure-valued Processes and Population Models*, Mathematisches Institut, Universität Erlangen, 2001

[DEF+02a] D. Dawson, A. Etheridge, K. Fleischmann, L. Mytnik, E. Perkins, J. Xiong: *Mutually catalytic branching in the plane: Finite measure states*, Ann. Probab. **30**(4), 1681-1762, 2002

[DEF+02b] D. Dawson, A. Etheridge, K. Fleischmann, L. Mytnik, E. Perkins, J. Xiong: *Mutually catalytic branching in the plane: Infinite measure states,* EJP **7**, paper 15

[DFM+02] D. Dawson, K. Fleischmann, L. Mytnik, E. Perkins, J. Xiong: *Mutually catalytic branching in the plane: Uniqueness,* WIAS Berlin, Preprint No. 641, Ann. Inst. Heri Poincaré Probab. Statist. (in print), 2002

[DF97a] D. Dawson, K. Fleischmann: *A continuous super-Brownian motion in a catalytic medium,* J. Theoret. Probab. **10(1)**, 213-276, 1997

[DF97b] D. Dawson, K. Fleischmann: *Longtime behavior of a branching process controlled by branching catalysts*, Stochastic processes and their applications, **71**, 241-257, 1997

[DGW01] D. Dawson, L. G. Gorostiza, A. Wakolbinger: *Occupation Time Fluctuations in Branching Systems*, J. Theoret. Prob. **14**(3), p. 729pp, 2001

[DG93a] D. Dawson, A. Greven: *Multiple time scale analysis of hierarchically interacting systems,* A Festschrift to honor G Kallianpur, 41-50, Springer-Verlag, 1993

[DG93b] D. Dawson, A. Greven: *Multiple time scale analysis of interacting diffusions.* Probab. Theory Rel. Fields **95**, 467-508, 1993

[DG93c] D. Dawson, A. Greven: *Hierarchical models of interacting diffusions: multiple time scale phenomena. Phase transition and pattern of cluster formation.* Probab. Theory Rel. Fields **96**, 435-473, 1993

[DG96] D. Dawson, A. Greven: *Multiple space-time scale analysis for interacting branching models*, EJP **1**/14, 1996

[DG03] D. Dawson, A. Greven: *State dependent multitype spatial branching processes and their longtime behavior,* EJP **8**/4, 2003

[Du79] R. Durrett: *An infinite particle system with additive interactions,* Adv. Appl. Prob. **11**, 355-383, 1979

[DP98] D. Dawson, E. Perkins: *Long-time behavior and coexistence in a mutually catalytic branching model*, Ann. Prob. **26**/3, 1088-1138, 1998

[E00] A. Etheridge: *An introduction to superprocesses*, American Mathematical Society, Providence (R. I.), 2000

[EF98] A. Etheridge, K. Fleischmann: *Persistence of a two-dimensional super-Brownian motion in a catalytic medium.* Probab. Theory Related Fields **110**, 1-12, 1998

[EK86] S. Ethier, T. Kurtz: *Marcov processes, Characterization and convergence*, Wiley, New York, 1986

[Fe] Feller: *An introduction to Probability Theory and Its applications, volume II*, Wiley, New York, 1971

[FG86] K. Fleischmann, J. Gärtner: *Occupation Time Processes at a Critical Point*, Math. Nachr. **125**, 275-290, 1986

[FX01] K. Fleischmann, J. Xiong: *A cyclically catalytic super-Brownian motion,* Ann. Prob. **29**(2), p. 820-861, 2001

[GKW99] A. Greven, A. Klenke, A. Wakolbinger: *The longtime behavior of branching random walk in a catalytic medium,* EJP **4**/12, 1999

[GKW02] A. Greven, A. Klenke, A. Wakolbinger: *Interacting diffusions in a random medium: comparison and longtime behavior,* Stochastic Processes and Appl., **98**, 23-41, 2002

[GW91] L. G. Gorostiza and A. Wakolbinger: *Persistence criteria for a class of branching particle systems in continuous time.* Ann. Prob. **19(1)**, 266-288, 1991

[Isc86] I. Iscoe: *A Weighted Occupation Time for a Class of Measure-Valued Branching Processes*, PTRF **71**, 85-116, 1986

[JM] A. Joffe, M. Métivier: *Weak convergence of sequences of semimartingales with applications to multitype branching processes,* Adv. Appl. Prob., **18**, 20-65, 1986

[Kal77] O. Kallenberg: *Stability of cluster fields,* Math. Nachr., **77**, 7-43, 1977

[K98] A. Klenke: *Clustering and invariant measures for spatial branching models,* Ann. Prob. **26(3)**: 1057-1087, 1998

[K00] A. Klenke: *Longtime Behavior of Stochastic Processes with Complex Interactions,* Habilitation thesis, University Erlangen, Germany, 2000

[KS] I. Karatzas, S. Shreve: *Brownian motion and stochastic calculus*, Springer, New York 1991

[LG83] J. F. Le Gall, *Applications du temps local aux equations differentielles stochastiques unidimensionelles*, Lecture Notes in Mathematics **986**, 15-31, Springer, Berlin, 1983

[Lig85] T. Liggett: *Interacting particle systems.* Springer-Verlag, New York 1985

[LS81] T. Liggett, F. Spitzer: *Ergodic theorems for coupled random walks and other systems with locally interacting components*, Z. Wahrsch. verw. Gebiete, **56**(4), 443-486, 1981

[Li92] T. Lindvall, *Lectures on the coupling method,* Wiley, 1992

[Myt96] L. Mytnik: *Superprocesses in random environments,* PhD Thesis, Technion, 1996

[Myt98] L. Mytnik: *Uniqueness for a mutually catalytic branching model*, Prob. Theory Related Fields, **112**, 245-253, 1998

[Ok] B. Øksendal: *Stochastic differential equations*, 5th edition, Springer, New York, 1991

[Pf03] P. Pfaffelhuber: *State-dependent interacting multitype branching systems,* Doktorarbeit, Universität Erlangen, 2003

[R79] R. Rebolledo: *La méthode des martingales appliquée à la convergence en loi des processus*, Mem. Soc. Math. France Suppl., 1-125, 1979

[Sh92] T. Shiga: *Ergodic theorems and exponential decay of sample paths for certain interacting diffusion systems,* Osaka J. Math **29**, 789-807, 1992

[Spi76] F. Spitzer: *Principles of random walk.* Springer, New York, second edition 1996. Translated from the first (1980) russian edition by R. P. Boas.

[Sz] A.-S. Sznitman: *Brownian Motion, Obstacles and Random Media*, Springer, 1998

[W99] A. Winter: *Multiple Scale Analysis of Spatial Branching Processes under the Palm Distribution,* Dissertation, Universität Erlangen-Nürnberg, 1999

[YW] T. Yamada, S. Watanabe: *On the uniqueness of solutions of stochastic differential equations,* J. Math. Kyoto Univ., **11**, 155-167, 1971

Index of Notation

$a(\cdot,\cdot)$	transition rates of a Markov chain
$\bar{a}(\cdot,\cdot)$	transposed matrix of the matrix a
$\hat{a}(\cdot,\cdot)$	symmetrization of a
$B_k(r)$	closed ball with center k and radius r
$B(E)$	bounded functions on E
$card(A)$	cardinality of the set A
$\mathcal{C}(A,B)$	continous mappings from A to B
$\mathcal{C}_b(A,B)$	bounded continous mappings from A to B
$\delta_{\{a\}}$	atomar probability distribution at a
$\delta_{\{\underline{a}\}}$	atomar probability distribution corresponding to constant configuration with intensity a
$\delta_{a,b}$	dirac delta: $\delta_{a,b}=1$ if $a=b$, else $\delta_{a,b}=0$
d_{Sk}	Skorohod distance
$\mathcal{D}(\mathbb{R},B)$	set of cadag functions from $\mathbb{R}$ to B
Δ_{RW}	discrete Laplacian
$\mathcal{L}\{X\}$	distribution of random variable X
$\mathcal{M}(\Lambda)$	set of measures on set Λ
$\mathcal{M}_{loc}(\Lambda)$	set of locally finite measures on Λ
$\mathcal{M}^1(\Lambda)$	set of probability measures on Λ
$\mathcal{N}(\Lambda)$	set of integer-valued measures on Λ
$\mathfrak{n}$	counting measure
$\mathbb{T}_N^d$	$\{-N,...,N-1\}^d$ equipped with torus topology
$\langle\cdot,\cdot\rangle$	dot product, integration
$\mathbb{I}_A(\cdot)$	indicator function: $\mathbb{I}_A(x)=1$ if $x\in A$, else 0
$\subset\subset$	"finite subset of"
$\Longrightarrow$	weak convergence

Lebenslauf

Name: Christian Penßel

Geburtsdatum: 16. Juli 1974

Geburtsort: Nürnberg

Schulbildung

1981-1985	Grundschule in Nürnberg
1985-1994	Hans-Sachs-Gymnasium, Nürnberg
	(mathematisch-naturwissenschaftlicher Zweig, Sprachenfolge Englisch, Latein)
1. 7. 1994	Abitur

Universität

1994-1999	Studium der Mathematik und Physik, Universität Erlangen
1996/1997	Vordiplome in Physik und Mathematik
1999	Diplom in Mathematik

Stellen

1999-2003	wissenschaftlicher Mitarbeiter an der Universität Erlangen
9/02-10/02	Praktikum bei imbus AG, Möhrendorf

Stipendien

1994-1999	Stipendium für besonders begabte Abiturienten des Freistaates Bayern
1998-2002	Förderung durch die Studienstiftung des deutschen Volkes